# TRACE ANALYSIS

### Volume 2

# Contributors to This Volume

Thomas V. Alfredson
R. F. Browner
J. B. Green
P. L. Grizzle
Ronald E. Majors
W. A. Saner
J. N. Thompson
R. C. Willoughby

# TRACE ANALYSIS

## Volume 2

*Edited by*

## James F. Lawrence

*Food Research Division*
*Health Protection Branch*
*Health and Welfare Canada*
*Ottawa, Ontario, Canada*

1982

**ACADEMIC PRESS**
*A Subsidiary of Harcourt Brace Jovanovich, Publishers*

New York   London
Paris   San Diego   San Francisco   São Paulo   Sydney   Tokyo   Toronto

COPYRIGHT © 1982, BY ACADEMIC PRESS, INC.
ALL RIGHTS RESERVED.
NO PART OF THIS PUBLICATION MAY BE REPRODUCED OR
TRANSMITTED IN ANY FORM OR BY ANY MEANS, ELECTRONIC
OR MECHANICAL, INCLUDING PHOTOCOPY, RECORDING, OR ANY
INFORMATION STORAGE AND RETRIEVAL SYSTEM, WITHOUT
PERMISSION IN WRITING FROM THE PUBLISHER.

ACADEMIC PRESS, INC.
111 Fifth Avenue, New York, New York 10003

*United Kingdom Edition published by*
ACADEMIC PRESS, INC. (LONDON) LTD.
24/28 Oval Road, London NW1 7DX

ISBN 0-12-682102-X

ISSN 0275-844X

This publication is not a periodical and is not subject to copying under CONTU guidelines.

PRINTED IN THE UNITED STATES OF AMERICA

82 83 84 85    9 8 7 6 5 4 3 2 1

# CONTENTS

CONTRIBUTORS . . . . . . . . . . . . . . . . . . . . . . vii
PREFACE . . . . . . . . . . . . . . . . . . . . . . . . . ix
CONTENTS OF VOLUME 1 . . . . . . . . . . . . . . . . . . xi

## Trace Analysis of Vitamins by Liquid Chromatography

### J. N. Thompson

I. Introduction . . . . . . . . . . . . . . . . . . . . 1
II. Fat-Soluble Vitamins . . . . . . . . . . . . . . . . 3
III. Water-Soluble Vitamins . . . . . . . . . . . . . . . 43
IV. Conclusion . . . . . . . . . . . . . . . . . . . . . 60
References . . . . . . . . . . . . . . . . . . . . . 61

## Combining Liquid Chromatography with Mass Spectrometry

### R. C. Willoughby and R. F. Browner

I. Introduction . . . . . . . . . . . . . . . . . . . . 69
II. The Mass Spectrometer as a Detector for Liquid Chromatography . . . . . . . . . . . . . . . . . . . 70
III. Off-Line Techniques . . . . . . . . . . . . . . . . . 74
IV. On-Line Techniques . . . . . . . . . . . . . . . . . 77
V. Alternative Methods of Vaporization and Ionization with Possible Applications to LC–MS . . . . . . . . . . . 97
VI. Applications and Conclusions . . . . . . . . . . . . 101
References . . . . . . . . . . . . . . . . . . . . . 105

## Applications of Steric Exclusion Chromatography in Trace Analysis

### Ronald E. Majors and Thomas V. Alfredson

I. Introduction . . . . . . . . . . . . . . . . . . . . 112
II. Basics of SEC . . . . . . . . . . . . . . . . . . . . 112
III. Advantages of the SEC Technique . . . . . . . . . . 114
IV. Disadvantages of the SEC Technique . . . . . . . . . 115
V. Recent Advances in SEC Columns . . . . . . . . . . . 116
VI. Utility of SEC in Trace Analysis . . . . . . . . . . 117
VII. Approaches for Use of SEC in Trace Analysis . . . . 118
VIII. Off-Line Multidimensional SEC Techniques . . . . . . 128
IX. On-Line Multidimensional SEC Techniques . . . . . . 134
X. Problems and Troubleshooting . . . . . . . . . . . . 143
References . . . . . . . . . . . . . . . . . . . . . 149

## Trace-Enrichment Techniques for Organic Trace Analysis

### W. A. Saner

|    |    |
|----|----|
| I. Trace Enrichment: Definition | 152 |
| II. Environmental Applications | 164 |
| III. Clinical Applications | 202 |
| IV. Pharmacological Applications | 214 |
| References | 220 |

## HPLC Analysis of Polar Substances on Unmodified Silica

### J. B. Green and P. L. Grizzle

|    |    |
|----|----|
| I. Introduction | 223 |
| II. Literature Survey | 224 |
| III. Nonaqueous Systems Containing Aliphatic Carboxylic Acids | 227 |
| IV. Nonaqueous Systems Containing Amines | 238 |
| V. Coordinating Amine- and Carboxylic Acid-Based Separations for Optimum Resolution of Complex Samples | 257 |
| VI. Conclusions | 264 |
| References | 265 |

INDEX . . . . . . . . . . . . . . . . . . . . 267

# CONTRIBUTORS

Numbers in parentheses indicate the pages on which the authors' contributions begin.

**Thomas V. Alfredson** (111), Varian Associates, Inc., Walnut Creek Instrument Division, Walnut Creek, California 94598

**R. F. Browner** (69), School of Chemistry, Georgia Institute of Technology, Atlanta, Georgia 30332

**J. B. Green** (223), U.S. Department of Energy, Bartlesville Energy Technology Center, Bartlesville, Oklahoma 74003

**P. L. Grizzle**[1] (223), U.S. Department of Energy, Bartlesville Energy Technology Center, Bartlesville, Oklahoma 74003

**Ronald E. Majors** (111), Varian Associates, Inc., Walnut Creek Instrument Division, Walnut Creek, California 94598

**W. A. Saner**[2] (151), U.S. Coast Guard Research and Development Center, Avery Point, Groton, Connecticut 06340

**J. N. Thompson** (1), Department of National Health and Welfare, Health Protection Branch, Tunney's Pasture, Ottawa, Ontario, Canada K1A OL2

**R. C. Willoughby** (69), School of Chemistry, Georgia Institute of Technology, Atlanta, Georgia 30332

---

[1]Present address: Sun Exploration and Production Co., P.O. Box 936, Richardson, Texas 75080.
[2]Present address: U.S. Army Corps of Engineers, Ohio River Division Laboratory, Mariemont, Ohio 45227.

# PREFACE

The publication of Volume 1 of *Trace Analysis* initiated a new multivolume publication devoted to state-of-the-art critical discussions of selected topics in organic and inorganic analytical chemistry. These include instrumentation, techniques, and applications to the detection, identification, and quantitation of trace quantities of substances in a large variety of sample materials.

Volume 1 dealt with high-performance liquid chromatography, including such subjects as the determination of trace organics in aqueous environmental samples, electrochemical detectors for chromatographic and flow analysis systems, determination of metal species by complexation and ion exchange, analysis of mycotoxins, and applications of ion chromatography. Volume 2 continues with selected topics in high-performance liquid chromatography, beginning with a comprehensive article on vitamin analysis in foods and tissues. This is followed by a discussion of liquid chromatography–mass spectrometry with emphasis on new approaches to interfacing. An article on gel-permeation (or steric exclusion) chromatography is included which discusses applications of high-efficiency columns for the removal of sample matrix interferences. Trace enrichment is the subject of the fourth article, where different approaches to the isolation and concentration of trace organics from aqueous samples are presented. The final article describes the advantages of unmodified silica gel for the chromatographic separation of polar substances. This volume is an important follow-up to Volume 1; together, the topics covered provide the reader with a great deal of information that will be extremely helpful for the development of trace analytical methodology by high-performance liquid chromatography.

I wish to express my thanks to J. N. Thompson, R. C. Willoughby, R. F. Browner, Ronald E. Majors, Thomas V. Alfredson, W. A. Saner, J. B. Green, and P. L. Grizzle for their diligent efforts in providing their comprehensive, informative articles.

JAMES F. LAWRENCE

# CONTENTS OF VOLUME 1

Determination of Trace Organic Compounds in Aqueous Environmental Samples by High-Performance Liquid Chromatography
  *Jeffrey A. Graham*

Electrochemical Detectors for High-Performance Liquid Chromatography and Flow Analysis Systems
  *K. Brunt*

The Separation and Determination of Metal Species by Modern Liquid Chromatography
  *R. M. Cassidy*

Liquid Chromatography in the Analysis of Mycotoxins
  *Peter M. Scott*

Applications of Ion Chromatography in Trace Analysis
  *Hamish Small*

Index

# TRACE ANALYSIS OF VITAMINS BY LIQUID CHROMATOGRAPHY

### J. N. Thompson

Department of National Health and Welfare
Health Protection Branch
Ottawa, Ontario, Canada

| | | | |
|---|---|---|---|
| I. | Introduction | | 1 |
| II. | Fat-Soluble Vitamins | | 3 |
| | A. | Vitamin A and Carotenoid Provitamins | 3 |
| | B. | Vitamin D | 17 |
| | C. | Vitamin E | 29 |
| | D. | Vitamin K | 37 |
| III. | Water-Soluble Vitamins | | 43 |
| | A. | Ascorbic Acid | 43 |
| | B. | Riboflavin and Thiamine | 47 |
| | C. | Niacin | 49 |
| | D. | Vitamin $B_6$ | 52 |
| | E. | Folacin | 55 |
| | F. | Vitamin $B_{12}$ | 60 |
| IV. | Conclusion | | 60 |
| | References | | 61 |

## I. INTRODUCTION

This review of methods of analysis is devoted entirely to the application of high-performance liquid chromatography (HPLC) to the measurement of vitamins in foods and tissues. The subject is still in its infancy, but progress has been rapid and the literature is already extensive.

It has been possible to cite most of the publications appearing before and during 1980, but this has left little room for references to background information. Readers wishing to compare other methods of analysis are therefore directed to the manuals of the Association of Vitamin Chemists (Freed, 1966) and the Association of Official Analytical Chemists (AOAC, 1980); Volumes VI and VII of "The Vitamins" (Gyorgy and Pearson, 1967); several volumes of "Methods in Enzymology" (McCormick and Wright, 1970–1980); and the books by Hashmi (1972), Strohecker and Henning (1965), and Osborne and Voogt (1978).

Although detailed background information must be sought elsewhere, each section in this article opens with a brief description of the properties of the vitamins, and although lacking depth and literature citations, these introductory sections at least give the chemical structures of substances mentioned later in the review. Without this information, many readers would probably find the subject unintelligible because of the widespread use of trivial names and ambiguous systems of nomenclature.

The vitamins are divided, as is the usual custom, into two groups according to their solubility in fat or water. More work has been done on the application of HPLC to the fat-soluble vitamins than to the water-soluble vitamins, and only a brief review of the latter group is possible at the present time. Four water-soluble vitamins—choline, inositol, pantothenic acid, and biotin—have had to be omitted altogether.

Few of the methods published so far will survive the era of rapid change that is now upon us; techniques and instrumentation are improving continuously, and much of what has been published in the last seven years is already obsolete. Most publications, in spite of their customary optimism and air of finality, are assumed to describe preliminary investigations, and criticism of these procedures as if they were finished methods would be not only pointless and unjust, but also extremely laborious. The author has therefore glossed over many obvious deficiencies on the assumption that most readers will use the information in the literature as the basis for the development of better and more convenient methods, and that few will expect to find, at this stage, an established protocol.

Although methods designed for the analysis of pharmaceuticals can rarely be applied directly to extracts of foods or tissues, publications on this subject are included in this review because useful information can often be obtained from them concerning the chromatographic properties of the vitamins.

The earliest publications on HPLC and vitamins described the separation of simple mixtures, such as might be found in pharmaceuticals (Williams *et al.*, 1972; Kissinger *et al.*, 1973; Vecchi *et al.*, 1973; Hatano *et al.*, 1973; Callmer and Davies, 1974; Matthews *et al.*, 1974; Stillman and Ma, 1974; Tomkins and Tscherne, 1974; Unger and Nyamah, 1974; Wittmer and Haney, 1974) and less often the analysis of foods (Van De Weerdhof *et al.*, 1973; Van Niekerk, 1973; Cavins and Inglett, 1974; Kissinger *et al.*, 1974). Recently, as the following review should indicate, more effort has been devoted to the separation of minute quantities of vitamins from the constituents of foods and tissues, i.e., trace analysis.

## II. FAT-SOLUBLE VITAMINS

### A. Vitamin A and Carotenoid Provitamins

#### 1. Introduction

Substances with vitamin A activity in foods are divisible into two distinct groups: the first consists of a number of carotenoids, known as provitamins, and the second includes the vitamin itself (retinol) and related biologically active "retinoids" (Fig. 1). The most important provitamin is β-carotene, which is structurally two molecules of retinol joined tail to tail. β-Carotene and other provitamins are biosynthesized in plants and then are split enzymatically in the intestines of animals to form retinol. The chemical and biological properties of the carotenoids are comprehensively reviewed in monographs by Zechmeister (1962), Karrer and Jucker (1950), Isler (1971), and Bauernfeind (1981).

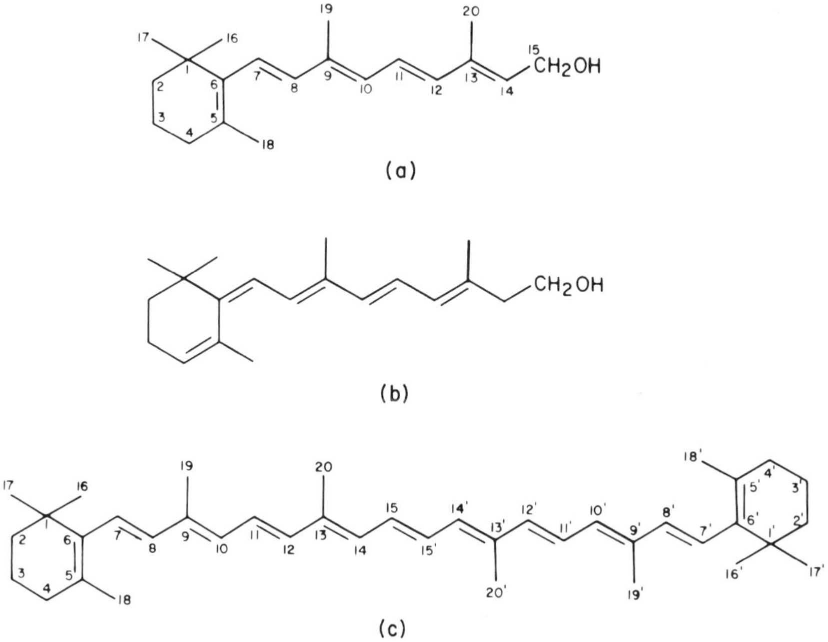

**Fig. 1.** Structures of (a) retinol, (b) retroretinol, and (c) β-carotene.

The measurement of carotenoids in foods is a difficult task because of their multiplicity, structural complexity, and instability. Each pigment can exist in a number of cis–trans forms, and isomerization often takes place during extraction and chromatography. They are readily oxidized when exposed to air.

Vitamin A is a primary alcohol (retinol) with five conjugated double bonds. Another form of the vitamin, formerly known as vitamin $A_2$, is found in some fish tissues and oils. It contains an extra double bond in the ring and is thus designated 3-dehydroretinol. Vitamin A occurs only in animal tissues. Small quantities of the free alcohol have been detected in blood, kidney, and liver, but the largest deposits of the vitamin are usually found in the liver in the form of the palmitate ester. The liver vitamin A in land animals often represents a substantial store which can be drawn upon during long periods of dietary deficiency. Liver as a food is an excellent source of the vitamin, and fish-liver oils were formerly commercially important in the manufacture of concentrates.

Oxidation of retinol yields the aldehyde, retinal, which functions in vision as the chromophore of the visual pigments. The corresponding acid, retinoic acid, is thought to be a metabolite of retinol in animals, but so far only trace amounts have been detected in tissues.

The chain of conjugated double bonds in retinol is readily oxidized and isomerized. It is therefore difficult to preserve the vitamin in pure form, and even specimens that are carefully stored under nitrogen will frequently yield many peaks during chromatography. Standards for quantitative analysis are unreliable when prepared gravimetrically, as a large part of the weight of dried specimens usually represents breakdown products. It is therefore important to verify the concentration of the vitamin in solutions by spectrophotometry and to reject preparations that contain substantial amounts of absorbing contaminants. When the results of analysis must be expressed in terms of equivalent amounts of retinol, it is convenient to calculate standards of retinol and retinol esters in hexane with an $E_{1cm}^{1\%} = 1830$; this assumes that the molecular extinction coefficient of the alcohol and esters is 52,200 (Boldingh et al., 1951).

Most of the substances produced during the breakdown of vitamin A have not been characterized. However, some have been identified, including 5,6-epoxy- and 5,8-oxyretinoids produced by oxidation and substances with retro structures (Fig. 1) produced by rearrangement of the double bonds. Retroretinol and its esters have negligible vitamin activity, but they are common contaminants of commercial preparations of the vitamin and they interfere in many methods of analysis. Retro ethers produced by irradiation are often present in solutions of retinyl esters, and the hydrocarbon anhydrovitamin A, which contains six double bonds in a retro configuration, is responsible for much of the yellow color of decomposing vitamin A.

Methods for measuring vitamin A have exploited its UV absorption, which is maximal near 325 nm, and its moderately strong fluorescence, which is emitted near 480 nm. Probably the most widely used procedures, however, have been based on the development of a transient blue color with antimony trichloride in the Carr–Price reaction. In critical work, some form of preliminary chromatography is usually necessary to eliminate interfering substances (Parrish, 1977).

## 2. Carotenoids in Foods

The separation of plant pigments on columns of calcium carbonate marked the birth of chromatography at the turn of the century. Column chromatography is still the method of choice for separating carotenoids, most of which are unstable substances prone to isomerization and decomposition. Gas–liquid chromatography (GLC) demands too high a temperature and thin-layer chromatography (TLC) involves too much exposure to oxidation, although the introduction of this technique was responsible for an increase in the number of identifiable carotenoids from about 80 in 1948 to over 350 at the present time. Column chromatography has continued to play an important role, however, in carotenoid analysis, and an unusual variety of adsorbents have been tested and recommended for special separations.

In the earliest applications of HPLC, carotenes and xanthophylls (hydroxylated carotenoids) from citrus fruits were separated on columns of magnesium oxide (Fig. 2), silica (Stewart and Wheaton, 1971; Stewart, 1977a,b), and basic alumina (Reeder and Park, 1975). Later, Botey Serra and Garcia Fite (1975) separated xanthophylls in eggs and chicken feeds with a reverse-phase system, and carotenes, xanthophylls, and other pigments in plants were separated on silica (De Jong and Woodlief, 1978) and reverse-phase columns (Eskins et al., 1977). Fiksdahl et al. (1978) published a systematic study of the separation of carotenes, carotenoid diols, cis–trans isomers, and diastereoisomers on modern high-efficiency silica packings and recommended a commercial 5-μm Spherisorb column.

A common practical problem is the measurement of β-carotene in fortified foods, where it makes a significant contribution to vitamin A activity. Sometimes β-carotene is the sole or the major low-polarity liposoluble pigment and there is little need for sophisticated chromatographic separations. In these circumstances, the ability of HPLC systems to provide quantitative measurements is exploited and the choice of column is of secondary importance. Often these measurements are undertaken as part of a vitamin A analysis that includes retinol and carotene. Thus β-carotene has been measured in fortified margarine using simple chromatographic systems (Maruyama et al., 1977; Thompson et al., 1977, 1980; Landen and Eitenmiller, 1979). Caution is needed in taking the same approach with foods containing plant material because carotenoids other than β-carotene may interfere.

Van De Weerdhof et al. (1973) measured β-carotene in spinach and vegetable soup using alumina columns, but the specificity and general applicability of this

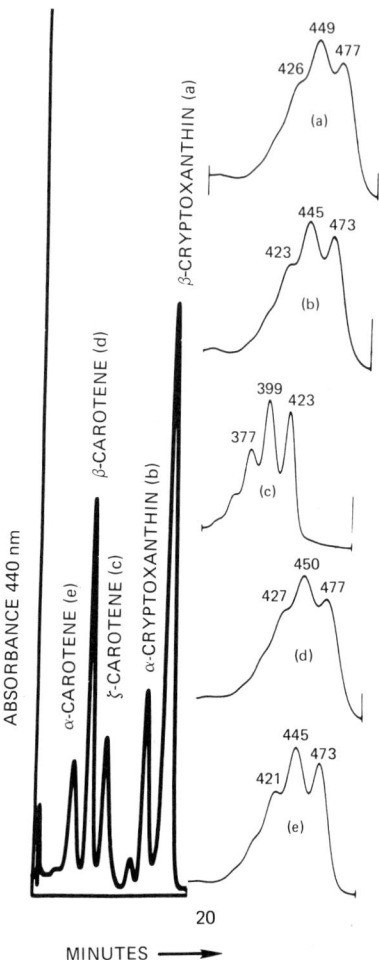

**Fig. 2.** HPLC on MgO column of the principal carotenes and cryptoxanthins in citrus juice. Absorption spectra of peaks shown at right. [Reprinted with permission from Stewart (1977b). *Agric. Food Chem.* **25**, 1132–1137. Copyright 1977, American Chemical Society.]

method are uncertain. In this author's experience, it is difficult to obtain stable retention times for carotenes during adsorption chromatography because of the need for low-polarity solvent systems, which are markedly affected by traces of water. Reproducible separations of carotenoids can be obtained on reverse-phase columns eluted with mixtures of organic solvents such as methanol, tetrahydrofuran, methylene chloride, and acetonitrile. Parris (1978) described this approach as "nonaqueous reversed-phase liquid chromatography." The present author has found it useful for the measurement of β-carotene in samples containing relatively large amounts of the pigment lycopene, such as extracts of human blood and foods containing tomatoes; a method applicable to tomatoes is described in detail by Zakaria et al. (1979).

## 3. Cis–trans Isomers of Retinol and Other Retinoids

If all 4 of the double bonds in the side chain of retinol (Fig. 1) could exist without restriction in either a cis or a trans configuration, there would be 16 possible combinations. There is steric hindrance, however, between the C-20 methyl group and the hydrogen on C-10 in the 11-cis isomer, and between methyl groups at C-19 and C-18 in the 7-cis isomer. Only the all-trans, 9-cis, 13-cis, and 9,13-di-cis isomers, therefore, are free from this type of steric hindrance, and these are thus "preferred" structures.

The occurrence of other isomers was at first considered to be improbable. Nevertheless, in spite of what was originally assumed, hindered isomers can exist, and 5 have been identified. Indeed, the 11-cis isomer of retinal has turned out to be important biologically because of a role in vision. Some of the hindered isomers have been synthesized chemically, the most recent being 9,11,13-tri-*cis*-retinol (Knudsen et al., 1980). Investigation of cis–trans isomers has been facilitated by HPLC techniques. This is a field of special interest to those investigating the biochemistry of vision, of which the cis–trans isomerization of retinal is an important feature. In addition, as isomerization is induced by irradiation and other means, cis isomers are common in food, and there is now an interest in the detailed investigation of the isomeric composition of vitamin A in foods and tissues.

The separations of isomers by HPLC achieved so far in various fields of research are listed in Table I. To facilitate comparisons among these results, the nine known isomers are coded a–i according to the order of elution of isomers of methyl retinoate on a reverse-phase column, which is the most sophisticated separation achieved so far (Halley and Nelson, 1979). Contrary to what one would expect, the isomers of retinol, retinyl esters, and retinal elute in straight-phase systems in the same general order as isomers of methyl retinoate in reverse-phase systems. The order of elution of isomers of retinoids on reverse-phase systems appears to vary. For example, the 13-cis isomer of retinol has been reported to elute before (Egberg et al., 1977) and after (McCormick et al., 1978a) the all-trans isomer. A systematic investigation of these reversals would seem to be worthwhile.

Most workers have reported good recoveries of isomers during HPLC, but the instability of vitamin A is such that tests for decomposition are advisable. Vecchi et al. (1973) noticed isomerization of vitamin A on some columns of Corasil. They overcame this problem by injecting small amounts of the antioxidant BHT every 15 min over a period of several hours. A deleterious effect of silica was

## TABLE I
### Separation of Geometric Isomers of Vitamin A

| Sample and column | Solvent system[a] | Isomers separated[b] | Reference |
|---|---|---|---|
| Retinol: | | | |
| Corasil II | 5 Dioxane | d, i | Vecci et al. (1973) |
| | 1 Isopropanol | f, i | Bridges (1976) |
| µPorasil | 20 Diethyl ether | f, i | Bridges (1976) |
| Zorbax SIL | 12 Ethyl acetate | (d + f), g, i | Tsukida et al. (1977) |
| | 20 Diethyl ether | (d + f), g, i | Tsukida et al. (1977) |
| | 7.5 Ethyl acetate + 9.3 dichloromethane | f, d, g, i | Tsukida et al. (1977) |
| | 6.2 Ethyl acetate + 7.7 dichloromethane | f, d, g, i | Tsukida et al. (1977) |
| Si60, 5 µm | 2.5–10 Dioxane | f, d, g, i | Paanakker and Groenendijk (1979) |
| Vydac, 10-µm ODS | 65 Acetonitrile + 35 water | d, i | Egberg et al. (1979) |
| µBondapak C-18 | 20 Methanol + 80 water | i, d | McCormick et al. (1978a) |
| Partisil | 3 Isobutanol | b, a, f | Knudsen et al. (1980) |
| Retinyl acetate: | | | |
| 3,3'-Oxydipropionitrile on Si60 | Hexane | b, f, d, g, i | Vecci et al. (1973) |
| Corasil II | 0.1 Dioxane | d, i | Vecci et al. (1973) |
| Si60, 5µm | 2 Dioxane | d, i | Vecci et al., (1973) |
| Alumina, 10 µm | 1 Diethyl ether | f, d, b, g, i | Vecci et al. (1973) |
| Alumina, 5 µm | 2 Diisopropyl ether | b, f, d, e, g, i | Steuerle (1981) |

| | | | |
|---|---|---|---|
| Retinyl palmitate: | | | |
| μPorasil | 2 Diethyl ether | f, i | Bridges (1976) |
| Si60, 5 μm | 0.1 Dioxane | d, f, g, i | Paanakker and Groenendijk (1979) |
| Alumina, 5 μm | 2 Diisopropyl ether | b, f, d, e, g, i | Steuerle (1981) |
| Retinal: | | | |
| μPorasil | 12.5 Diethyl ether | f, i | Bridges (1976) |
| | 2 Diethyl ether | d, f, g, i | Rotmans and Kropf (1975) |
| | 2 Diethyl ether | d, e, f, g, i | Waddell et al. (1976) |
| | 6 Diethyl ether | d, f, g, h, i | Tsukida et al. (1980) |
| | 1 Diethyl ether | f, d, g, i | Pilkiewicz et al. (1977); Adams and Nakanishi (1979) |
| μBondapak CN | 12 Ethyl acetate | d, f, g, i | Tsukida et al. (1977) |
| Zorbax SIL | 1.25–5 Dioxane | d, f, g, i | Paanakker and Groenendijk (1979) |
| Si60, 5 μm | | | |
| Retinoic acid: | | | |
| μBondapak C-18 | 75 Methanol + 25 water | d, i | McCormick et al. (1978a) |
| Partisil 10 ODS | 65 Methanol + 35 water | a, b, (d + f + e), g, i | McKenzie et al. (1978) |
| | 65 Acetonitrile + 35 water | d, i | Frolik et al. (1978a) |
| Methyl retinoate: | | | |
| Partisil PXS 10/25 ODS-2 | 85 Methanol + 15 water | a, b, c, d, e, f, g, h, i | Halley and Nelson (1979) |

[a] Percent in hexane, except in reverse-phase systems.
[b] In order of elution; coded as follows:

a  9,11,13-tri-cis   d  13-cis        g  9-cis
b  11,13-di-cis      e  9,13-di-cis   h  7-cis
c  7,13-di-cis       f  11-cis        i  all-trans

also noticed by Tsukida *et al.* (1980), and it was especially pronounced with 3-dehydroretinols.

Isomers of retinol have been well separated on columns of silica. The pioneering work by Vecchi *et al.* (1973) with retinyl acetate is of special interest in that it included a separation of retro forms on columns of alumina (Fig. 3). It was also discovered that retinyl esters dissolved in peanut oil could be injected directly onto columns of silica, a finding of considerable practical importance.

Isomers of retinal can also be separated on silica. Tsukida *et al.* (1980) separated isomers of 3-dehydroretinal. It will be noted that 13-cis and 11-cis isomers (d and f in Table I) of retinal are eluted in the opposite order of the corresponding isomers of retinol and retinyl esters.

The 11-cis isomer of retinal, which is important in the study of vision, can be conveniently isolated on a CN-bonded column (Pilkiewicz *et al.*, 1977), which holds other isomers more strongly. The syn and anti isomers of the oximes of all-*trans*-retinal and 13-*cis*-retinal have been separated on silica (Groenendijk *et al.*, 1979). Conversion of the aldehyde to the oxime seems to reduce artifactual isomerization (Groenendijk *et al.*, 1980).

### 4. Vitamin A Esters

Retinyl esters used commercially (palmitate, propionate, and acetate) can be readily separated from each other on silica (Aitzetmuller *et al.*, 1979; Thompson *et al.*, 1980). De Ruyter and De Leenheer (1979) have described the separation of long-chain esters in methanol on reverse-phase columns. When silver ions were added to the solvent system, unsaturated and saturated esters were separated by argentation reverse-phase chromatography. Silver ions did not affect the chromatography of saturated esters of vitamin A in spite of the fact that there are five double bonds in the vitamin itself. Steuerle (1981) has described the separation of cis–trans isomers of retinyl acetate and palmitate on alumina.

### 5. Vitamin A Esters in Foods

Vitamin A in foods, both added and naturally occurring, is mainly in the form of retinyl esters. The most common naturally occurring ester is the palmitate; it is accompanied by smaller amounts of other long-chain esters. Synthetic retinyl palmitate and, less frequently, the acetate or propionate are added to foods. Although oily preparations are used to fortify fats and oils, most of the vitamin added to foods is in a stabilized form, such as in a matrix of gelatin and sugar, and is not readily extracted by organic solvents.

Saponification is an obvious approach to the analysis of foods for vitamin A. It has three obvious advantages: it eliminates most of the lipid; it overcomes difficulties of extraction; and it converts the various esters of the vitamin to free retinol. This approach has been used in HPLC methods for various foods (Van De Weerdhof *et al.*, 1973; Egberg *et al.*, 1977; Osborne and Voogt, 1978; Head

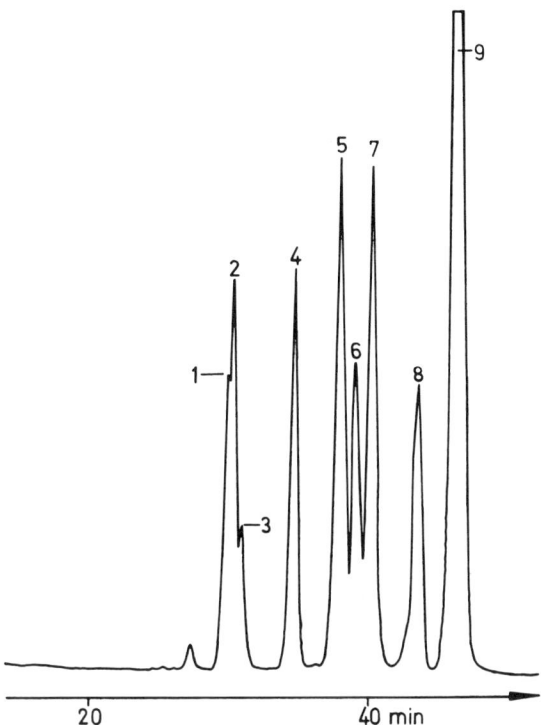

**Fig. 3.** Separation of *retro*-vitamin A acetate isomers from vitamin A acetate isomers on $Al_2O_3$. Mobile phase: hexane plus 1% diethyl ether. (1) 11-*cis*-Vitamin A acetate; (2) 13-*cis*-vitamin A acetate; (3) 11,13-di-*cis*-vitamin A acetate; (4) 9-*cis*-vitamin A acetate; (6) all-*trans*-vitamin A acetate; (5), (7), (8), and (9) *retro*-vitamin A acetate isomers. [With permission, from Vecchi *et al.* (1973).]

and Gibbs, 1977; Ruckemann, 1981); cereals (Dennison and Kirk, 1977); margarine, milk, and infant formula (Thompson and Maxwell, 1977; Mankel, 1979); dairy products (Bui-Nguyen and Blanc, 1980); and concentrates and premixes (Ranfft and Ruckemann, 1978a). Various columns have been used with these methods, including reverse-phase, silica, and alumina.

The extraction of unsaponifiable lipids is laborious, hazardous, and expensive, especially when large volumes of solvent are involved. As HPLC measurements can be made on relatively small amounts of extract, there is a temptation to reduce the scale of all steps in the analysis and thus the cost and effort. A problem then arises in ensuring that the small portion of the sample used in the analysis is truly representative of the whole. Liquids such as milk are easily mixed, and fats such as margarine can be sampled after melting or dissolving in hexane. The analysis of other foods, however, is more difficult. Egberg *et al.*

(1977) and Dennison and Kirk (1977) mixed or ground their samples and then took small portions for analysis. Because of the instability of vitamin A, this approach can involve losses; moreover, moist samples are difficult to mix at room temperature. A safer method is that proposed by Van De Weerdhof et al. (1973); they homogenized foods in liquid nitrogen and then removed 2.5-g portions for analysis.

In this author's opinion, it is better to saponify a reasonably large sample (10–20 g), extract the unsaponifiable lipid with hexane, and then evaporate small portions of the extract for HPLC on a silica column.

A different approach can be used when the HPLC involves a reverse-phase system. Egberg et al. (1977) neutralized the digest after saponification with acetic acid and then diluted the product with acetonitrile. Fatty acids and soaps precipitated, and after centrifugation, filtration, and dilution, a clear solution was obtained which could be injected directly into the HPLC system. An example of the analysis of butter by this technique is shown in Fig. 4.

Although most early workers with HPLC took the orthodox approach to vitamin A analysis and performed chromatography after saponification, it has recently been realized that it is possible, with some samples at least, to measure the vitamin directly in the form of the ester. Widicus and Kirk (1979) measured retinyl palmitate in ethanol–chloroform extracts of fortified cereal products. The ester was separated on a silica column using either an absorbance or fluorescence detector. Similar methods have been proposed for margarine and milk. Aitzetmuller et al. (1979) prepared a solution of margarine in hexane; after passing it through a small column of sodium sulfate and sodium chloride, they applied it to a HPLC column of silica and measured the vitamin A ester. It is also possible to prepare hexane extracts of fortified milk and measure the vitamin A ester directly by HPLC on silica (Thompson et al., 1980; Woollard and Woollard, 1981). A minor advantage of this approach is that the different esters in commercial use can be identified; margarine, for example, may contain palmitate, acetate, or propionate. A major advantage is that with careful design evaporation can be avoided. Often during vitamin A analysis, some of the vitamin is destroyed when extracts are evaporated to dryness. A silica column is preferable to a reverse-phase system for the direct estimation of vitamin A esters because the solvent system, usually mainly hexane, readily dissolves triglycerides. Moreover, silica columns seem to tolerate relatively large amounts of triglycerides.

Foppa (1981), however, has measured retinyl palmitate in vitaminized oils by reverse-phase HPLC using a mixture of chloroform and methanol as solvent system. Previously, Landen and Eitenmiller (1979) described a method for measuring vitamin A esters in margarine by reverse-phase chromatography. They found that it was necessary to remove triglycerides from their extracts by preliminary gel-permeation chromatography, and consequently their method is relatively laborious.

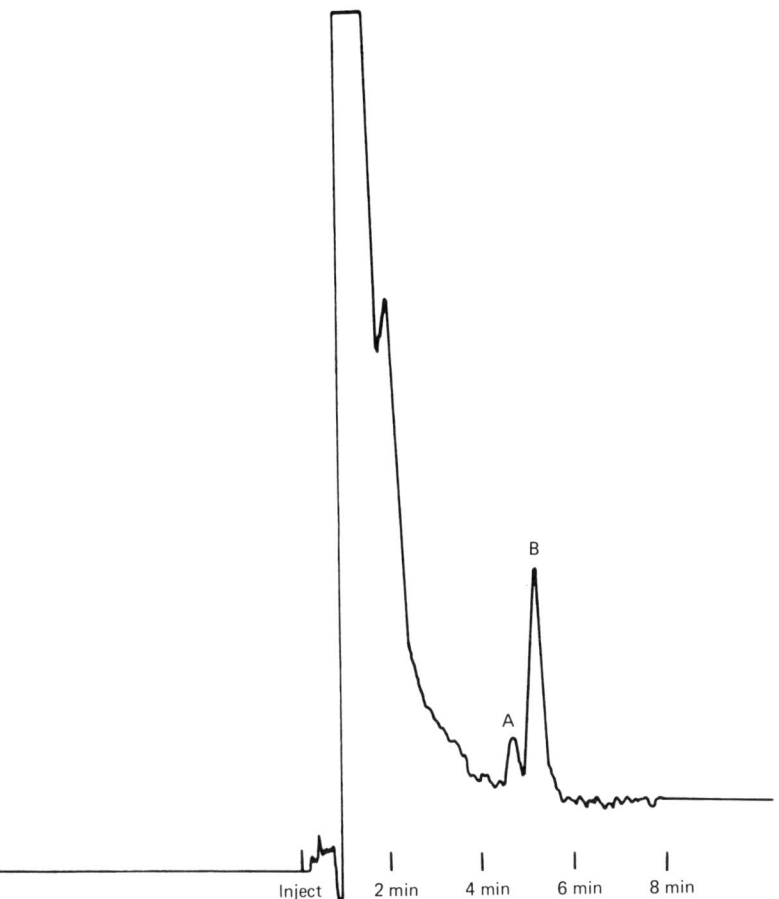

**Fig. 4.** Chromatogram of butter extract on ODS column, 0.01 AUFS, in acetonitrile–water, flow rate 1.5 ml/min. Peak A, 0.002 µg 13-*cis*-retinol; peak B, 0.011 µg of all-*trans*-retinol injected. [Reprinted with permission from Egberg *et al.* (1977). *J. Agric. Food Chem.* **25**, 1127–1132. Copyright 1977, American Chemical Society.]

## 6. Retinol, Retinyl Esters, and Retinoic Acid in Blood and Tissues

Retinol and its esters can be determined in tissues by a wide variety of HPLC procedures employing straight-phase or reverse-phase columns. Trace amounts of the vitamin can be measured by absorbance or fluorometric detectors. Extraction can be performed with or without saponification.

The routine analysis of serum for vitamin A can be performed by HPLC, and this is potentially the most reliable and convenient procedure, especially if automated. As in all types of vitamin A analysis, evaporation should be avoided whenever possible; this can easily be achieved in HPLC methods. Retinol is conveniently extracted from serum (denatured with alcohol) in hexane; thus silica columns, rather than reverse-phase columns, are compatible. An absorbance detector at 325 nm (or 313 nm) provides sufficient sensitivity and specificity. This procedure is basically that described by De Ruyter and De Leenheer (1976). Abe et al. (1977) proposed the use of a fluorescence detector, but this seems to have been an unnecessary refinement, especially as they failed to exploit the increased sensitivity to avoid evaporation. Bieri et al. (1979) also evaporated extracts and then subjected them to reverse-phase chromatography in spite of problems in dissolving the sample. The lipids were in fact dissolved in a mixture of diethyl ether and methanol, which, because of its volatility, required the use of an internal standard. As tocopherols were determined at the same time, a detector at 280 nm was used, with a loss of specificity for vitamin A.

Vahlquist (1980) has measured retinol in human skin after saponification and reverse-phase HPLC. A substance with some of the properties of 3-dehydroretinol was isolated from individuals with psoriasis.

The determination of retinoic acid is a problem of considerably greater magnitude; it had never been satisfactorily solved prior to the development of HPLC. Puglisi and De Silva (1978) describe a method for measuring 13-*cis*-retinoic acid, all-*trans*-retinoic acid, and other retinoids in blood after these substances have been administered orally. Whole blood was extracted with ether or ethyl acetate; after evaporation, the lipids were chromatographed on Partisil (10 μm) in methylene chloride containing 0.5–1% glacial acetic acid (the author cautions that this mixture forms HCl, which will damage equipment). It was possible to measure 10–20 ng/ml retinoic acid in blood.

Frolik et al. (1978a) found silica columns inconvenient for the analysis of common forms of vitamin A and their metabolites because of the wide range of polarity exhibited by these substances. Gradients were needed for satisfactory separations and then time-consuming reconditioning of the adsorbents. Alumina columns held the retinoic acid so strongly that it could not be eluted with common solvents. Better results were obtained with reverse-phase columns. Radioactively labeled retinoic acid, retinol, retinal, and retinyl acetate were separated on Spherisorb-ODS in acetonitrile–1% aqueous ammonium acetate (60:40). A Partisil ODS-2 column was preferred for retinoic acid and polar metabolites, and good separation of 13-*cis*- and all-*trans*-retinoic acid was achieved. Ammonium acetate prevented broadening of retinoic acid peaks; a 1% solution was necessary with Partisil ODS-2, but a 0.2% solution was adequate with Spherisorb-ODS. The same system was used to measure retinol and retinyl esters in liver and intestine and to investigate the metabolism of two synthetic retinoids, *N*-acetylretinylamine and retinylacetylhydrazone (Roberts et al., 1978). Several solvent systems were investigated for the extraction of lyophilized tissues. It was discovered that a serious disadvantage of the reverse-phase mode was the difficulty with which retinyl esters dissolved in methanol or acetonitrile. Besner *et al.* (1980) have also measured 13-*cis*-retinoic acid in plasma on a reverse-phase column.

Frolik *et al.* (1978b) applied their procedure to the measurement of 13-*cis*- and all-*trans*-retinoic acid in blood from humans given oral doses of these substances. Ascorbic acid and EDTA were added to plasma which was then lyophilized. The dried material was extracted with methanol containing 50 µg/ml BHT and then chromatographed directly on Partisil 10 ODS-2 in acetonitrile–1% ammonium acetate solution (65:35). It was possible to measure 50 ng/ml retinoic acid by this method, which was thus satisfactory for the investigation of subjects given oral doses of the acid but not for measuring normal circulating levels—which according to selected-peak-monitoring–GLC–mass-spectroscopy measurements of De Ruyter *et al.* (1979) are only 1–3 ng/ml.

De Ruyter *et al.* (1979) describe an HPLC method for measuring endogenous retinoic acid. They extracted plasma with hexane and discarded the extracted lipids. The residue was acidified and again extracted with hexane. After evaporation, the second hexane extract was chromatographed on 5-µm silica in hexane–acetonitrile–acetic acid (99.5:0.2:0.3) or on a DIOL column in hexane–dichloromethane–isopropanol (97.5:2.0:0.5) containing 0.4 or 0.6% acetic acid. It would be unwise, however, to assume that the small peaks sometimes observed in this simple method really represent endogenous retinoic acid.

## 7. *Metabolites of Retinol, Retinoic Acid, and Other Retinoids*

In spite of decades of research, only the first steps in the metabolism of vitamin A are understood. Briefly described, it has been established that vitamin A ingested in the diet is stored in the liver as retinyl esters. A small amount of retinol circulates in the blood, bound to special carrier proteins, and a trace of the aldehyde retinal is formed in the retina where it functions in vision. Retinal is probably oxidized to the acid, retinoic acid, but as the latter substance is difficult to detect in tissues, the significance of this step is unclear. Even when retinoic acid is administered to animals, it rapidly disappears; some is excreted in the bile as retinoyl glucuronides, but the fate of the remainder is obscure.

Early experiments with labeled retinol and retinoic acid indicated that a number of polar metabolites were formed in animals, but these substances could not be adequately purified by TLC or simple-column chromatography. The development of HPLC should undoubtedly rekindle interest in the study of vitamin A metabolism. Important clues have already been provided by Hanni *et al.* (1976), who identified several cyclohexanone derivatives as metabolites of radioactivity-labeled retinoic acid in rats. These workers, like many others in this field, were motivated by an interest in the use of retinoids, such as 13-*cis*-retinoic acid, in the treatment of cancer. They acidified urine and passed it through a column of Amberlite XAD-2. After washing the column with water, they eluted labeled

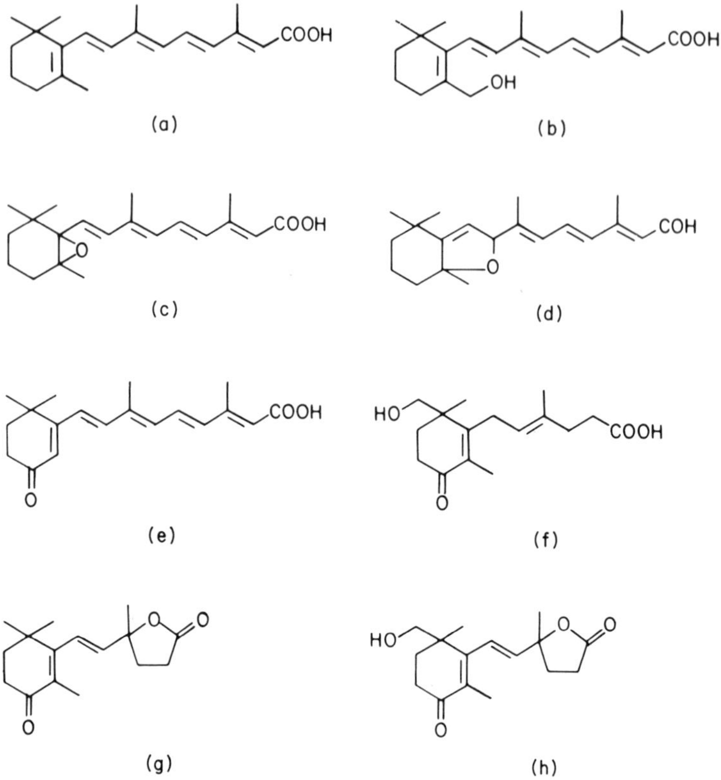

**Fig. 5.** Structures of substances suspected to be metabolites of vitamin A: (a) retinoic acid; (b) 18-hydroxyretinoic acid; (c) 5,6-epoxyretinoic acid; (d) 5,8-oxyretinoic acid; (e) 4-oxoretinoic acid; (f) metabolite 4; (g) metabolite 2; (h) metabolite 3.

metabolites with ethyl acetate and then purified them chromatographically on ordinary columns of silica and by TLC. The metabolites were finally separated by HPLC on Partisil-5 silica in pentane containing small amounts of acetonitrile, dichloromethane, or tetrahydrofuran. The structures of these metabolites, designated by the numerals 2, 3, and 4, are shown in Fig. 5. Three other metabolites, 4-oxoretinoic acid, 18-hydroxyretinoic acid, and 9-*cis*-18-hydroxyretinoic acid, have been isolated from feces (Hanni and Bigler, 1977). McCormick *et al.* (1978a) also described the HPLC separation of 4-hydroxy- and 4-oxoretinoic acid on a reverse-phase column in methanol containing 0.01 $M$ ammonium acetate. More recently, HPLC methods have been used to demonstrate the formation of 4-hydroxy- and 4-oxoretinoic acid from tritiated retinoic acid in tracheal organ cultures (Frolick *et al.*, 1979) and hamster-liver microsomes (Roberts *et al.*, 1980). Two polar metabolites of retinoic acid in rat intestine have been identified as 5,6-epoxyretinoic acid and 5,8-oxyretinoic acid (McCormick *et al.*, 1978b, 1979).

The possibilities for research in this area are indicated by the successes achieved in pharmaceutical laboratories in studies of the metabolism of drugs structurally related to vitamin A: 18 metabolites of an aromatic analog of retinoic acid have been isolated from human urine (Hanni *et al.*, 1977), and an assay has been developed which is applicable to plasma (Hanni *et al.*, 1979).

## B. Vitamin D

### 1. Introduction

The structure of the most important form of vitamin D, known as cholecalciferol or vitamin $D_3$, is shown in Fig. 6. Another common form, produced industrially and perhaps sometimes naturally, is ergocalciferol, or vitamin $D_2$; it differs structurally from cholecalciferol in that there is a double bond between carbons 22 and 23 and there is an additional methyl group attached to carbon 24.

Related substances of interest to the analyst include the provitamins 7-dehydrocholesterol and ergosterol, in which carbons 10 and 9 are linked as in sterols. The provitamins are converted to vitamins when exposed to UV light; this conversion takes place both *in vitro* and *in vivo* and it is exploited commercially during the production of vitamin concentrates or "resins." Intermediary products, such as lumisterols, tachysterols, and previtamins, may therefore contaminate commercial products. The vitamins themselves form toxisterols, lumicalciferols, and suprasterols when irradiated (Fig. 7).

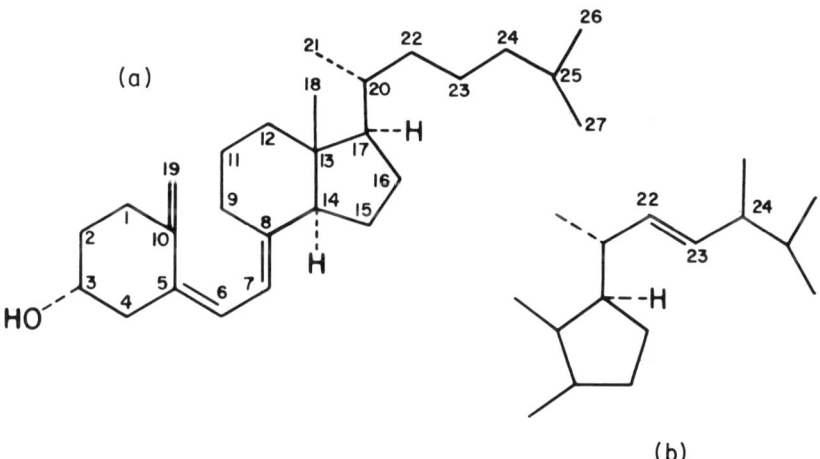

**Fig. 6.** Structures of (a) vitamin $D_3$ and (b) vitamin $D_2$.

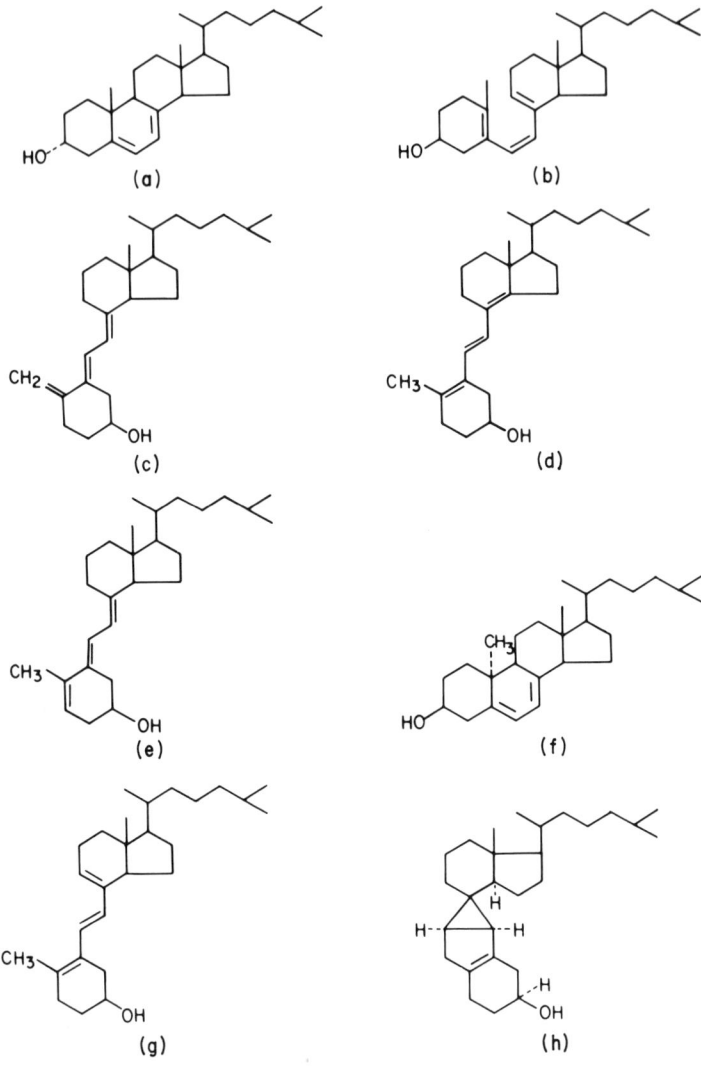

**Fig. 7.** Structures of isomers of cholecalciferol (vitamin $D_3$) and related substances: (a) 7-dehydrocholesterol (provitamin $D_3$); (b) previtamin $D_3$; (c) 5,6-*trans*-vitamin $D_3$; (d) isotachysterol$_3$; (e) isovitamin $D_3$; (f) lumisterol; (g) tachysterol$_3$; (h) suprasterol II.

Vitamin D can be converted by the action of light, heat, or acidity to isomers with differing biological and chemical properties. Traces of acid destroy the vitamin by promoting, among other changes, isomerization to 5,6-*trans*-vitamin D, isotachysterol, and isovitamin D. Heat promotes the formation of the isomer *pre*vitamin D (which should not be confused with *pro*vitamin D).

In many respects, vitamin D is similar to common, naturally occurring sterols from which it must often be separated during analysis. A characteristic feature—in the absence of which analysis by HPLC would be extremely difficult—is the cis-triene structure which is responsible for the absorption at 265 nm. The absorption spectrum is atypical; it differs markedly from even the 5,6-trans isomer, which has a single peak at 273 nm. The spectra of isotachysterol, isovitamin D, and tachysterol are similar to spectra of other trienes and have fine structure with maxima at longer wavelengths (270–300 nm). The previtamins have single maxima at 260 nm (Shaw *et al.*, 1957).

The UV absorption has been used for decades to measure the levels of the vitamin in potent sources, but it was considered to be far too nonspecific for the analysis of foods. It is now almost always the basis of detection in HPLC. So far, most analytical work has been done with an absorbance detector set at 254 nm or 265 nm. However, the absorption spectrum of vitamin D extends to 290 nm, and there is less interference—and little reduction in sensitivity—when the HPLC detector is set at 280 nm (Lofty *et al.*, 1981).

Seamark *et al.* (1980) converted vitamin D and its metabolites to isotachysterol derivatives before HPLC. The vitamin can also be converted to colored products (Nair, 1966; Kodicek and Lawson, 1967) by treatment with antimony trichloride, mercuric *p*-chlorobenzoate, iodine, or glycerol dichlorohydrin; and, although native fluorescence has not been reported, fluorescent derivatives are formed in strong acids. These reactions have not been exploited so far for detection during HPLC.

According to early bioassays, few foods contain more than trace amounts of naturally occurring vitamin D. Fish tissues are an exception, containing typically 0.02–0.1 μg/g, and oils derived from fish livers are very potent sources. Otherwise, the main foods containing vitamin D are egg yolk (0.07 μg/g), butter (0.02 μg/g), cheese (0.01 μg/g), and liver (0.01 μg/g). Although meats are usually considered to contain little vitamin D (0.002 μg/g), bioassays (Kummerow *et al.*, 1976) have revealed significant amounts in lean beef (0.005 μg/g), beef fat (0.015 μg/g), chicken breast (0.02 μg/g), chicken liver (0.095 μg/g), and liver sausage (0.03 μg/g). Naturally occurring vitamin D is often believed to contribute little to the total vitamin D intake, but rough calculations indicate that in North America typical adults obtain up to 40% (1.5–4 μg/day) from this source. A larger quantity of vitamin D is obtained from fortified foods, the most important of which is milk. Milk is usually fortified with 0.01 μg/ml, which represents a tenfold increase over the natural level. Other foods have also been fortified from time to time, but fears of excessive intakes have led to limitation of this practice. Infant formulas and other special products that are intended to provide all essential nutrients must also contain vitamin D. Margarine is another common food to which vitamin D is sometimes added.

As animal feeds are often fortified with vitamin D, methods are needed for concentrates, premixes, and finished feeds. The relatively large quantities of interfering substances in foods and feeds have until recently prevented the use of physicochemical methods. In fact, the only reliable method for measuring vitamin D in foods, before the use of HPLC, was the expensive and time-consuming bioassay with rats or chicks. Color reactions and column chromatography have been investigated frequently but have proved to be barely adequate for the measurement in vitamin D in concentrates and pharmaceuticals. Application of these procedures to foods has required many steps of purification, too laborious for application to routine work (Parrish, 1979).

## 2. Cholecalciferol and Ergocalciferol

These two important forms of vitamin D differ only in the structure of their side chains. They have slightly different retention times during HPLC on silica, but they are well separated in reverse-phase systems. This otherwise difficult separation was first demonstrated by Williams *et al.* (1972) on a Permaphase-ODS column with methanol–water (78:22). Equivalent or better separations have since been achieved in methanol–water mixtures on Vydac reverse-phase (Osadca and Araujo, 1977) or Zorbax-ODS (Jones, 1978) columns and in acetonitrile–methanol mixtures on a Spherisorb 10 ODS column (unpublished data). Excellent separation can be obtained in methanol on a Waters Radial-Pak 5 μm C-18 cartridge.

## 3. Isomers of Cholecalciferol

*a. Previtamins.* Solutions of vitamin D at room temperature always contain the isomer previtamin D. When equilibrated at 20°C, the ratio of vitamin to previtamin is 93:7. More previtamin is formed at higher temperatures, and the interconversion is faster: At 100°C equilibrium is attained in less than 30 min, and 28% of the mixture is previtamin; whereas at 0°C no more than 4% previtamin is formed, and conversion is protracted over many months. The kinetics are discussed by Keverling Buisman *et al.* (1968), but as their data were obtained by spectrophotometry it would be worthwhile to reexamine the reaction using HPLC. This isomerization is important to the analyst because the previtamin is surprisingly easily separated from the vitamin during chromatography. Moreover, the previtamin is almost always present in standards and samples, and its existence must somehow be allowed for during the calculation of the results of an analysis.

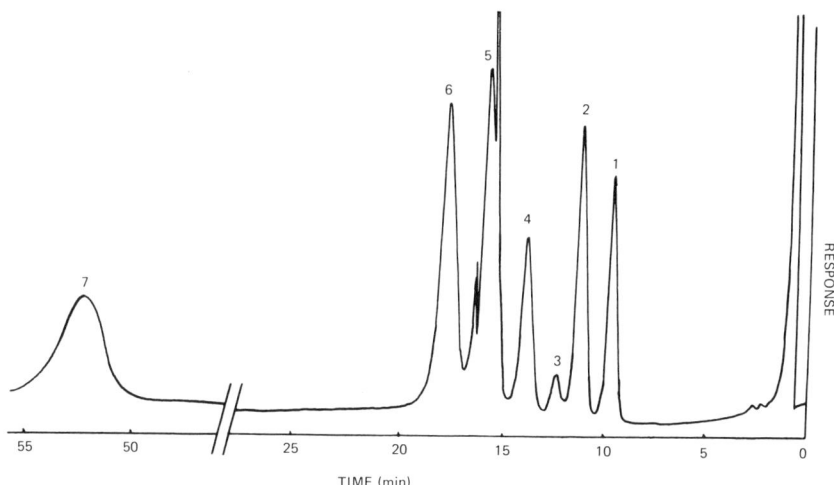

**Fig. 8.** Reverse-phase HPLC separation in acetonitrile–propionitrile–water of (1) *trans*-vitamin D$_3$; (2) isotachysterol$_3$; (3) previtamin D$_3$; (4) tachysterol$_3$; (5) vitamin D$_3$; (6) lumisterol$_3$; (7) 7-dehydrocholesterol. [With permission, from De Vries *et al.* (1979).]

Vitamin D and previtamin D separate readily during chromatography in spite of their structural similarity. The separation was formerly performed by TLC (Hanewald *et al.*, 1968) and later by chromatography on columns of HAPS* (Krol *et al.*, 1972). Early work with HPLC revealed that the vitamin and previtamin were widely separated on silica (Vydac) in hydrocarbon solvents containing small amounts of an ether (tetrahydrofuran), the retention time of the previtamin being less than two-thirds of that of the vitamin (Krol *et al.*, 1972). The vitamin is also eluted after the previtamin on reverse-phase columns (Fig. 8, peaks 3 and 5).

Similar thermal isomerization occurs in hydroxylated metabolites. The separation of prehydroxyvitamins has been investigated by Vanhaelen-Fastré and Vanhaelen (1979).

*b. Isomers in Resins.* Colorimetric methods for measuring the potency of resins (i.e., commercial sources of vitamin D produced by irradiation of provitamins) often fail to discriminate between active and inactive isomers; costly biological assays were therefore needed during the manufacture of vitamin D. There has thus been little hesitation in exploiting HPLC techniques. Steuerle (1975) reported the complete separation of the important constituents of vitamin D resins; thus, previtamin D$_3$, lumisterol, tachysterol, vitamin D$_3$, and 7-de-

---

*Hydroxyalkoxypropyl Sephadex, a lipophilic gel (Nystrom and Sjovall, 1975).

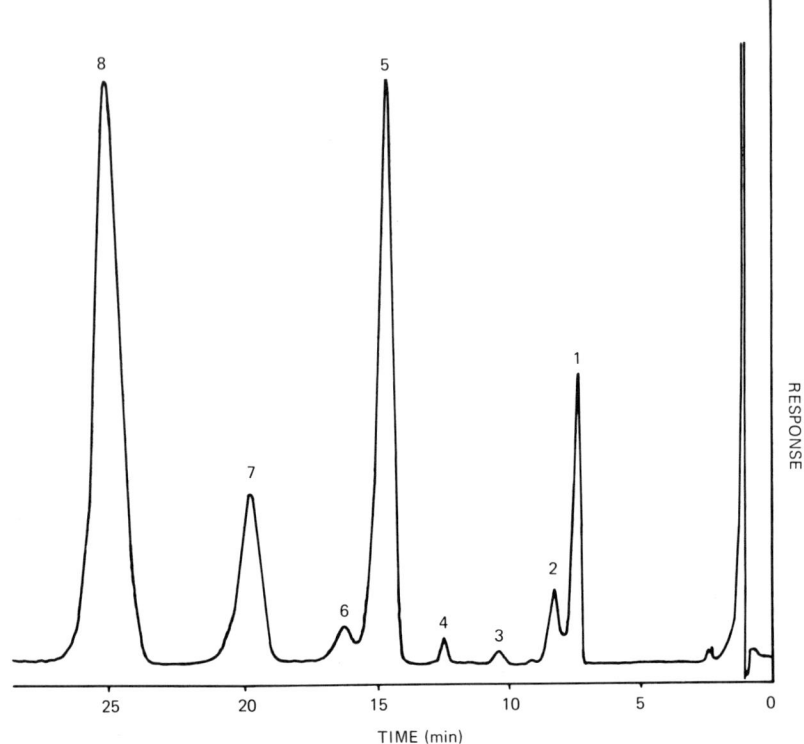

**Fig. 9.** Straight-phase HPLC separation of: (1) previtamin $D_3$; (2) *trans*-vitamin $D_3$; (3) lumisterol$_3$; (4) isotachysterol$_3$; (5) vitamin $D_3$; (6) tachysterol$_3$; (7) 4,6-cholestadienol; (8) 7-dehydrocholesterol. [With permission, from De Vries *et al.* (1979).]

hydrocholesterol were separated on alumina (Merck Alox T, 5 μm) using chloroform as solvent. The separation of tachysterol was a special achievement, as it is difficult by other procedures. Hofsass *et al.* (1976) separated in addition pyrocalciferol, isotachysterol, and isopyrocalciferol on silica (Partisil, 10 μm) in isooctane–ethanol (98.4 : 1.6). Small adjustments to the solvent systems and the use of finer packings (LiChrosorb Si60, 5 μm) have allowed the separation of *trans*-vitamin D from previtamin D. Thus De Vries *et al.* (1979) recommend hexane–amyl alcohol (99.7 : 0.3) as solvent system (Fig. 9). The application of this methodology to the analysis of resins has been studied collaboratively (Mulder *et al.*, 1979). Hofsass (private communication) has found that the addition of traces of water to a solvent system containing chloroform–hexane–ethyl acetate (70 : 30 : 2) has a dramatic effect on the separation of *trans*-vitamin D, previtamin D, and lumisterol, the optimum addition being 0.015%.

Although the isomers of vitamin D are usually separated on silica columns, it is interesting to note that they can also be separated on reverse-phase columns, but the order of elution is quite different and is not merely reversed. De Vries *et al.* (1979) eluted in sequence *trans*-vitamin D, isotachysterol, previtamin D, lumisterol, and 7-dehydrocholesterol (Fig. 8) from LiChrosorb RP18, 10 μm with acetonitrile–propionitrile–water (79:15:6).

## 4. Hydroxylated Metabolites

Hydroxylated metabolites of biological and medical interest include 25-hydroxycholecalciferol, 1α,25-dihydroxycholecalciferol, 24R,25-dihydroxycholecalciferol, 1α,24R,25-trihydroxycholecalciferol, and corresponding structures of the ergocalciferol (vitamin $D_2$) series. The metabolites of cholecalciferol (vitamin $D_3$) are usually abbreviated as follows: 25-$(OH)D_3$; 1,25-$(OH)_2D_3$; 24,25-$(OH)_2D_3$; and 1,24,25-$(OH)_3D_3$. Similar abbreviations are used for the ergocalciferol (vitamin $D_2$) series: e.g., 1,24,25-$(OH)_3D_2$. It should be noted that the calciferols already contain a hydroxyl group; therefore there are two hydroxyls in "hydroxy" metabolites, three in "dihydroxy," and four in "trihydroxy."

Matthews *et al.* (1974) reported the separation of 1,25-$(OH)_2D_3$, 25,26-$(OH)_2D_3$, 25-$(OH)D_3$, and 1α-$(OH)D_3$ in reverse-phase systems with water–methanol gradients. The use of silica columns has been pioneered by Jones and DeLuca (1975) and Jones (1980). Jones and DeLuca found that it was difficult to separate isocratically all the metabolites of interest in a reasonable time because of the large differences in polarity. A useful compromise was obtained with two 25-cm columns of Zorbax-SIL eluted with isopropanol–hexane (1:9), but different proportions of isopropanol were recommended for particular separations. The order of elution was: $D_2$ + $D_3$; 25-$(OH)D_2$; 25-$(OH)D_3$; 24,25-$(OH)_2D_2$; 24,25-$(OH)_2D_3$; 1-$(OH)D_2$ + 1-$(OH)D_3$; 25,26-$(OH)_2D_2$; 25,26-$(OH)_2D_3$; 1,25-$(OH)_2D_2$; and 1,25-$(OH)_2D_3$. It is interesting to note that 1-$(OH)D_2$ and 1-$(OH)D_3$ were held tenaciously by the column and that when eluted they were not separated from each other. It appears that a hydroxyl group must be introduced into the side chain before separation of the vitamin $D_2$ and vitamin $D_3$ series occurs on silica. Jones (1980) has reported that ternary solvent systems (hexane–isopropanol–methanol) give better separations because there is less tailing of the peaks. Ikekawa and Koizumi (1976) describe methods for separating C-24 epimers of 1,24-$(OH)_2D_3$ and 1,24,25-$(OH)_3D_3$ on Zorbax-SIL. Epimers of 24-$(OH)D_3$ and 24,25-$(OH)_2D_3$ were separated after conversion to trimethylsilyl derivatives.

Vanhaelen-Fastré and Vanhaelen (1979) have compared the separation of hydroxyvitamins from the corresponding prehydroxyvitamins in straight-phase (LiChrosorb Si60) and reverse-phase (μBondapak C-18) systems. Good separation was achieved only on silica columns.

## 5. Foods and Feeds

*a. Methods Not Involving Saponification.* Most HPLC methods for measuring vitamin D in foods have a saponification step which not only reduces the bulk of the sample by eliminating glycerides but also ensures complete dissolution of the sample, liberates the vitamin from its protective coating, and destroys pigments such as the chlorophylls which otherwise complicate the chromatography. The value of saponification is illustrated by the problems that are encountered in methods which omit this step.

Ray *et al.* (1977) extracted feed supplements with chloroform–methanol (9 : 1) and, after filtration of the extracts, performed HPLC on silica. They discovered that samples containing gelatin-coated beadlets had to be treated with alkali to release the vitamin. In spite of the high levels of vitamin in the samples, their published chromatograms indicate that separation from interfering substances was incomplete and accurate quantitation would be difficult. Better results were obtained after a second chromatography on a reverse-phase column (Fig. 10).

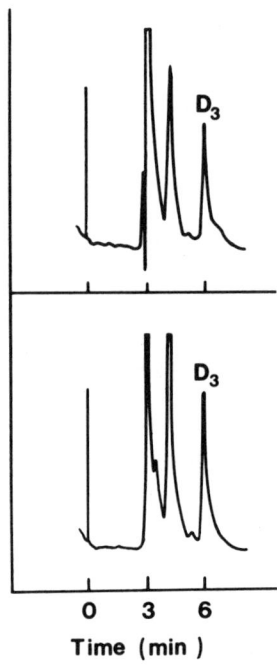

**Fig. 10.** HPLC analysis of feed supplement. Rechromatography of fraction from μPorasil column on μBondapak C-18. Detectors set at two wavelengths: 280 nm (top) and 254 nm (bottom). [With permission, from Ray *et al.* (1977).]

Cohen and Lapointe (1978) devised a method of extraction and cleanup for the measurement of fat-soluble vitamins in animal feeds and premixes. The sample was first extracted with dioxane–isooctane (1:4) for 90 min. As other workers (Tomkins and Tscherne, 1974; Eriksen, 1980) were unable to extract vitamins from certain commercial preparations, such as gelatin-coated beadlets, with organic solvents, the applicability of the method of Cohen and Lapointe would appear to be limited. Moreover, as recovery tests were reported for feeds spiked with vitamin already in solution, the efficiency of extraction is in any case undemonstrated. Chlorophyll was removed from the extract by treatment with an amine (tetraethylene pentamine), and the preparation was then extracted with acetonitrile–isooctane, washed with sodium phosphate solution (to remove minerals and residual amine), and filtered. HPLC was performed on a LiChrosorb $NH_2$ column after preliminary chromatography on a column of silica (Cohen and Lapointe, 1979).

More recently, Cohen and Lapointe (1980) described a new method for feeds, in which the sample was first treated with sodium phosphate solution and dichloromethane to release and dissolve the vitamins. The extract was then passed through a Waters Sep-Pak (silica) and then chromatographed first on Sephadex-LH20 and second on Partisil-10; finally, the vitamin was measured during HPLC on a LiChrosorb $NH_2$ column. Similar procedures have been described for premixes (Lein et al., 1980) and nonfat dried milk (Cohen and Wakeford, 1980).

Lofty et al. (1981) employed Waters Sep-Pak cartridges for precolumn cleanup of extracts of oils, concentrates, and gelatin-protected beadlets.

*b. Methods Involving Saponification.* When considering the use of saponification, it is important to remember that heat isomerizes vitamin D to previtamin. Recoveries from samples refluxed in ethanol, as in hot saponification, are consistently 10–20% low (Thompson et al., 1977; Takada et al., 1979) because the previtamin is separated from the vitamin in most chromatography systems and is then usually discarded or ignored. Some authors have overlooked this source of error in the design of their methods, and their failure to notice its effects later in tests is puzzling. Others have obtained correction factors by passing standards through the entire procedure (Egaas and Lambertsen, 1979; Ali, 1978) or by analysis of heated solutions of the vitamin (Knapstein et al., 1979; Takada et al., 1979). One way of eliminating the need for factors in calculations and other extra work would be to heat the standard solutions used in HPLC. The author prefers, however, to avoid the problem in the first place by saponifying at room temperature (Thompson et al., 1977; Henderson and Wickroski, 1978).

Various cleanup procedures have been used after saponification before one or more steps of HPLC. Antalick et al. (1977) saponified foods, concentrates, and premixes, and extracted the unsaponifiable lipids with petroleum ether and passed them through columns of magnesia and alumina before HLPC on silica. A lower limit of 2 IU/g was claimed for the method.

Chromatography of unsaponifiable lipids from fortified milk on a column of HAPS yields a fraction containing vitamin D which can be resolved by HPLC on silica (Thompson et al., 1977). In another method for fortified milk, unsaponifiable lipids were purified by chromatography on a column of alumina before HPLC in a reverse-phase system (Henderson and Wickroski, 1978).

Ruckemann and Ranfft (1977) also employed a reverse-phase system in the analysis of premixes, but they used columns of Celite and Florex for cleanup. Knapstein et al. (1979) saponified poultry feeds and purified the unsaponifiable lipids by chromatography on alumina, TLC, and reverse-phase HPLC. The vitamin D was measured in a final HPLC on silica.

Egaas and Lambertsen (1979) saponified fish tissues and oils spiked with vitamin $D_2$ as an internal standard and removed sterols from the unsaponifiable lipids by crystallization from methanol. They purified the residue in several batches by HPLC on silica and pooled fractions containing vitamin D. A final reverse-phase chromatography completed the purification and in addition separated vitamin $D_3$ from the internal standard, vitamin $D_2$.

Ali (1978) saponified fish oils and removed sterols from the unsaponifiable lipids with digitonin. HPLC was performed on silica after filtration through Florex. Digitonin was also used by Adachi and Kobayashi (1979) in a method for measuring natural levels of vitamin D in milk. Lipid was extracted, saponified, treated with digitonin–Celite and subjected to "preparative TLC" before HPLC on silica.

The miscellaneous, time-consuming cleanup procedures described above will probably be abandoned in the near future in favor of short purification steps followed by two or more steps of HPLC. Van Niekerk and Smit (1980), for example, chromatographed the unsaponifiable lipids from fortified margarine twice on silica and once on a reverse-phase column. At the present time, the author measures vitamin D in fortified foods in four steps: (1) saponification, (2) filtration through a short column of alumina, (3) HPLC on a silica column (LiChrosorb Si60, 5 μm), and (4) HPLC on a reverse-phase column (Waters Radial-Pak).

*c. Calculation of Potency.* The analyst is actually faced with a number of decisions when calculating vitamin D potency. For example, it is debatable whether "vitamin D" should be considered to be pure calciferol, the equilibrium mixture at ambient temperature containing the previtamin, or some other combination of vitamin and previtamin. In the analysis of resins and concentrates, the levels of previtamin are often added to those of the vitamin to give the "potential vitamin D" (Hanewald et al., 1968), whereas in the analysis of foods, the previtamin is usually ignored.

Another decision to be made concerns the relative potencies of vitamins $D_2$ and $D_3$. Recent research has indicated that the metabolism of vitamin $D_2$ differs from that of vitamin $D_3$ in most species, including humans. At the present time, however, it is assumed in analytical work that the two vitamins are equivalent, and this is the basis of the use of international units. It is preferable to define the international unit as 65 pmol rather than 0.025 µg of the calciferols, which is the official definition, because there is certainly no evidence that molecule for molecule, vitamin $D_3$ (MW 385; 0.025 µg = 65 pmol) is less potent than vitamin $D_2$ (MW 397; 0.025 µg = 63 pmol). This new definition of the international unit was proposed by Norman (1972), but it has not been officially sanctioned or recognized. The effect of Norman's definition on analytical results may at first seem trivial (a mere 3%), but actually it can be combined conveniently with spectroscopic measurements, which are the basis of HPLC determinations, to simplify calculations and eliminate illogical correction factors. If it is assumed that the molecular extinction coefficients $\epsilon$ of vitamins $D_2$ and $D_3$ are both equal to 18,500 (Shaw et al., 1957; Havinga et al., 1955; Kodicek and Lawson, 1967), calculations will reveal that solutions in ethanol containing 100 IU/ml will always have an absorbance at 265 nm of 0.12. Similarly, it will be noticed during HPLC that only one factor is needed for both vitamins in the calculation of international units from the areas beneath peaks.

## 6. Vitamin D and Its Metabolites in Tissues

The metabolism of vitamin D (DeLuca, 1979) involves hydroxylation in the liver to 25-(OH)D, which then circulates in blood. This metabolite is converted in the kidneys to 1,25-$(OH)_2$D or 24,25-$(OH)_2$D.

Horst and Littledike (1979) measured vitamins $D_2$ and $D_3$ in methanol–methylene chloride extracts of bovine plasma by reverse-phase HPLC after cleanup on columns of LH20 and HAPS. Vitamin $D_3$ formed in skin exposed to light has been measured by HPLC on silica after saponification (Takada et al., 1979).

Although research with animals often involves the measurement of vitamin D in tissues, there is probably more interest, from a clinical point of view, in the levels of 25-(OH)D in blood. Until recently, 25-(OH)D was measured in blood and tissues by competitive-binding assays, but the levels (1–20 ng/ml) are high enough for direct measurement by HPLC.

Koshy and VanDerSlik (1976) obtained the first successful results with bovine serum, but the volume of sample needed for analysis (25 ml) was inconveniently large. The method has since been improved to measure 25-$(OH)D_2$ and 25-$(OH)D_3$ separately (Koshy and VanDerSlik, 1978) in only 2.5 ml cow serum (Koshy and VanDerSlik, 1980) or human serum (Koshy and VanDerSlik, 1977a; Koshy, 1980). A similar procedure has been described for 50-g samples of bovine liver, kidney, and muscle (Koshy and VanDerSlik, 1977b) and 20-g samples of chicken egg yolk (Koshy and VanDerSlik, 1979). In these methods, extracts were obtained and partitioned with various organic solvents. The remaining lipids were then purified on a short column of silica gel prepared in a Pasteur or serological pipette. After rinsing of the column with hexane–ether (1 : 1), the metabolites were eluted with ether–ethyl acetate (9 : 1). Finally, the extracts were purified by partition chromatography on Celite with a stationary phase of methanol–water (8 : 2) and a mobile phase of pentane. This column, in addition to purifying the extracts, separated 25-$(OH)D_3$ from 25-$(OH)D_2$. Preparations from some samples, such as egg yolk, had to be further purified at this stage by passage through a column of microparticulate silica in isopropanol–hexane (2.5 : 97.5). HPLC was finally performed on Zorbax-ODS in acetonitrile–methanol–water (95 : 5 : 5) or Zorbax-SIL in isopropanol–hexane (3 : 97).

Jones (1978) measured cholecalciferol ($D_3$), ergocalciferol ($D_2$), and the corresponding 25-hydroxy derivatives in human blood in two steps of HPLC. Plasma was extracted with a mixture of chloroform and methanol, and the evaporated

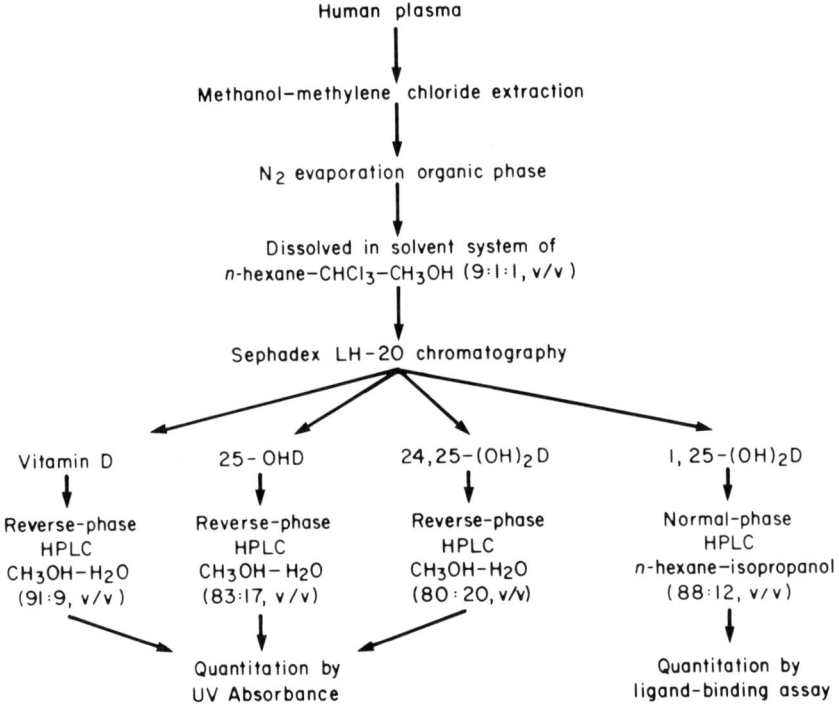

**Fig. 11.** Summary of the quantitation of vitamin D and three of its physiologically important metabolites. [With permission, from Lambert *et al.* (1977). Copyright 1977, Pergamon Press, Ltd.]

extract was chromatographed on Zorbax-SIL in isopropanol–hexane (5.5 : 94.5). Two fractions were collected containing the unchanged vitamins and the metabolites, respectively. Each was then rechromatographed on Zorbax-ODS to separate the $D_2$ and $D_3$ forms. Several groups have reported the use of Waters Sep-Pak cartridges in the measurement of 25-$(OH)D_2$ and 25-$(OH)D_3$ in blood (Kohl and Schaefer, 1981; Dabek *et al.*, 1981; Adams *et al.*, 1981).

A sensitive HPLC method for measuring 25-$(OH)D_3$ in rat blood and liver was developed during research on the formation of vitamin D during the exposure of animals to UV irradiation (Okano *et al.*, 1978). The method is of special interest in that it involves saponification. The unsaponifiable lipids were purified on a column of HAPS and by TLC before HPLC on Zorbax-SIL. Saponification was also used in the isolation of 25-$(OH)D_2$ from tissues of rats given ergocalciferol (Okano *et al.*, 1979).

HPLC has also been used to measure metabolites beyond 25-(OH)D, although the levels of these substances are extremely low, and after chromatography some can be detected and measured only with the assistance of competitive protein-binding assays. An outline of a procedure developed for human plasma by Lambert *et al.* (1977) is shown in Fig. 11. Another multiple assay for plasma, described by Shephard *et al.* (1979), involves purification on HAPS and two steps of HPLC. It has recently been used to investigate the levels of vitamin D and its metabolites in the tissues of rats given various quantities of the vitamin in the diet (Shephard and DeLuca, 1980).

## C. Vitamin E

### 1. Structure and Chemical Properties

Vitamin E is a naturally occurring antioxidant. According to "The National Formulary" (APA, 1975) it is free or esterified α-tocopherol (Fig. 12). Its structure includes three asymmetric carbon atoms (C-2, C-4', and C-8') and thus theoretically there are many sterioisomeric forms. Tocopherol molecules formed by nature are all of one configuration, probably $2R,4'R,8'R$, whereas those prepared by chemists are partially or entirely racemic, depending upon the origin of the phytol used in the synthesis.

Although changes of configuration at C-2 have a large effect on biological activity, the orientations at C-4' and C-8' seem to be unimportant. In practice, therefore, only two forms are distinguished: *d*-α-tocopherol, of natural origin, and *dl*-α-tocopherol, which is synthetic. The optical activity of *d*-α-tocopherol is actually very small, and this isomer is usually identified after oxidation with alkaline potassium ferricyanide to a product with higher specific rotation (Ames and Drury, 1975).

Vitamin E is often measured in international units, which take into account the different biological activities of the isomers and their esters. One unit is equiv-

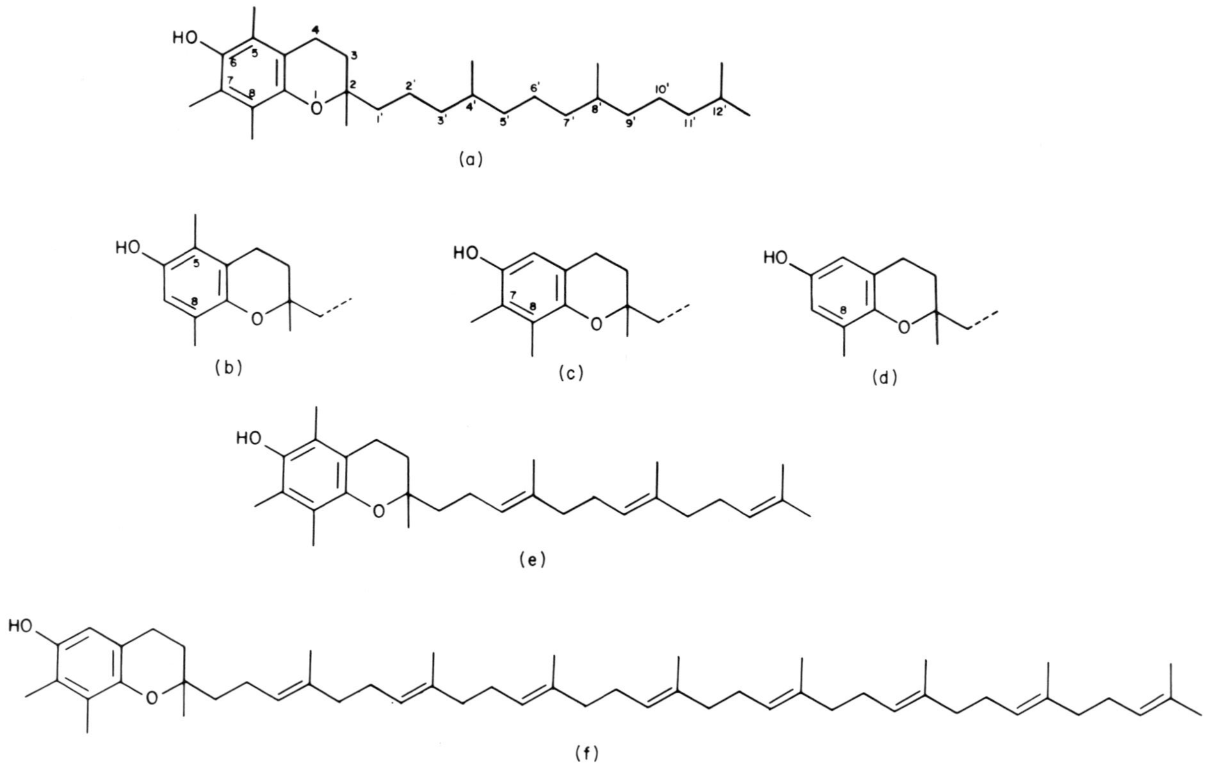

**Fig. 12.** Structures of tocopherols and related substances: (a)–(d) α-, β-, γ-, and δ-tocopherol, respectively; (e) α-tocotrienol; (f) plastochromanol-8.

alent to 0.67 mg $d$-α-tocopherol or 0.91 mg $dl$-α-tocopherol; factors for the esters are listed in "The National Formulary" (APA, 1975).

Some analysts are obliged to use the National Formulary definition of vitamin E (i.e., α-tocopherol), and many more adopt it because it is unambiguous and practical. Most nutritionists, however, extend the definition to include several naturally occurring substances which differ structurally from α-tocopherol because they lack one or more of the methyl groups attached to the ring or because they possess double bonds in the side chain (Fig. 12). Thus vitamin E is considered to be a family of four tocopherols, designated α, β, γ, and δ (abbreviated α-T, β-T, γ-T, and δ-T), and four corresponding tocotrienols (α-$T_3$, β-$T_3$, γ-$T_3$, and δ-$T_3$). These additional forms of the vitamin have lower biological activity than α-tocopherol, but some, such as γ-tocopherol, are sometimes present in foods in larger quantities. For this reason there is interest in measurement of all the secondary forms of vitamin E, which must in any case be separated from α-tocopherol during analysis.

Hitherto, the analysis for vitamin E was difficult, mainly because of its tendency to oxidize. The problem was compounded by the lack of a specific color reaction or other property which could be used for detection and quantitation. Most often vitamin E was measured in unsaponifiable lipids, before or after chromatography, using the Emmerie–Engel reaction which detects the presence of a liposoluble reducing agent by the rapid reduction of ferric ions to ferrous. The vitamins absorb, albeit weakly, UV light maximally at 292–295 nm. This region of the spectrum was considered to be too deep in the UV to be useful in trace analysis. They also fluoresce, but the emission at 310–330 nm is close to the excitation, and it is difficult to measure with simple filter fluorometers; there was therefore little interest in this property before the development of the variable-wavelength spectrophotofluorometer.

There are a number of naturally occurring isoprenoid lipids with structural similarities to the tocopherols that might be encountered during analysis, such as ubichromenols and solanachromene (Morton, 1965). Plastochromanol-8 (Fig. 12) is of special interest as it can be considered to be γ-tocotrienol with an elongated side chain; it occurs in plants and vegetable oils (Dunphy *et al.*, 1966), but it appears to be of no value to animals as a vitamin. Some form of chromatography is therefore usually necessary during analysis for vitamin E except with samples such as serum, in which it has been established that simple tests provide a reasonably accurate answer. In early work, tocopherols were separated by column chromatography using adsorbents such as alumina or secondary magnesium phosphate. Better separations were obtained by paper chromatography, using paper impregnated with zinc carbonate or coated with liquid paraffin (Green *et al.*, 1955), and later by TLC (Pennock *et al.*, 1964). Quantitation was difficult, however, because of losses which in TLC usually exceeded 20% (Hjarde *et al.*, 1973). The most recently developed technique prior to HPLC involved GLC, and it appears to have been the most successful (Slover et al., 1969); however, it required several steps of sample cleanup (Parrish, 1980a).

## 2. General Chromatographic Properties

The chromatographic properties of the tocopherols are related to the number and, to a lesser extent, the position of the methyl groups attached to the aromatic ring. Unsaturation in the side chain, as in the tocotrienols, is associated with an increase in polarity. These relationships emerged in early work with columns of alumina when three fractions were usually eluted: first, $\alpha$-T and $\alpha$-$T_3$; second, $\beta$-T, $\gamma$-T, and $\beta$-$T_3$; and finally, $\delta$-T, $\gamma$-$T_3$, and $\delta$-$T_3$. Thin-layer chromatography of the tocopherols and some synthetic positional isomers on silica revealed the following order of polarity (Kofler et al., 1962): 5,7,8-trimethyl- (i.e., $\alpha$-T); 5,7-dimethyl-; 5,8-dimethyl- (i.e., $\beta$-T); 7,8-dimethyl- (i.e., $\gamma$-T); 7-methyl-; 5-methyl-; and finally 8-methyltocol (i.e., $\delta$-T). The tocotrienols are more polar than the corresponding tocopherols, and differences in mobility were especially marked during reverse-phase paper chromatography (Green et al., 1955).

During HPLC on silica the sequence of elution of naturally occurring tocopherols and tocotrienols is $\alpha$-T, $\alpha$-$T_3$, $\beta$-T, $\gamma$-T, $\beta$-$T_3$, $\gamma$-$T_3$, $\delta$-T, and $\delta$-$T_3$ (Cavins and Inglett, 1974; Thompson and Hatina, 1979). The order of elution of tocopherols is reversed during HPLC with reverse-phase columns, but the separation is inferior (Vatassery et al., 1978); the behavior of tocotrienols has not been investigated.

There are no reports of separation of d and dl isomers during chromatography and it seems unlikely that it will be achieved. The separation of cis–trans isomers of tocotrienols may, however, be possible. Fujitani (1976) has investigated the HPLC separation of a number of dimers of $\gamma$- and $\delta$-tocopherols.

## 3. Tocopherols and Tocotrienols in Foods

An early example of vitamin E analysis with a closed-column chromatography system is that described by Hjarde et al. (1973). They connected a proportioning pump to a column of calcium phosphate, mixed the effluent with reagents needed for the Emmerie–Engel reaction, and measured the color continuously with a colorimeter equipped with a flow cell. Lipids extracted from foods were saponified, and the unsaponifiable residue was applied to the column. A gradient of diethyl ether in petroleum ether was used as eluting solvent, and common forms of vitamin E were separated over a period of 3 hr. Other substances observed during chromatography included plastochromanol, dimeric tocopherols, and the antioxidant ethoxyquin. The method was applicable to foods containing as little as 0.5 µg/g $\alpha$-tocopherol.

Absorbance detectors appear to be adequate for some samples, but they have insufficient specificity for general use. Cavins and Inglett (1974) used detectors set at 254 and 280 nm and separated tocopherols and tocotrienols in 2 hr on a 2-m

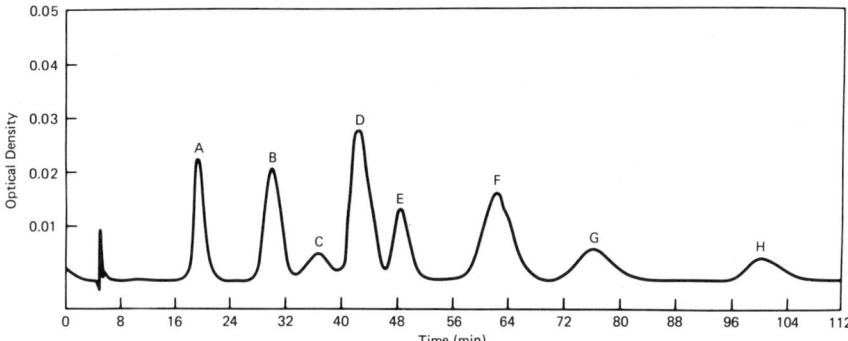

**Fig. 13.** Separation of tocopherol and tocotrienol standards. Absorbance detector at 254 nm; column, Corasil II, 2 m × 2 mm; mobile phase, 0.5% tetrahydrofuran in hexane. Peaks A, C, D, G: α-, β-, γ-, δ-tocopherol, respectively. Peaks B, E, F, H: α-, β-, γ-, δ-tocotrienol, respectively. [With permission, from Cavins and Inglett (1974).]

column of Corasil II using 0.5% tetrahydrofuran in hexane as solvent system (Fig. 13). They were able to measure the vitamin directly in some oils extracted from grains, but according to their published chromatograms, separation from interfering substances in wheat bran lipids, for example, was incomplete. Carpenter (1979) reported a similar method for vegetable oils using a 30-cm μPorasil column on which tocopherols were separated—but not to baseline—in 5–10 min. Pickston (1978) saponified milk substitutes and Mankel (1979) saponified margarines and vegetable oils before HPLC analysis in a reverse-phase system with an absorption detector. Hung et al. (1980) measured α-tocopherol in fish liver with an absorbance detector, but they did not confirm that their method of extraction was adequate. In any case, as the method involves the use of dioxane (a potential carcinogen) as extractant, it cannot be recommended.

Although the absorbance detector is satisfactory for the direct analysis of some samples, such as extracts of serum (Bieri *et al.*, 1979) and brain (Westerberg *et al.*, 1981) and solutions of vegetable oils, most samples of foods and tissue require one or more steps of purification before chromatography. Manipulation of small amounts of vitamin E is inevitably accompanied by oxidative destruction, and thus full advantage cannot be taken of the excellent recovery of HPLC techniques.

Fortunately, it is not necessary to undertake much purification of extracts before HPLC. A dramatic increase in specificity and sensitivity can be obtained with a fluorescence detector. There appear to be surprisingly few lipids in foods and tissues with the chromatographic properties and fluorescence characteristics of the tocopherols and tocotrienols. A fluorometric detector was first used in column chromatography of vitamin E with columns of HAPS (Thompson *et al.*,

1972). A specific method was developed for tocopherols in a variety of foods, but separation on the gel was time consuming. Van Niekerk (1973) used a fluorometer fitted with an interference filter as a detector in an early HPLC system and measured tocopherols in vegetable oils on a column of Corasil. Abe et al. (1975) also separated fluorescent peaks in vegetable oils by HPLC, but they used a true spectrofluorometer, rather than a simple filter fluorometer, as a detector. When a suitable spectrofluorometer is used as a detector, the sensitivity and specificity are such that tocopherols can be detected and measured directly in lipids extracted from foods (Thompson et al., 1979; Thompson and Hatina, 1979). This simple approach is possible not only with oils from grains, which are rich sources of vitamin E, but also with meats, vegetables, and complete meals (Fig. 14). As little as 2 µg tocopherol can be measured in 1 g fat. In this procedure, it is first necessary to extract lipids from foods with solvents such as ethanol or acetone. Polar lipids should be removed by dissolving the extracts in

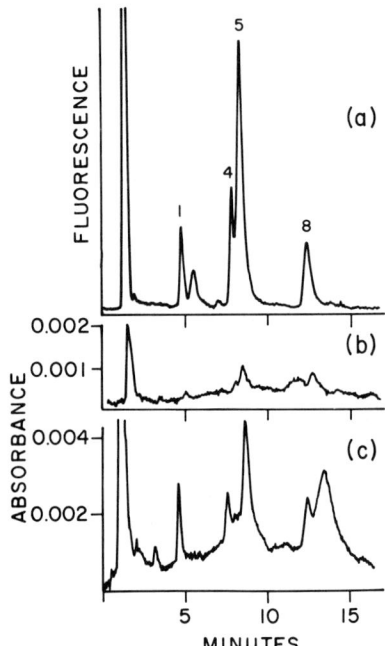

**Fig. 14.** HPLC of extract of TV dinner passed first through fluorescence detector (a) and second through absorbance detector (b). Fluorescence measured at 290 nm excitation and 330 nm emission; absorbance measured at 295 nm. (c) HPLC of concentrated extract using absorbance detector; note presence of interfering substances. Identification of peaks: (1) α-tocopherol; (4) antioxidant BHA; (5) γ-tocopherol; (8) δ-tocopherol. [From Thompson and Hatina (1979), p. 338, by courtesy of Marcel Dekker, Inc.]

hexane and washing this solution with aqueous ethanol or acetone. These steps can be combined to give an extraction procedure which is applicable to a wide variety of foods (Thompson and Hatina, 1979). The chromatography system must separate individual tocopherols and must also tolerate samples containing large amounts of triglyceride. Columns of silica, such as 5-μm LiChrosorb Si60, are recommended (1) because they provide excellent separation and (2) because the associated solvent systems are good solvents for the samples. Reverse-phase systems are inferior in both respects.

Prospective users of a fluorometric detector for vitamin E determinations should, however, be aware of the following:

1. The fluorescence detector must have sufficient energy and selectivity for the measurement of fluorescence at low wavelengths. Unfortunately, the performance of many simple fluorescence detectors is inadequate. So far, the best results have been obtained with modified spectrofluorometers (Abe et al., 1975; Thompson and Hatina, 1979).

2. The fluorescence of the tocopherols is markedly affected by solvents; it is enhanced by small amounts of alcohols and abolished by even traces of chlorinated hydrocarbons. Each tocopherol is affected differently (Thompson et al., 1972). These effects must be considered during the selection of solvent systems for HPLC, and problems in quantitation can be expected with gradients.

3. If substances with strong absorption at 280–320 nm are eluted during the chromatography at the same time as the tocopherols, fluorescence may be partially or completely quenched even though such substances are otherwise "invisible" to the detector. The effects are sometimes bizarre, and single peaks may appear to split into two. In the development of new methods or the examination of unfamiliar samples, it is advisable to monitor the effluent with an in-line absorbance detector set near 290 nm.

*4. Determination of Tocopheryl Acetate*

Esters of tocopherols and tocotrienols are rarely found in nature. When tocopherol is added to foods and feeds, however, it is almost always in the form of the acetate. The free tocopherols are less resistant to oxidation than the esters, and they are added to foods only when they are intended to serve as antioxidants. There is therefore a need to measure vitamin E acetate, which is added to foods in the form of a gelatin-coated beadlet. If the equipment is available, the vitamin added to foods can be determined by HPLC using a fluorescence detector as described already for naturally occurring tocopherols. The levels in fortified foods are high (5–500 μg/g), and measurement is therefore relatively straightforward. It is merely necessary to saponify the sample, or an extract, before chromatography to convert the weakly fluorescent esters to free tocopherol; an established procedure can be followed (Bunnell, 1967). This approach was recommended by Soderhjelm and Andersson (1978) and McMurray and Blanchflower (1979a). However, both groups employed reverse-phase HPLC; better results would probably have been obtained with a silica column. Soderhjelm and Andersson described the simultaneous determination of vitamins A and E, but the recovery of both vitamins was low. There seems to be a need for the systematic study of techniques for the saponification and extraction of tocopherol esters.

There are a few reports of the use of an absorbance detector. Shaikh et al. (1977a) published a method for three feeds which contained 1–3 mg/g tocopherol acetate, which is far above usual levels of fortification. Ruckemann and Ranfft (1978) saponified concentrates and supplements containing 300 µg/g tocopherol acetate and presumably measured the tocopherol at 254 nm.

Saponification does not seem to be necessary with cereal products fortified with 700 µg/g tocopherol acetate; the acetate and vitamin A palmitate can be measured simultaneously in extracts by HPLC on silica (Widicus and Kirk, 1979). It is important to verify that stabilized preparations of the vitamins are completely extracted. Eriksen (1980) recommended that feeds containing gelatin-coated beadlets be treated with water and alcohol before extraction with perchlorethylene. He filtered the extract and then applied it directly to a reverse-phase column. Levels as low as 50 µg/g were measurable. The accuracy of the method was calculated to be ±5%, but this optimistic figure is not supported by the appearance of the published chromatograms.

Foppa (1981) has measured tocopheryl acetate in vitaminized oils by direct chromatography on a reverse-phase column.

## 5. Measurement of Vitamin E in Serum and Tissues

Little is known about the metabolism of vitamin E, and analytical work has hitherto been confined to the measurement of tocopherol in blood serum or plasma, and much less frequently, tissues. Although the vitamin can be measured in the blood by simple colorimetric and fluorimetric methods, HPLC is advantageous because it eliminates doubts about interfering substances and allows the individual tocopherols to be discriminated. De Leenheer et al. (1978) used an absorbance detector and a reverse-phase column for serum analysis. Tocol was added as an internal standard. It was claimed that 0.6 µg/ml tocopherol serum could be measured by this method. Bieri et al. (1979) developed a procedure for the simultaneous determination of vitamins A and E that also utilized a reverse-phase column. Nilsson et al. (1978) used silica and reported a high limit of detection near 2.5 µg/ml.

A fluorescence detector provides much greater sensitivity. This was first demonstrated by Abe and Katsui (1975), who measured $\alpha$-, $\beta$-, $\gamma$-, and $\delta$-tocopherol in human and horse serum, the values being as low as 0.1 µg/ml. Jansson et al. (1980) reported that with a similar method, the minimum detectable level, which gave a response double that of the detector noise, was 40 ng/ml tocopherol.

Detector response can be increased in some instruments by using a shorter excitation wavelength, such as 205 nm (Hatam and Kayden, 1979). This involves a marked decrease in specificity, however, and an aggravation of quenching effects.

McMurray and Blanchflower (1979b) put high sensitivity to good use by eliminating the usual evaporation steps and measured tocopherol after direct injection of hexane extracts of bovine and porcine plasma. This approach not only saves time but also reduces losses.

The fluorescence detector is also ideal for the analysis of tissues. Reports have already appeared describing its use in the analysis of red blood cells (Mino *et al.*, 1979), platelets, and liver (Tangney *et al.*, 1981).

## D. Vitamin K

Vitamin K functions in mammals and birds in the formation of proteins involved in blood clotting. There are three forms of vitamin K. The first and most important is phylloquinone (vitamin $K_1$), which occurs in plants. It is a naphthoquinone with an aliphatic side chain containing 4 isoprenoid units, all except 1 of which are saturated (Fig. 15). In the side chain of natural phylloquinone, the double bond is in the trans configuration and the optically active centers (seventh and eleventh carbons) have the same configuration as natural phytol. The second form (vitamin $K_2$) is produced by microorganisms; it is thought to consist of a series of menaquinones with side chains containing 6–13 unsaturated isoprenoid units. A third group of vitamins includes synthetic substances with no side chains, such as menadione (vitamin $K_3$) and water-soluble derivatives, that are added to animal feeds. Phylloquinone and the menaquinones absorb UV light between 240 and 280 nm and thus are easily detected by conventional absorbance detectors during HPLC.

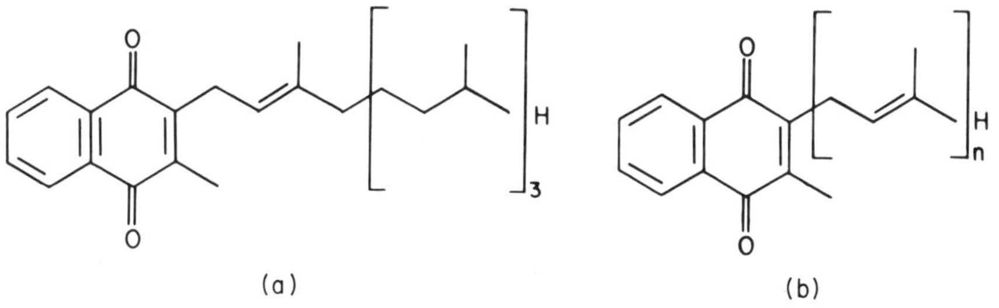

**Fig. 15.** Structures of (a) phylloquinone and (b) menaquinones.

Vitamin K may be accompanied in plant and animal tissues by a number of related naturally occurring quinones such as ubiquinones, plastoquinones, and various benzoquinones. The properties of these substances are reviewed in a monograph by Morton (1965). The small amount of information available concerning the levels of vitamin K in foods has been obtained by bioassays with chicks (Parrish, 1980b). Chemical analysis of foods rich in lipids is difficult because of the instability of the quinones. They are rapidly decomposed in alkali and thus it is not possible to isolate them from fat by saponification. A major problem in HPLC analysis is thus the presence of triglycerides, which require columns of substantial size for chromatography. Triglycerides can be removed by preliminary enzymatic hydrolysis in neutral solutions, a technique which has been applied in the analysis of infant formula (Barnett et al., 1980), but this method requires more study. Barnett and co-workers used a complicated HPLC system with programmed changes in solvent composition, flow rate, and detector wavelength to measure several vitamins simultaneously (Figs. 16 and 17).

Ranfft and Ruckermann (1978b) have measured menadione bisulfite in premixes by HPLC on a reverse-phase column. The measurement of phylloquinones and menaquinones in plants and other foods, however, is a more difficult problem. This analysis requires HPLC on two columns: first, silica and second,

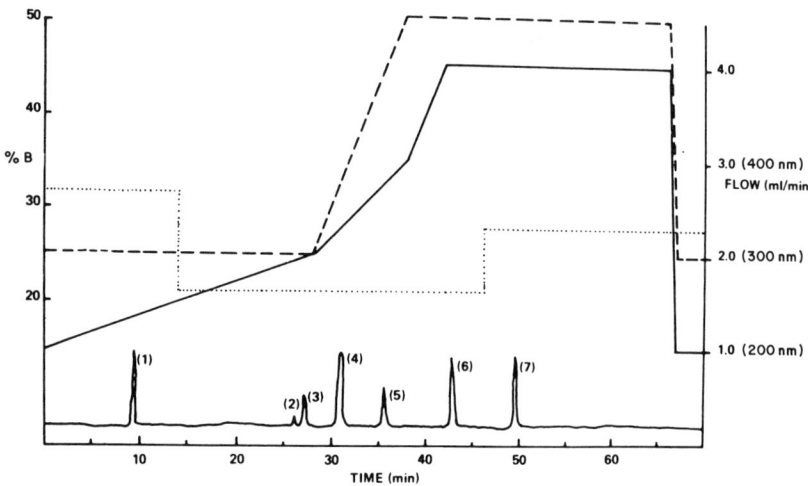

**Fig. 16.** Analysis of infant formula for fat-soluble vitamins. Examples of solvent system (- - -), flow (—), and wavelength (···) programs used with two Zorbax ODS columns connected in series. Solvent system: A, methanol–ethyl acetate (86:14); B, acetonitrile. Peak identification of standards: (1) retinol; (2) vitamin $D_3$; (3) α-tocopherol; (4) tocopheryl acetate; (5) phylloquinone; (6) cholesterol phenyl acetate internal standard; (7) retinyl palmitate. [Reprinted with permission from Barnett et al. (1980). Anal. Chem. **52**, 610–614. Copyright 1980, American Chemical Society.]

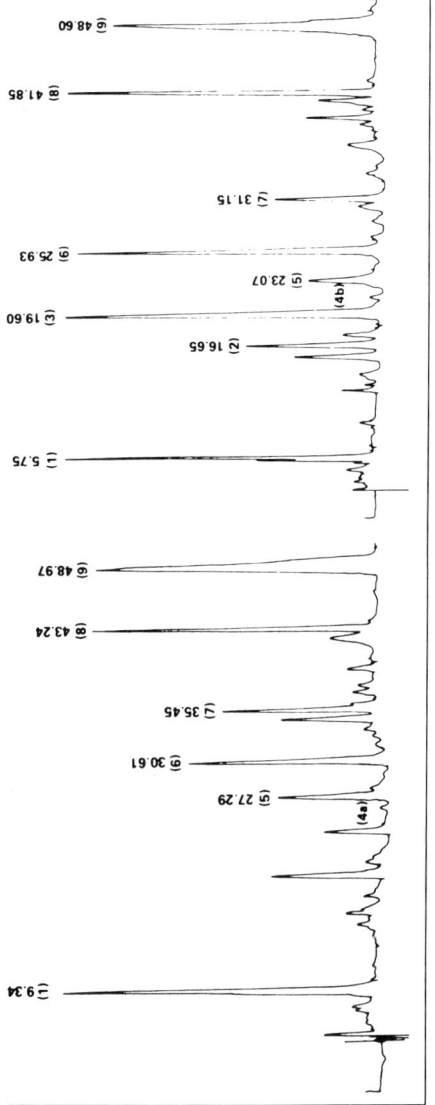

**Fig. 17.** HPLC of soy-base infant-formula products (after enzymatic hydrolysis) using systems similar to that shown in Fig. 16. Peak identification: (1) retinol; (2) δ-tocopherol; (3) γ-tocopherol; (4a) vitamin $D_3$; (4b) vitamin $D_2$; (5) α-tocopherol; (6) tocopheryl acetate; (7) phylloquinone; (8) cholesterol phenyl acetate; (9) retinyl palmitate. Retention times given in minutes. [With permission, from Earnett et al. (1980). *Anal. Chem.* **52**, 610–614. Copyright 1980, American Chemical Society.]

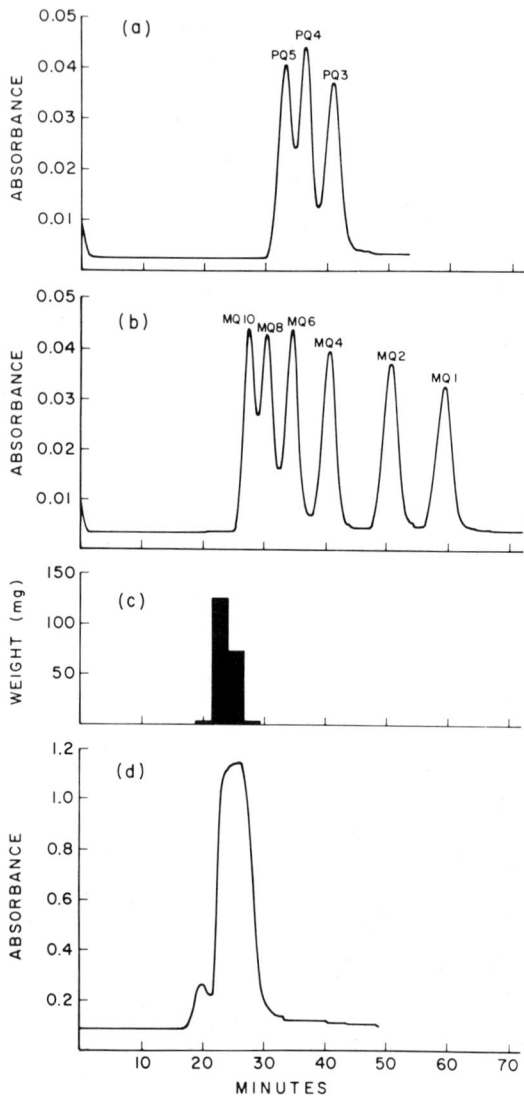

**Fig. 18.** Separation of phylloquinones, menaquinones, and triglycerides on 120-cm column HAPS in hexane (1 ml/min): (a) phylloquinones 3, 4, and 5; (b) menaquinones 1, 2, 4, 6, 8, and 10; (c) and (d) soybean oil (250 mg). Absorbance detector at 254 nm. [With permission, from Thompson et al. (1979).]

reverse-phase. Lipids can be extracted from plants with organic solvent, filtered through a short column of HAPS to remove polar lipids and chlorophylls, and then subjected to HPLC first on silica and then on reverse-phase columns for quantitative measurement of phylloquinone (Thompson *et al.*, 1979). When larger amounts of lipid are present in the sample, the vitamin can be purified by more careful chromatography on HAPS. The separation of a number of synthetic phylloquinones and menaquinones from triglycerides along a column of HAPS is shown in Fig. 18. The HAPS column may be used to reduce both the weight and the absorbance of lipid extracts when isolating phylloquinones and menaquinones 1, 2, 4, 6, and 8. Unfortunately, menaquinones 10 and higher are less well separated from interfering materials. The vitamins are separated from other impurities and from the remaining traces of triglycerides by HPLC on silica. The menaquinones can also be separated from the phylloquinones at the same time (Fig. 19). Individual phylloquinones and menaquinones can then be separated in order of decreasing length of the side chain on a reverse-phase column (Fig. 20). This procedure was used to measure phylloquinone in plant tissues and milk, but

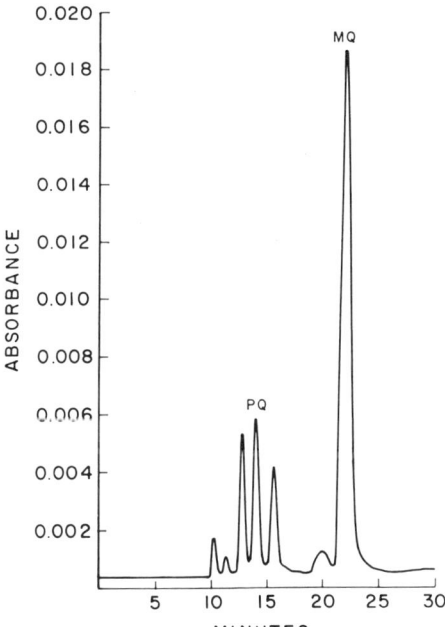

**Fig. 19.** HPLC of phylloquinones (PQ) and menaquinones (MQ) on 25 cm × 1 cm column LiChrosorb Si60, 5 μm. Solvent system: 0.03% isopropanol in hexane; 10 ml/min; absorbance detector at 262 nm. [With permission, from Thompson *et al.* (1979).]

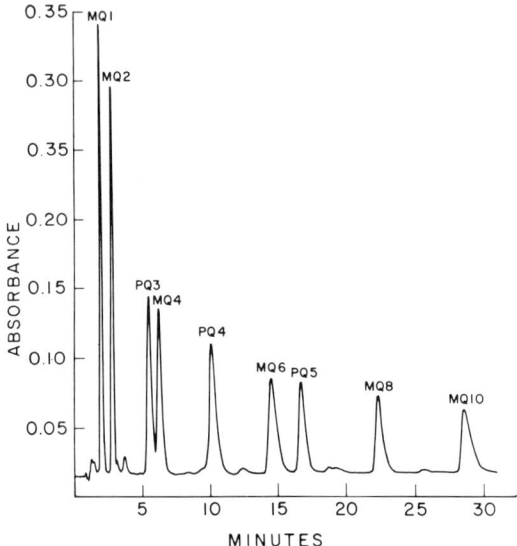

**Fig. 20.** HPLC of phylloquinones (PQ) and menaquinones (MQ) on 25 cm × 3.2 mm Li-Chrosorb reverse-phase column. Solvent system: 30-min gradient 85% aqueous methanol to methanol; 1 ml/min; absorbance detector at 262 nm. [With permission, from Thompson et al. (1979).]

when it was applied to liver the expected menaquinones were not detected (Thompson et al., 1979).

Shearer et al. (1980) measured phylloquinone in a wider variety of foods by a similar approach, except that the initial purification of the lipid extracts was achieved on a gravity column of silica instead of HAPS. Two steps of HPLC were performed on Partisil-5 and Zorbax-ODS. Phylloquinone was also measured in human and cow's milk (Haroon et al., 1980a).

Similar methods can be used for the investigation of the metabolism of vitamin K in animals. Shearer et al. (1980) used their procedure for foods to measure phylloquinone in plasma from humans given injections of vitamin K. The levels in the fasting state could not be measured accurately, but they were estimated to be less than 1 ng/ml. Lefevere et al. (1979), however, measured endogenous phylloquinone in serum by direct application of extracts first to silica and then to reverse-phase columns, and the values obtained with normal, healthy adults ranged from 5 to 30 ng/ml. Abe et al. (1979) measured phylloquinone and menaquinone 4 in plasma and liver from rats given these vitamins by a single HPLC chromatography on a reverse-phase column. They mixed the effluent with a solution of sodium borohydride and then measured the fluorescence of the reduced quinones.

Phylloquinone epoxide is an important metabolite which accumulates in liver and plasma of animals given anticoagulants which are antagonists of vitamin K. 2,3-Epoxides of phylloquinones and menaquinones can be separated from the corresponding vitamins by HPLC on reverse-phase columns (Donnahey *et al.,* 1979); this procedure has been used to measure radioactively labeled 2,3-epoxides in plasma (Bjornsson *et al.,* 1978). Haroon *et al.* (1980b) have investigated the separation of cis–trans isomers of phylloquinone and phylloquinone epoxide on several silica and reverse-phase columns and, in addition to providing a number of potentially useful systems, have strongly recommended reverse-phase columns with nonaqueous solvent systems.

Mack (1980) has made a thorough investigation of the separation of menaquinones on reverse-phase columns in chloroform–methanol mixtures. When silver ions were included in the solvent system, menaquinone 3 and higher menaquinones were eluted earlier. Haroon *et al.* (1981) have described the separation of menaquinones on reverse-phase, silica, and cyano bonded-phase columns.

## III. WATER-SOLUBLE VITAMINS

### A. Ascorbic Acid

Ascorbic acid (vitamin C, L-threohexono-1,4-lactono-2-ene) is an optically active monobasic acid with powerful reducing properties (Fig. 21). It occurs in most fruits and vegetables at levels between 20 μg and 3 mg/g; less important sources are some foods derived from animals, such as kidney (100 μg/g), liver (300 μg/g), and milk (10 μg/ml).

During analysis, it is often necessary to include a measurement of the oxidized form, dehydroascorbic acid, which has about 80% of the biological activity of the true vitamin. Only traces of dehydroascorbic acid are found in fresh foods, but larger amounts are formed during processing or prolonged storage.

Many isomers of L-ascorbic acid and other structurally related substances have been synthesized and investigated biologically, and most have been found to be

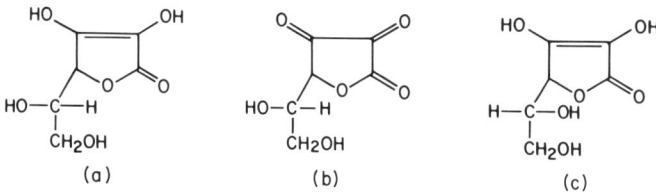

**Fig. 21.** Structures of vitamin C and related substances: (a) ascorbic acid; (b) dehydroascorbic acid; (c) isoascorbic acid.

inactive. The isomer D-erythrohexono-1,4-lactono-2-ene is of concern to the analyst because it is a common preservative, better known as isovitamin C, araboascorbic acid, or erythorbic acid. At the present time, it is difficult to distinguish between ascorbic acid and isoascorbic acid in foods by chemical methods.

Several colorimetric and electrometric procedures have been developed for measuring vitamin C, and most foods and tissues can be analyzed by established methods. There are, however, exceptions that yield highly colored extracts or contain other interfering substances.

The methods are too numerous to review comprehensively or even list, but the principles of three in common use can be mentioned. In one, the reducing capacity of an extract is measured with an oxidizing indicator such as iodine, ferricyanide, methylene blue, or more often 2,6-dichlorophenolindophenol. Another is based on the reaction of ascorbic acid with diazotized 4-methoxy-2-nitroaniline to form an adduct which has an intense blue color when made alkaline. In the third—and perhaps most important—method, ascorbic acid is oxidized to dehydroascorbic acid and then condensed with 2,4-dinitrophenylhydrazine to form an osazone which is red in sulfuric acid. This method can be used to measure separately ascorbic acid, dehydroascorbic acid, and other substances producing osazones by comparing the intensities of the colors produced after oxidation, reduction, or other preliminary treatment of sample extracts. Spectrophotometric methods based on the measurement of the absorption of UV light have poor specificity, and their usefulness is limited to the assay of relatively pure specimens of the vitamin. The absorption is obviously important, however, in the application of HPLC. The maximum is at 245 nm at pH values below 1.5 and shifts to 265 nm at pH 6.8. The molecular extinction coefficient has been determined with difficulty to be approximately 16,500 in alkaline solutions.

Ascorbic acid rapidly decomposes in alkali, but it is relatively stable when in acid or when dry. Considerable effort has been expended on the development of nondestructive extraction techniques. Special precautions are needed to prevent oxidation caused by exposure to air, oxidative enzymes, or metal ions, which in addition to occurring naturally in foods, can be produced accidentally from worn metal parts in blenders. In addition to careful homogenizing under nitrogen, it is usual to add a protective agent such as metaphosphoric acid, oxalic acid, EDTA, or perchloric acid. This information, which is reviewed elsewhere (Freed, 1966; Gyorgy and Pearson, 1967), should not be overlooked by those designing HPLC methods.

Early work on the measurement of vitamin C in foods by HPLC involved the use of the electrochemical detector. The use of this detector in advance of other approaches is accidental; the value of more orthodox detectors has not yet been fully appraised, and thus the real advantages of the electrochemical detector are unknown, and those claimed to date are hypothetical.

When Kissinger *et al.* (1973) pioneered the development of electrochemical detectors for HPLC almost a decade ago, the measurement of ascorbic acid was an obvious application. A method was developed for the analysis of urine using a column of SAX resin eluted with 0.05 $M$ acetate buffer at pH 4.6 (Kissinger *et al.*, 1974). Later, procedures were described for infant foods, fruits, fruit juices, fruit drinks, fortified cereals, milk products, and vitamin premixes (Pachla and Kissinger, 1976). The samples were extracted with 3% metaphosphoric acid in 8% acetic acid. As the ionic strength of this reagent was too high for the chromatography system, the extracts were diluted with cold 0.05 $M$ perchloric acid, which protected ascorbic acid against oxidation. Pretreatment of the SAX columns with ascorbic acid was necessary to prevent losses during chromatography. The method was sensitive enough for the analysis of serum, which typically contains 5–15 $\mu$g/ml ascorbic acid, and white blood cells (Tsao and Salimi, 1981). However, as pointed out by Rouseff (1979), the values obtained with citrus juices, at least, were surprisingly high.

Thrivikraman *et al.* (1975) used a similar detector to measure ascorbic acid in extracts of brain containing 10–20 $\mu$g/ml vitamin. The tissue was extracted with 0.05 $M$ perchloric acid, 0.002 $M$ thiourea, and 0.001 $M$ EDTA. A column of strong cation-exchange resin, Bondapak AX/Corasil in the phosphate form, was eluted with 0.01 $M$ monosodium phosphate.

Mason *et al.* (1980) have described a method for measuring ascorbic acid in plasma and urine that is suitable for routine repetitive work. An electrochemical detector was used with a Bondapak $NH_2$ bonded-phase column.

Carr and Neff (1980) measured ascorbic acid in tissues from various marine invertebrates using an anion-exchange column and an electrochemical detector. The samples were homogenized in cold perchloric acid and centrifuged, and then an aliquot of the supernatant was applied to the column.

The author's experience with an early electrochemical detector did not confirm a hope that it might be useful for the routine analysis of foods. The results were difficult to reproduce and the detector was troublesome to maintain. There have been great improvements in electrochemical detectors, however, and they will probably be used extensively for the measurement of ascorbic acid in the future (Stulik and Pacakova, 1980).

Meanwhile, the application of the absorbance detector requires investigation. Sood *et al.* (1976) obtained encouraging results with a detector operating at 254 nm. Various foods were extracted with 6% metaphosphoric acid, and ascorbic acid was separated from other detectable substances by ion-pair chromatography. The method could be used to measure levels as low as 5 $\mu$g vitamin/ml extract. A reverse-phase column was eluted with quaternary ammonium hydroxides or formates in aqueous methanol at pH 5. Several bases, including tetramethyl, tetraethyl, tetrapropyl, tetrabutyl, and tetrahexyl ammonia, were investigated, and the best results were obtained with tridecylammonium formate. The chromatography of ascorbic acid from tomato juice is shown in Fig. 22. The validity

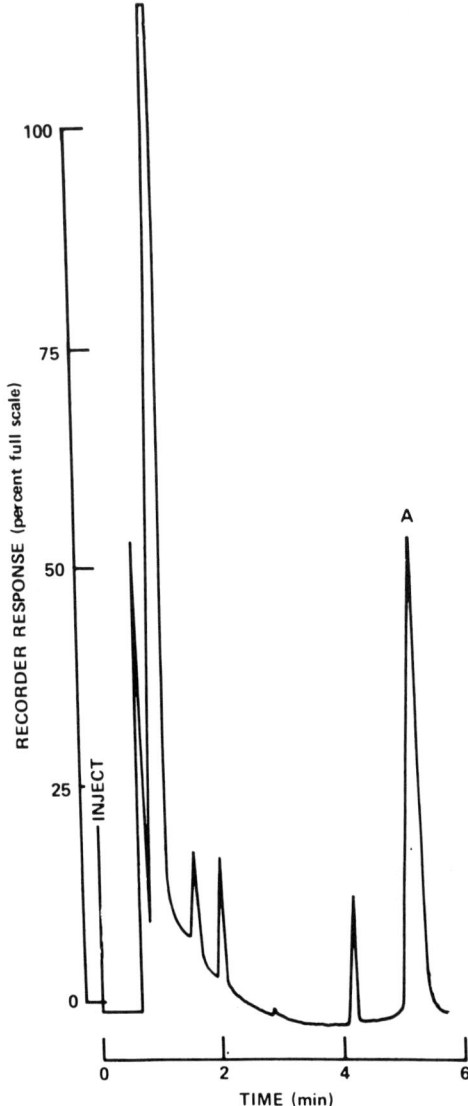

**Fig. 22.** Ion-pair HPLC of an extract of tomato juice containing 17.7 mg ascorbic acid (A) per 100 g. Column, μBondapak C-18; mobile phase, $1.0 \times 10^{-3}$ $M$ tridecylammonium formate in 50% methanol. [With permission, from Sood et al. (1976). *Anal. Chem.* **48**, 796–798. Copyright 1976, American Chemical Society.]

of the method was confirmed by comparing the results with those of established titrimetric and turbimetric procedures. Moreover, the ascorbic acid peak was reduced when the extracts were treated with the enzyme ascorbic oxidase before HPLC. Ruckemann (1980) has also used ion-pair chromatography to measure ascorbic acid in fruits and vegetables. He used tetrabutylammonium sulfate in aqueous methanol as solvent system. In contrast, Rouseff (1979) obtained disappointing results with ion-pair chromatography, and he recommended an amino–cyano bonded-phase column (Whatman PAC) for the analysis of citrus juices.

Methods for vitamin C should be applicable to a wide variety of samples, should measure ascorbic acid and dehydroascorbic acid separately, and ideally should discriminate between ascorbic and isoascorbic acids. Dehydroascorbic acid can be converted to ascorbic acid by treatment with homocysteine, and then analysis for ascorbic acid yields "total vitamin C." The direct measurement of dehydroascorbic acid is difficult because the absorbance is greatest below 220 nm and it is low at 254 nm.

Rose and Nahrwold (1981) attempted the direct measurement of ascorbic acid and dehydroascorbic acid in extracts of tissues and foods using two absorbance detectors set at 254 nm and 210 nm, respectively. The sensitivity of the method for dehydroascorbic acid was poor, however, and conversion to ascorbic acid before chromatography was recommended.

Ascorbic acid and isoascorbic acid are diastereoisomers (Fig. 21), and their separation has challenged the chromatographer. Archer *et al.* (1980) noticed a small difference in retention times when the two isomers were chromatographed on $NH_2$ bonded-phase columns (Wills *et al.*, 1977). Almost complete separation, however, has been obtained on a weak anion-exchange column (Waters carbohydrate analysis column) eluted with acetonitrile–water mixtures (Geigert *et al.*, 1981).

It thus seems probable that reproducible and practical liquid chromatographic methods will soon be developed for measuring vitamin C and related substances in foods. Although the electrochemical detector has received much attention in recent years, use of the absorbance detector seems equally attractive. Derivatization before or after chromatography has yet to be explored. Derivatives used previously in vitamin C determinations, such as osazones, might also prove to be useful in HPLC methods.

## B. Riboflavin and Thiamine

Riboflavin (Fig. 23) is resistant to acids, air, and common oxidizing agents but is destroyed by exposure to strong light. The absorption spectrum of neutral solutions has maxima at 475, 446, 359–375, 268, and 223 nm. The molecular extinction coefficient is 76,000 at 268 nm. Neutral solutions are fluorescent,

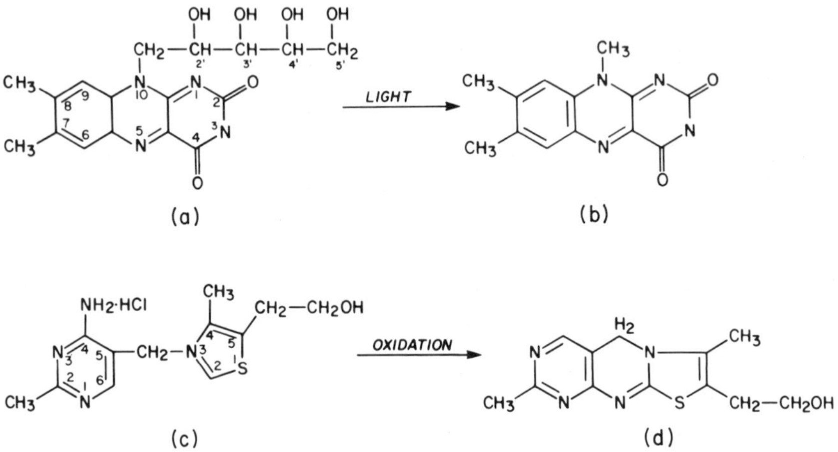

Fig. 23. Structures of (a) riboflavin, (b) lumiflavin, (c) thiamine hydrochloride, and (d) thiochrome.

emitting at 530 nm. In foods and tissues, riboflavin is combined mainly as the phosphate ester or as flavin–adenine dinucleotide. Free riboflavin can be liberated enzymatically or by treatment with acids.

Irradiation of alkaline solutions of riboflavin yields lumiflavin (Fig. 23), a highly fluorescent substance that is soluble in chloroform. Although the formation of lumiflavin is the basis of many analytical methods, including some of the recently developed HPLC procedures, it should be noted that conversion of riboflavin to lumiflavin is usually incomplete; special care is needed to perform it reproducibly.

Thiamine (Fig. 23) absorbs light below 300 nm, the maximum being near 232 nm in neutral solutions. In acid, a maximum appears at 246 nm. Thiamine is readily oxidized to a fluorescent substance, thiochrome, and this property is often exploited in analytical methods. Thiamine occurs in foods and tissues in the free form or combined as phosphate esters.

Callmer and Davies (1974) described the separation of thiamine, riboflavin, and riboflavin phosphate in pharmaceutical preparations by ion-exchange chromatography on HS Pellionex SCX. Wittmer and Haney (1974) used a fluorometric detector and a LiChrosorb Si60 column eluted with chloroform–methanol–acetate buffer, pH 4, (60:28:4.5) to measure riboflavin in multivitamin preparations. The behavior of riboflavin and thiamine on reverse-phase columns with ion-pair systems has been investigated by Wills *et al.* (1977).

Riboflavin and thiamine have been measured in hydrolyzed extracts of foods by HPLC on silica eluted with alcohol–buffer mixtures (Van De Weerdhof *et al.*,

1973). A fluorescence detector was used, and thiamine was oxidized after chromatography to thiochrome by mixing with alkaline potassium hexacyanoferrate. Riboflavin was measured from its native fluorescence. A similar method, but partly automated, has been described by Osborne and Voogt (1978).

Toma and Tabekhia (1979) and Kamman et al. (1980) measured riboflavin and thiamine in hydrolyzed extracts of rice and fortified cereals by HPLC on reverse-phase columns with ion-pair reagents and an absorbance detector at 254 nm. Ang and Moseley (1980) reported that this method does not have sufficient sensitivity or specificity for meat products. They converted riboflavin to lumiflavin by irradiation with UV light and oxidized thiamine to thiochrome. Lumiflavin was extracted with chloroform, and thiochrome was extracted with isobutanol. These fluorescent products were chromatographed on silica (Spherisorb) in chloroform–methanol (9:1), and a fluorescence detector was used for quantitation.

Riboflavin has been measured in urine by HPLC on a reverse-phase column in phosphate buffer, pH 5–methanol (65:35) using a fluorescence detector (Smith, 1980). The results were higher than those obtained by a manual fluorometric method. Riboflavin has also been measured in human and canine serum by HPLC. Filtered specimens were chromatographed on reverse-phase columns using a gradient of increasing amounts of methanol in phosphate buffer, pH 5.6 (Assenza and Brown, 1980). In addition, to riboflavin, 15 low-molecular-weight substances were identified with absorbance and fluorescence detectors.

Ishii et al. (1979a) converted thiamine and its mono-, di-, and triphosphates to the corresponding thiochrome derivatives with cyanogen bromide before chromatography on LiChrosorb-NH$_2$ in acetonitrile–phosphate buffer, pH 8.4, (6:4). This procedure was applied to extracts of brain, liver, heart, and kidney (Ishii et al., 1979b). Better results were obtained later with a C-18 column (Sanemori et al., 1980).

Thiamine in tissues has been separated from its phosphate esters by ion-exchange chromatography and by ion-pair systems (Gubler and Hemming, 1980; Hemming and Gubler, 1980). Although some problems were encountered in the determination of thiamine triphosphate, satisfactory results were obtained with thiamine and thiamine pyrophosphate.

## C. Niacin

Niacin refers to the vitamin pyridine-3-carboxylic acid (nicotinic acid, Fig. 24) and the corresponding amide (nicotinamide), which are precursors of the phosphopyridine nucleotide coenzymes (DPN and TPN). Some mammals satisfy their needs for niacin by converting the amino acid tryptophan to nicotinic acid. Others, such as humans and dogs, are unable to obtain sufficient niacin by this route and must include the vitamin in their diet.

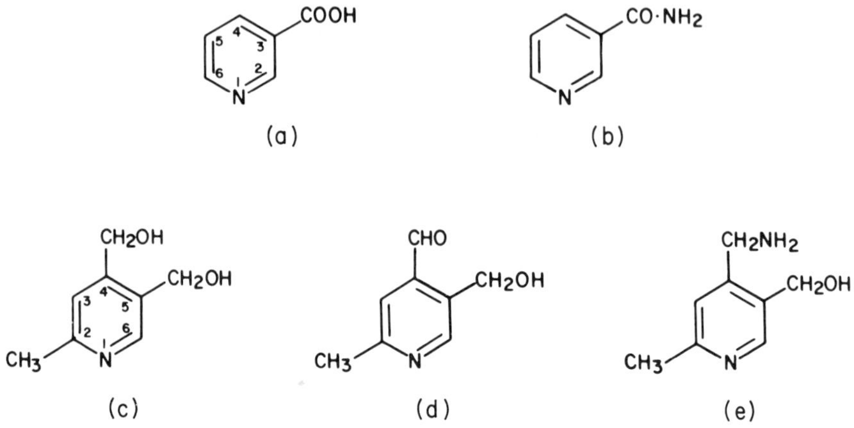

**Fig. 24.** Structures of niacin and vitamin $B_6$: (a) nicotinic acid; (b) nicotinamide; (c) pyridoxine; (d) pyridoxal; (e) pyridoxamine.

Nutritionists usually allow for the conversion of tryptophan to niacin even in man, and reference is sometimes made to "niacin equivalents." This term refers to niacin plus 1/60 of the tryptophan. It implies that analysis for the amino acid is needed in addition to that for the vitamin, but this can be avoided by assuming that vegetable and animal proteins contain 1 and 1.4% tryptophan, respectively. The levels of niacin in foods vary from 1 μg/g (milk) to 400 μg/g (yeast).

Nicotinic acid is stable in dilute acid and alkali even when heated. The absorption spectrum has a maximum at 261 nm that is independent of pH. The molecular extinction coefficient is higher in acidic than in alkaline solutions.

Nicotinamide can be autoclaved unchanged in neutral solutions, but in the presence of acids or alkalis it is hydrolyzed to nicotinic acid. Its absorption maximum is at 261 nm, and the extinction coefficient is higher at low pH values. Nicotinic acid and nicotinamide have the properties of bases and form quaternary ammonium salts with acids. As nicotinic acid also forms salts with bases, its properties are similar in many respects to those of an amino acid.

Nicotinic acid is converted by treatment with cyanogen bromide to substances that couple with aromatic amines to form colored products. This is the basis of the chemical determination of niacin in foods (Freed, 1966).

Osborne and Voogt (1978) described an automated HPLC method that was essentially a refinement of the manual chemical method. Niacin was extracted from foods by hydrolysis and detached from proteins by treatment with papain; phosphates were split with diastase. After filtration, the treated extracts were subjected to HPLC on a silica column eluted with buffer. The effluent was mixed automatically with cyanogen bromide and *p*-aminoacetophenone and passed through a fluorescence detector.

Tyler and Shrago (1980) developed a method for cereals in which niacin was detected from the absorbance at 254 nm. Samples were autoclaved with calcium hydroxide suspension and centrifuged. The extract was passed through a short column of AG1-X8 anion-exchange resin, which retained the vitamin. The vitamin was then eluted with 5% acetic acid, and the effluent was treated with dilute potassium permanganate solution to remove additional interfering substances. Nicotinic acid was measured in the extract by ion-pair chromatography. The method was applied to samples of semolina and bread (Fig. 25), and the results were similar to those obtained in microbiological assays.

Nicotinamide has been measured in serum and urine by ion-pair chromatography after preliminary purification on a short column of reverse-phase packing (De Vries *et al.*, 1980). This procedure was designed for pharmaceutical research. Nicotinamide was separated from nicotinic acid and related substances, and isonicotinamide was used as an internal standard. An end product of nicotinamide metabolism, *N*-methylnicotinamide, has been determined in urine by

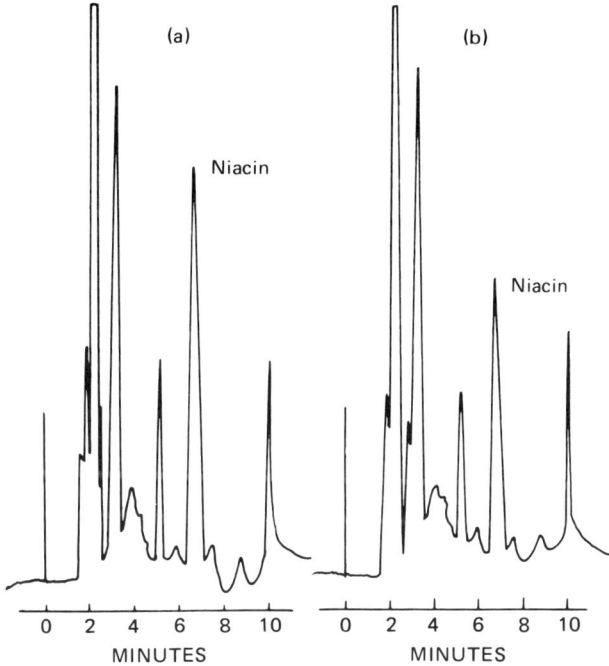

**Fig. 25.** Ion-pair HPLC determination of niacin in (a) semolina and (b) bread samples after ion-exchange and permanganate cleanup. Flow rate, 2 ml/min; mobile phase, 5% methanol in water with PIC-A. [From Tyler and Shrago (1980), p. 276, by courtesy of Marcel Dekker, Inc.]

HPLC on a Partisil SCX column after cleanup on strong cation-exchange resin, AG50 (Shaikh *et al.*, 1977b). Hengen *et al.* (1978) describe a method using ion-pair chromatography.

## D. Vitamin $B_6$

The vitamin occurs in many common foods such as cereals, fruits, meats, eggs, and dairy products. The levels range from 0.5–10 µg/g.

The term vitamin $B_6$ refers to three structures: a hydroxylated weak base, pyridoxine; a corresponding aldehyde, pyridoxal; and a stronger base, pyridoxamine (Fig. 24). These substances also occur in foods and tissues as the 5-phosphates. Pyridoxic acid is a metabolite of vitamin $B_6$ and is probably formed by liver aldehyde oxidase; it is the major excretory product in urine.

The vitamins absorb UV light below 350 nm, and one, two, or three maxima are present depending upon the pH. Aqueous solutions of free pyridoxine, pyridoxal, and pyridoxamine—but not the phosphates—are strongly fluorescent, emitting at 345–400 nm. The variations in the absorption and fluorescence spectra at different pH values indicate the formation and disappearance of numerous ionic species, each of which appears to have characteristic optical properties (Johnson and Metzlar, 1970; Udenfriend, 1969). These complicated changes can confuse quantitative work with HPLC detectors if careful attention is not given to the control of pH.

Hitherto, the vitamin was measured in foods by microbiological methods. The growth of some test organisms is stimulated equally by all three forms of the vitamin, whereas the response of others is greater to either pyridoxal or pyridoxamine. It is thus possible, in theory, to estimate the levels of the different forms of the vitamin by combining the results of several selective microbiological assays; however, this procedure is rarely straightforward in practice. More reliable results can be obtained by separating the three forms chromatographically before microbiological assay, but this is time consuming and laborious. As reliable chemical procedures for measuring vitamin $B_6$ in foods are not available, the only convenient method of analysis remains the microbiological assay of total vitamin $B_6$ activity. There is thus a need for more rapid and specific methods, and this has encouraged the application of HPLC to $B_6$ analysis.

In the differential microbiological assay, the three vitamins were released from foods by acid hydrolysis and the free forms separated by ion-exchange chromatography on resins such as Dowex AG50W. Yasumoto *et al.* (1975) used more rigid Aminex A5 resin in an early example of the application of HPLC techniques to ion-exchange resins for vitamin $B_6$ analysis. The three forms were separated in phosphate buffers and then detected and measured after chromatography by reaction with 5-chloroaniline 2,4-disulfonyl chloride, which yields orange products. It is not clear how successful this method was with foods, as many unidentified peaks appeared to interfere with those produced by the vitamins.

Williams and Cole (1975) and Williams (1980) described the use of Aminex A5 resin with an absorbance detector set at 260 nm. The pure vitamins were separated in 0.7 $M$ ammonium formate buffer, pH 5.6. This author has obtained better results with a column of Aminex A9 using a fluorescence detector. The rapid assay of pyridoxine in hydrolyzed infant formula with an isocratic system is shown in Fig. 26. The fluorescence detector improves the selectivity of the analysis; it has also been used with conventional resins such as Dowex AG50W (Gregory and Kirk, 1977).

Wong (1978) used a column of Zipax SCX to measure the vitamins in extracts of fruits and vegetables after enzymatic hydrolysis of the phosphate esters and preliminary purification on a column of Dowex AG50. An absorbance detector was used at 210 nm, which thus operated with relatively high sensitivity but low specificity.

Yoshida et al. (1978) used a fluorescence detector for the analysis of blood extracts on a Japanese TSK-Gel LS-160 column; pyridoxal phosphate was determined after treatment of the extracts with acid phosphatase. The method was applied to blood from dogs and rabbits dosed with pyridoxal phosphate. The detection limits were in the range 15–50 ng/ml.

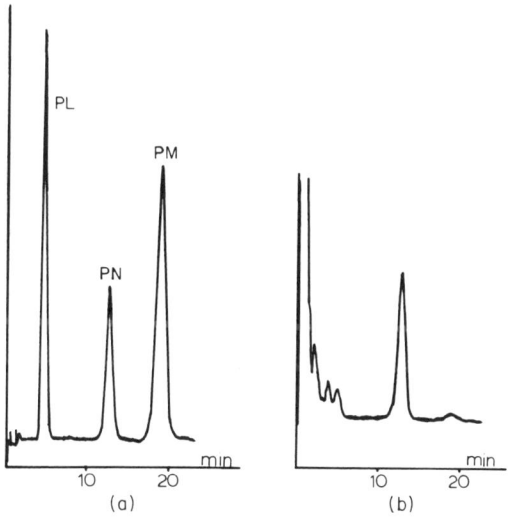

**Fig. 26.** (a) HPLC of pyridoxine (PN), pyridoxal (PL), and pyridoxamine (PM) on Aminex A9 in 0.5 $M$ ammonium formate buffer, pH 5.5. (b) HPLC of extract of hydrolyzed infant formula. Fluorometric detector.

Although the $B_6$ vitamins have usually been separated as cations in most ion-exchange chromatography, Vanderslice et al. (1979) employed alkaline buffers (pH 10) and achieved useful separations by anion-exchange chromatography on Bio-Rad A25 resin. One advantage of this approach is that it is possible to separate and measure the unhydrolyzed phosphates in addition to the free vitamins (Fig. 27). The sensitivity of the method has been improved by converting pyridoxal and pyridoxal phosphates to the more fluorescent oximes. This conversion was achieved by adding $0.005\ M$ semicarbazide to the buffer system. When a spectrofluorometer was used as detector, the limits of detection were better

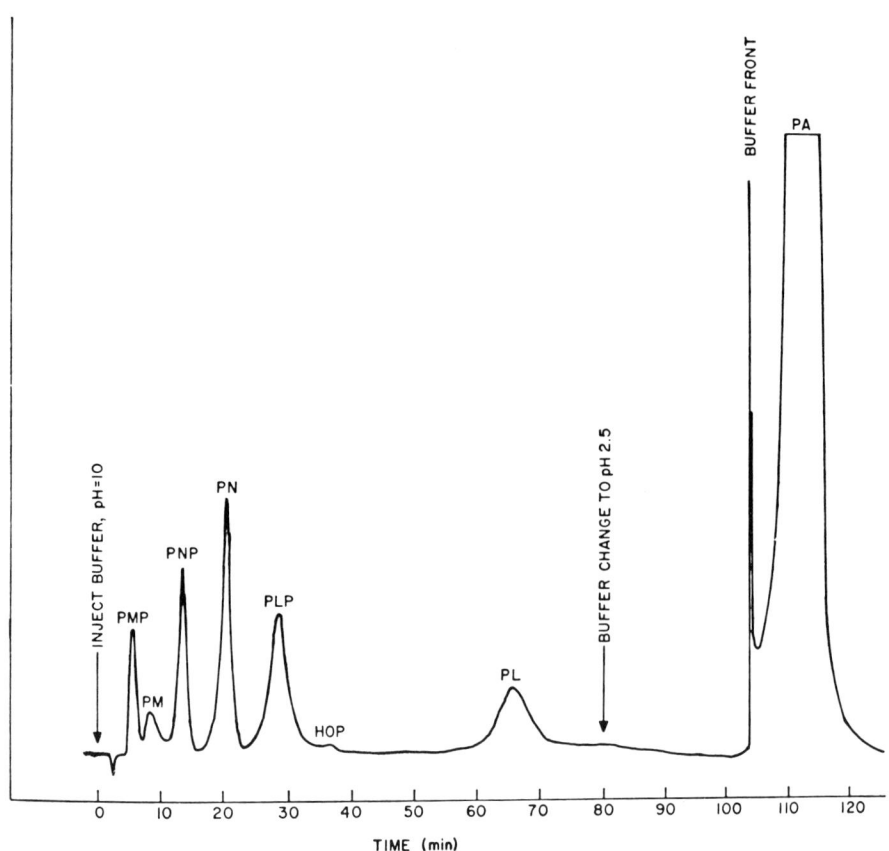

**Fig. 27.** HPLC of pyridoxine (PN); pyridoxal (PL); pyridoxamine (PM); the corresponding phosphates (PNP, PLP, PMP); pyridoxic acid (PA); and hydroxypyridine (HOP) on Bio-Rad A-25 resin in NaCl–glycine buffers. [With permission, from Vanderslice et al. (1979).]

than 0.5 ng/ml. The application of the method to the analysis of foods is presently being studied (Vanderslice and Maire, 1980), and improved extraction procedures utilizing sulfosalicylic acid have been developed (Vanderslice et al., 1980).

The vitamins may also be separated on reverse-phase columns. Gregory and Kirk (1978) reported the chromatography of hydrolyzed extracts of dehydrated foods on μBondapak C-18 columns eluted with 0.033 $M$ potassium phosphate buffer, pH 2.2, using a fluorescence detector. The results agreed with rat bioassays and microbiological assays of total vitamin $B_6$ activity. The method was also applicable to the measurement of pyridoxine in fortified breakfast cereals (Gregory, 1980a) and after addition of 5% methanol to the solvent system, to the measurement of pyridoxic acid in urine (Gregory and Kirk, 1979). This approach has also been used in the analysis of milk (Lim et al., 1980). Pyridoxal phosphate can be measured in perchloric acid extracts of tissues by conversion to the semicarbazone and chromatography on LiChrosorb RP-8 in 0.033 $M$ phosphate buffer containing 2.5% acetonitrile (Gregory, 1980b). O'Reilly et al. (1980) measured pyridoxine, pyridoxal, and pyridoxic acid in blood and urine using a reverse-phase column eluted with potassium phosphate buffer.

The first reported use of ion-pair chromatography to separate pyridoxine from related substances concerned drug analysis (Stewart et al. 1976). Systematic studies of the behavior of vitamin $B_6$ metabolites in various ion-pair systems have recently been published (Morita and Mizuno, 1980; Tryfiates and Sattsangi, 1982).

## E. Folacin

The term folacin encompasses a large group of substances which are all different forms of one vitamin required by animals for growth, reproduction, and the prevention of megaloblastic anemia. Vitamins in this group are often called collectively "folic acid" or "folates," but it is preferable to reserve these terms for the substance pteroylmonoglutamic acid (Fig. 28) and related polyglutamates (pteroyl-oligo-γ-L-glutamates).

Foods are fortified with the monoglutamate, but most of the naturally occurring vitamin is in the form of polyglutamates containing up to eight glutamyl residues. Microbiological methods have been used for many years to measure the levels of the vitamin in foods. The most commonly used organism, a folacin-requiring *Lactobacillus casei,* utilizes forms of the vitamin with up to three glutamyl residues for growth far more efficiently than higher polyglutamates. Values obtained after incubation of extracts with a conjugase enzyme, which splits the links in the side chain, are therefore higher than those obtained with untreated extracts; the former are referred to as "total-folate" values, whereas the latter represent "free folate." As the availability of different polyglutamates

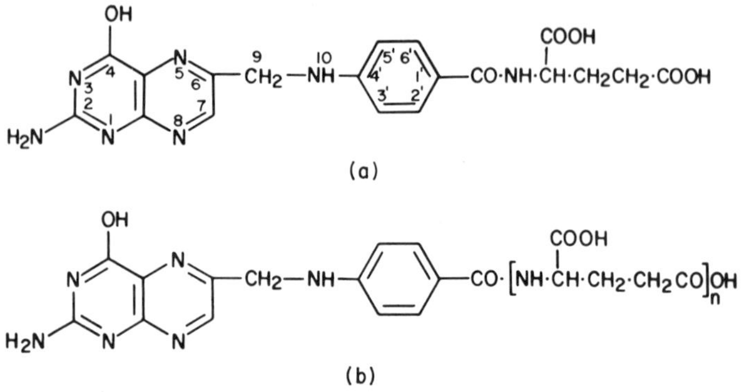

Fig. 28. Structure of (a) folic acid and (b) polyglutamates (folates).

to animals is still disputed, it is undecided whether "total" or "free" values should be used in nutritional calculations. In practice, however, the measurement of free folate is often complicated by uncontrolled deconjugation caused by processing, cooking, or the action of naturally occurring conjugase enzymes. For this and other reasons, the total folate values are now often preferred. The literature on the role of folacin in human nutrition has been reviewed by Rodriguez (1978).

In tissues, the pteridine nucleus of folacin may exist in the oxidized form (e.g., folic acid), in a 5,6-dihydro form and in a 5,6,7,8-tetrahydro form. Thus folic acid can be reduced to 5,6-dihydrofolic acid (DHF) and 5,6,7,8-tetrahydrofolic acid (THF). Furthermore, the N-5 and N-10 positions may be substituted with one or more of six different one-carbon structures. For example, derivatives of THF with physiological significance include 5-methyl-, 5-formyl-, 5,10-methylene-, 5,10-methenyl-, and 5-formiminotetrahydrofolic acid.

The measurement of the many forms of folacin in foods and tissues is thus potentially a complicated task. The problem is made more difficult by the susceptibility of the vitamin to destruction or change when exposed to heat, light, oxygen, acidity, alkalinity, or endogenous enzymes. Chemical methods are not sufficiently sensitive for the determination of naturally occurring folacin, and most analyses have been based on differential microbiological assays which exploit the unequal responses of different organisms to different forms of the vitamin.

Microbiological assays have provided useful data on total folates in foods, but the results in most other areas are unsatisfactory. The analysis of serum, for example, yields results which vary from laboratory to laboratory, and although the results of extensive surveys have been used to assess nutritional status, the absolute level of folacin in serum is still disputed. Some of the folacins have been separated chromatographically on paper and on columns of DEAE- and TEAEcellulose. This process is time consuming, especially when combined with microbiological assays, and the separation of the vitamins is in any case incomplete. As all forms of folacin absorb UV light, their detection and measurement during HPLC is straightforward with absorbance detectors, providing that the effects of acidity and alkalinity are taken into account; some forms of the vitamin are also fluorescent.

The first reported tests with HPLC involved weak anion-exchange packings which were equivalent to the cellulose columns used previously. Reed and Archer (1976) separated THF, 5-methylTHF, 5-formylTHF, and 5,10-methenylTHF on AL-Pellionex-WAX in 0.006 $M$ phosphate buffer. They added mercaptoethanol to their standard solutions to protect the vitamin against oxidation, but as this substance absorbed light at the detector wavelength (254 nm), it could not be added to the column eluent. Instead, they bubbled nitrogen through the buffer, a technique adopted by many later workers.

In similar work, Stout *et al.* (1976) first tried Tris-HClO$_4$ buffers at pH 7–7.5 and discovered that their Partisil 10 SAX columns were rapidly destroyed. They subsequently recommended a gradient of sodium chloride in phosphate buffer, pH 6.5, which was claimed to extend the life of the column to 8 weeks (sodium chloride probably caused more expensive damage to the equipment, but this was apparently not noticed). In addition to separating 5- and 10-substituted derivatives, they separated THF and corresponding tri-, penta-, and heptaglutamates.

Clifford and Clifford (1977) also used a sodium chloride gradient and an anion-exchange column with similar results except that the separations were inferior. Surprisingly, polyglutamates were not separated from each other or from monoglutamates.

Anion-exchange chromatography has also been studied by Cashmore *et al.* (1980), who described the separation of polyglutamates on a Partisil SAX column in phosphate buffers. They also compared the behavior of polyglutamates on reverse-phase columns. The retention of polyglutamates on anion-exchange columns was stronger in proportion to the number of carboxyl groups and thus increasing chain length. In reverse-phase chromatography at high pH values, ionization of the carboxyl groups increased the electrostatic interactions with the mobile solvent, and the polyglutamates were eluted in the reverse order, i.e., the longer chain polyglutamates were eluted first (Fig. 29). In contrast, at low pH values (i.e., <3), the carboxyl groups were not ionized, and retention was proportional to the number of glutamyl residues as in anion-exchange chromatography (Bush *et al.*, 1979). At pH 3.5 the opposing tendencies balanced on reverse-phase columns, and the polyglutamates did not separate.

Ion-pair chromatography (Fig. 30) with Waters Pic Reagent A has been used to separate folic acid, DHF, and THF (Branfman and McComish, 1978) and 5-formylTHF, DHF, and 5-methylTHF (Allen and Newman, 1980).

Unconjugated pteridines (2-amino-4-hydroxy substituted pteridines) occurring

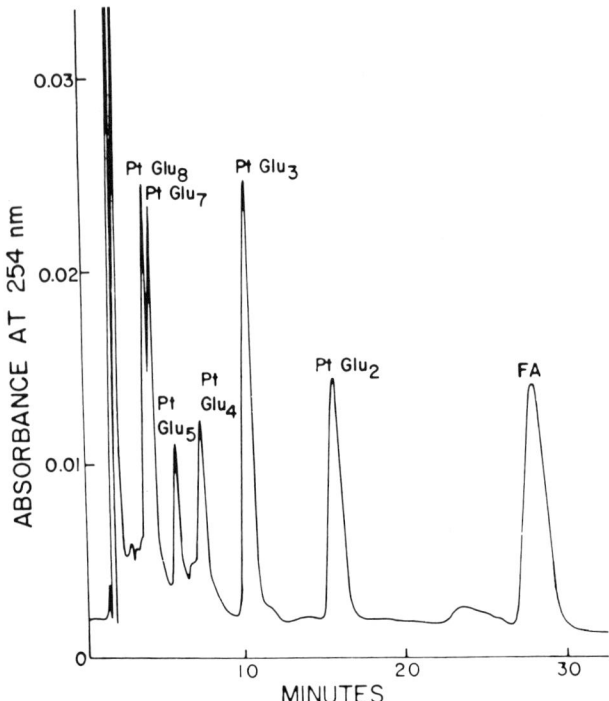

**Fig. 29.** Reverse-phase HPLC of pteroyl-oligo-γ-L-glutamates (Pt Glu$_n$) and folic acid (FA) on Spherisorb ODS column in 0.1 $M$ phosphate buffer, pH 6, containing 1% acetonitrile. [With permission, from Bush et al. (1979).]

in tissues have been separated on cation-exchange columns (Archer and Reed, 1980; Stea et al., 1979).

In the detection and measurement of folacin in extracts of foods and tissues, the different forms of the vitamin must be separated not only from each other but also from relatively enormous amounts of interfering material. Reingold and Picciano (1982) separated folate derivatives by ion-pair chromatography on C-18 and phenyl columns and on Radial-Pak C-8 cartridges. The *in situ* formation of folate ion pairs facilitated the removal of interfering substances during sample preparation, but the method was not completely successful when applied to infant formulas. Day and Gregory (1980), however, claim to have measured folate derivatives in liver, fortified cereals, and infant formula using coupled C-18 and phenyl columns. The extracts of the samples were purified on a hydrophobic resin (Bio-Beads SM-2) before HPLC. As the vitamins were not separated cleanly from interfering substances, quantitative measurements with this system would have dubious validity. Clifford and Clifford (1977) claim to have measured folic acid, DHF, THF, and 5-methylTHF in citrus juices and nuts

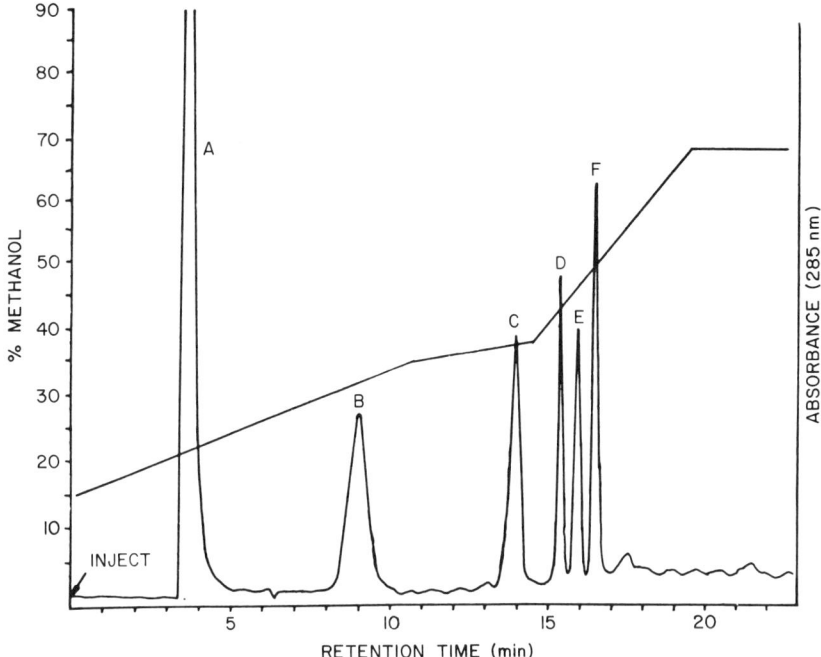

**Fig. 30.** Ion-pair separation of folic acid compounds and *p*-aminobenzoyl-L-glutamic acid (PABG) on reverse-phase column in methanol–water gradient containing PIC-A. Peaks: A, ascorbic acid; B, PABG; C, 5-formylTHF; D, DHF; E, 5-methylTHF; F, folic acid. [With permission, from Allen and Newman (1980).]

using an absorbance detector at 280 nm and an anion-exchange column, but convincing evidence that the observed peaks were correctly identified is lacking. A similar technique was applied to the analysis of rat liver (Clifford, 1976) and blood, kidney, and brain (Clifford and Clifford, 1976).

There is considerable interest in the analysis of blood not only from normal individuals but also from those given oral doses of folic acid. There are unresolved controversies concerning the nature of the metabolites in serum and plasma; some claim that even after dosing with folic acid, circulating folacin is mainly in the form of 5-methylTHF. Chapman *et al.* (1978) chromatographed serum, from volunteers given folic acid, on an anion-exchange column (Permaphase AAX) using a potassium perchlorate gradient. Two peaks of interest were observed. The first appeared to be 5-formylTHF, but this could not be confirmed by microbiological assay of the eluted material. The second peak was identified as either DHF or unchanged folic acid, both of which eluted at the same time in this system. Ion-pair chromatography with Waters Pic Reagent A confirmed the absence of significant amounts of DHF, and a peak corresponding to folic acid was observed. As standards of folic acid and 5-methyl folic acid eluted at the same time in the ion-pair system, a selective microbiological assay was used to confirm that the peak was indeed folic acid.

A more sensitive method for measuring folacin in serum and spinal fluid has been reported by Lankelma et al. (1980). It was based on a procedure for the measurement of the drug methoxytrate (Lankelma and Poppe, 1978). They stabilized extracts of plasma with ascorbic acid and removed protein with trichloracetic acid. The vitamin was first concentrated on a reverse-phase column and then back-flushed by means of a switching valve into an anion-exchange system (Partisil SAX). An electrochemical detector was used to measure 5-methylTHF, and identity of the peak was confirmed by microbiological assay. It is interesting to note that the HPLC analysis of serum gave lower results than were obtained with the usual microbiological method. It was suggested that the results of the microbiological assay were probably elevated because of the use of folic acid as a standard instead of 5-methylTHF.

## F. Vitamin $B_{12}$

Vitamin $B_{12}$ (cyanocobalamin) is a six-coordinate cobalt(III) complex with an impressive chemical structure corresponding to the empirical formula $C_{63}H_{88}O_{14}N_{14}PCo \cdot 25H_2O$. The properties of the vitamin and related substances, known collectively as cobalt corrinoids, are described in a monograph by Pratt (1972). The vitamin is relatively stable in solution, especially at pH values of 4.5–5, but it is destroyed by exposure to oxidizing or reducing agents or light. It absorbs light of all wavelengths below 550 nm. The absorption spectrum has peaks at 278, 361, and 550 nm and is altered little by changes in pH. Vitamin $B_{12}$ is present in dairy products, meats, eggs, and seafoods, but at very low concentrations (0.01–1 µg/g).

Only two publications have appeared concerning HPLC and vitamin $B_{12}$. Mourot et al. (1979) separated the vitamin from cyano-, methyl- and hydroxycobalamins in veterinary medical products on a reverse-phase column eluted with methanol–water gradients; an absorbance detector was used at 350 nm. Frenkel et al. (1979) in similar work used an absorbance detector at 254 nm and collected fractions for the measurement of cobalamins by a radioisotopic competitive-inhibition assay. They investigated the separation of a number of cobalamins on several reverse-phase systems.

## IV. CONCLUSION

HPLC has been applied successfully to the measurement of all the fat-soluble vitamins and many water-soluble vitamins in foods and tissues. It is clear from the results obtained so far that HPLC not only has improved the accuracy of quantitative measurements by eliminating interfering materials but also has permitted the measurement of different forms of each vitamin, isomers, and derivatives. However, if these advances are to be fully exploited, they will require improvement in other areas such as the preparation of standards and the definition of units.

Many analyses can be performed in a single HPLC step, provided that a suitable extraction and "cleanup" can be devised. More effort must be devoted to the design of the extraction and preliminary purification steps, but the goal must be simplicity with economy. The design of HPLC methods is suspect when it calls for many steps of extraction, or partitioning of extracts between numerous solvents, or preliminary chromatography on a variety of adsorbants. There should be a trend toward simplification and efficiency in all of the steps of analysis, especially when methods are intended for routine work.

## REFERENCES

Abe, K., and Katsui, G. (1975). *Bitamin* **49**, 259–263.
Abe, K., Yuguchi, Y., and Katsui, G. (1975). *J. Nutr. Sci. Vitaminol.* **21**, 183–188.
Abe, K., Ishibashi, K., Ohmae, M., Kawabe, K., and Katsui, G. (1977). *Bitamin* **51**, 275–280.
Abe, K., Hiroshima, O., Ishibashi, K., Ohmae, M., Kawabe, K., and Katsui, G. (1979). *Yakugaku Zasshi* **99**, 192–200.
Adachi, A., and Kobayashi, T. (1979). *J. Nutr. Sci. Vitaminol.* **25**, 67–78.
Adams, J. S., Clemens, T. L., and Holick, M. F. (1981). *J. Chromatogr.* **226**, 198–201.
Adams, M. A., and Nakanishi, K. (1979). *J. Liq. Chromatogr.* **2**, 1097–1136.
Aitzetmuller, K., Pilz, J., and Tasche, R. (1979). *Fette, Seifen, Anstrichm.* **81**, 40–43.
Ali, S. L. (1978). *Fresenius' Z. Anal. Chem.* **293**, 131–134.
Allen, B. A., and Newman, R. A. (1980). *J. Chromatogr.* **190**, 241–245.
Ames, S. R., and Drury, E. E. (1975). *J. Assoc. Off. Anal. Chem.* **58**, 585–594.
Ang, C. Y. W., and Moseley, F. A. (1980). *J. Agric. Food Chem.* **28**, 483–489.
Antalick, J. P., Debruyne, H., and Faugere, J. G. (1977). *Ann. Falsif. Expert. Chim.* **70**, 497–502.
AOAC (1980). "Official Methods of Analysis," 13th ed. Association of Official Analytical Chemists, Washington, D.C.
APA (1975). "The National Formulary," 14th ed. American Pharmaceutical Association, Washington, D.C.
Archer, M. C., and Reed, L. S. (1980). *In* "Methods in Enzymology" (D. B. McCormick and L. D. Wright, eds.), Vol. 66, pp. 452–459. Academic Press, New York.
Assenza, S. P., and Brown, P. R. (1980). *J. Chromatogr.* **181**, 169–176.
Barnett, S. A., Frick, L. W., and Baine, H. M. (1980). *Anal. Chem.* **52**, 610–614.
Bauernfeind, J. C. (1981). "Carotenoids as Colorants and Vitamin A Precursors." Academic Press, New York.
Besner, J.-G., LeClaire, R., and Band, P. R. (1980). *J. Chromatogr.* **183**, 346–351.
Bieri, J. G., Tolliver, T. J., and Catignani, G. L. (1979). *Am. J. Clin. Nutr.* **32**, 2143–2149.
Bjornsson, T. D., Swezey, S. E., Meffin, P. J., and Blaschke, T. F. (1978). *Thromb. Haemostasis* **39**, 466–473.
Boldingh, J., Cama, H. R., Collins, F. D., Morton, R. A., Gridgeman, N. T., Isler, O., Kofler, M., Taylor, R. J., Welland, A. S., and Bradbury, T. (1951). *Nature (London)* **168**, 598.

Botey Serra, J., and Garcia Fite, D. (1975). *Afinidad* **32**, 249–255.
Branfman, A. R., and McComish, M. (1978). *J. Chromatogr.* **151**, 87–89.
Bridges, C. D. B. (1976). *Exp. Eye Res.* **22**, 435–455.
Bui-Nguyen, M.-H., and Blanc, B. (1980). *Experientia* **36**, 374–375.
Bunnell, R. H. (1967). *In* "The Vitamins" (P. Gyorgy and W. N. Pearson, eds.), 2nd ed., Vol. VI, pp. 261–316. Academic Press, New York.
Bush, B. T., Frenz, J. H., Melander, W. R., Horvath, C., Cashmore, A. R., Dryer, R. N., Knipe, J. O., Coward, J. K., and Bertino, J. R. (1979). *J. Chromatogr.* **168**, 343–353.
Callmer, K., and Davies, L. (1974). *Chromatographia.* **7**, 643–650.
Carpenter, A. P. (1979). *J. Am. Oil Chem. Soc.* **56**, 668–671.
Carr, R. C., and Neff, J. M. (1980) *Anal. Chem.* **52**, 2428–2430.
Cashmore, A. R., Dreyer, R. N., Horvath, C., Knipe, J. O., Coward, J. K., and Bertino, J. R. (1980). *In* "Methods in Enzymology" (D. B. McCormick and L. D. Wright, eds), Vol. 66, pp. 459–468.
Cavins, J. F., and Inglett, G. E. (1974). *Cereal Chem.* **51**, 605–609.
Chapman, S. K., Greene, B. C., and Streiff, R. R. (1978). *J. Chromatogr.* **145**, 302–306.
Clifford, A. J. (1976). *Adv. Chromatogr. (N.Y.)* **13**, 1–36.
Clifford, A. J., and Clifford, C. K. (1976). *Nutr., Proc. Int. Congr., 10th, 1975*, 127.
Clifford, C. K., and Clifford, A. J. (1977). *J. Assoc. Off. Anal. Chem.* **60**, 1248–1251.
Cohen, H., and Lapointe, M. (1978). *J. Agric. Food Chem.* **26**, 1210–1213.
Cohen, H., and Lapointe, M. (1979). *J. Chromatogr. Sci.* **17**, 510–513.
Cohen, H., and Lapointe, M. (1980). *J. Assoc. Off. Anal. Chem.* **63**, 1158–1162.
Cohen, H., and Wakeford, B. (1980). *J. Assoc. Off. Anal. Chem.* **63**, 1163–1167.
Dabek, J. T., Harkonen, M., Wahlroos, O., and Adlercreutz, H. (1981). *Clin. Chem.* **27**, 1346–1351.
Day, B. P., and Gregory, J. F. (1980). *J. Agric. Food Chem.* **29**, 374–377.
De Jong, D. W., and Woodlief, W. G. (1978). *J. Agric. Food Chem.* **26**, 1281–1288.
De Leenheer, A. P., De Bevere, V. O., Cruyl, A. A., and Claeys, A. E. (1978). *Clin. Chem. (Winston-Salem, N.C.)* **24**, 585–590.
DeLuca, H. F. (1979). "Vitamin D: Metabolism and Function." Springer-Verlag, Berlin and New York.
Dennison, D. B., and Kirk, J. R. (1977). *J. Food Sci.* **42**, 1376–1379.
De Ruyter, M. G. M., and De Leenheer, A. P. (1976). *Clin. Chem.* **22**, 1593–1595.
De Ruyter, M. G. M., and De Leenheer, A. P. (1979). *Anal. Chem.* **51**, 43–46.
De Ruyter, M. G. M., Lambert, W. E., and De Leenheer, A. P. (1979). *Anal. Biochem.* **98**, 402–409.
De Vries, E. J., Zeeman, J., Esser, R. J. E., Borsje, B., and Mulder, F. J. (1979). *J. Assoc. Off. Anal. Chem.* **62**, 129–135.
De Vries, J. X., Gunthert, W., and Ding, R. (1980). *J. Chromatogr.* **221**, 161–165.
Donnahey, P. L., Burt, V. T., Rees, H. H., and Pennock, J. F. (1979). *J. Chromatogr.* **170**, 272–277.
Dunphy, P. J., Whittle, K. J., and Pennock, J. F. (1966). *Biochem. Chloroplasts, Proc., 1965* **1**, 165–171.
Egaas, E., and Lambertsen, G. (1979). *Int. J. Vitam. Nutr. Res.* **49**, 35–42.
Egberg, D. C., Heroff, J. C., and Potter, R. H. (1977). *J. Agric. Food Chem.* **25**, 1127–1132.
Eriksen, S. (1980). *J. Assoc. Off. Anal. Chem.* **63**, 1154–1157.
Eskins, K., Scholfield, C. R., and Dutton, H. J. (1977). *J. Chromatogr.* **135**, 217–220.
Fiksdahl, A., Mortensen, J. T., and Liaaen-Jensen, S. (1978). *J. Chromatogr.* **157**, 111–117.
Foppa, G. F. V. (1981). *Riv. Ital. Sostanze Grasse* **58**, 296–298.
Freed, M. (1966). "Methods of Vitamin Assay," 3rd ed. Wiley, New York.

Frenkel, E. P., Kitchens, R. L., and Prough, R. (1979). *J. Chromatogr.* **174**, 393–400.
Frolik, C. A., Tavela, T. E., and Sporn, M. B. (1978a). *J. Lipid Res.* **19**, 32–37.
Frolik, C. A., Tavela, T. E., Peck, G. L., and Sporn, M. B. (1978b). *Anal. Biochem.* **86**, 743–750.
Frolik, C. A., Roberts, A. B., Tavela, T. E., Roller, P. P., Newton, D. L., and Sporn, M. B. (1979). *Biochemistry* **18**, 2092–2097.
Fujitani, T. (1976). *Yukagaku* **25**, 46–48.
Geigert, J., Hirano, D. S., and Neidleman, S. L. (1981). *J. Chromatogr.* **206**, 396–399.
Green, J., Marcinkiewicz, S., and Watt, P. R. (1955). *J. Sci. Food Agric.* **6**, 274–282.
Gregory, J. F. (1980a). *J. Agric. Food Chem.* **28**, 486–489.
Gregory, J. F. (1980b). *Anal. Biochem.* **102**, 374–379.
Gregory, J. F., and Kirk, J. R. (1977). *J. Food Sci.* **42**, 1073–1076.
Gregory, J. F., and Kirk, J. R. (1978). *J. Food Sci.* **43**, 1801–1815.
Gregory, J. F., and Kirk, J. R. (1979). *Am. J. Clin. Nutr.* **32**, 879–883.
Groenendijk, G. W. T., De Grip, W. J., and Daemen, F. J. M. (1979). *Anal. Biochem.* **99**, 304–310.
Groenenddijk, G. W. T., De Grip, W. J., and Daemen, F. J. M. (1980). *Biochim. Biophys. Acta* **617**, 430–438.
Gubler, C. J., and Hemming, B. C. (1980). In "Methods in Enzymology" (D. B. McCormick and L. D. Wright, eds.), Vol. 62, pp. 63–68. Academic Press, New York.
Gyorgy, P., and Pearson, W. N. (1967). "The Vitamins," 2nd ed., Vols. VI and VII. Academic Press, New York.
Halley, B. A., and Nelson, E. C. (1979). *J. Chromatogr.* **175**, 113–123.
Hanewald, K. H., Mulder, F. J., and Keuning, K. J. (1968). *J. Pharm. Sci.* **57**, 1308–1312.
Hanni, R., and Bigler, F. (1977). *Helv. Chim. Acta* **60**, 881–887.
Hanni, R., Bigler, F., Meister, W., and Englert, G. (1976). *Helv. Chim. Acta* **59**, 2221–2227.
Hanni, R., Bigler, F., Vetter, W., Englert, G., and Loeliger, P, (1977). *Helv. Chim. Acta* **60**, 2309–2325.
Hanni, R., Hervouet, D., and Busslinger, A. (1979). *J. Chromatogr.* **162**, 615–621.
Haroon, Y., Shearer, M. J., McEnery, G., Allan, V. E., and Barkham, P. (1980a). *Proc. Nutr. Soc.* **39**, 49A.
Haroon, Y., Shearer, M. J., and Barkhan, P. (1980b). *J. Chromatogr.* **200**, 293–299.
Haroon, Y., Shearer, M. J., and Barkhan, P. (1981). *J. Chromatogr.* **206**, 333–342.
Hashmi, M. (1972). "Assay of Vitamins in Pharmaceutical Preparations." Wiley, New York.
Hatam, L. J., and Kayden, H. J. (1979). *J. Lipid Res.* **20**, 639–645.
Hatano, H., Yamamoto, Y., Saito, M., Mochido, E., and Watanabe, S. (1973). *J. Chromatogr.* **83**, 373–380.
Havinga, E., Koevoet, A. L., and Verloop, A. (1955). *Recl. Trav. Chim. Pays-Bas* **74**, 1230–1242.
Head, M. K., and Gibbs, E. (1977). *J. Food Sci.* **42**, 395.
Hemming, B. C., and Gubler, C. J. (1980). *J. Liq. Chromatogr.* **3**, 1697–1712.
Henderson, S. K., and Wickroski, A. F. (1978). *J. Assoc. Off. Anal. Chem.* **61**, 1130–1134.
Hengen, N., Seiberth, V., and Hengen, M. (1978). *Clin. Chem. (Winston-Salem, N.C.)* **24**, 1740–1743.
Hjarde, W., Leerbeck, E., and Leth, T. (1973). *Acta Agric. Scand. Suppl.* **19**, 87–96.
Hofsass, H., Grant, A., Alcino, N. J., and Greenbaum, S. B. (1976). *J. Assoc. Off. Anal. Chem.* **59**, 251–260.
Horst, R. L., and Littledike, E. T. (1979). *J. Dairy Sci.* **62**, 1746–1751.
Hung, S. S. O., Young, C. C., and Slinger, S. J. (1980). *J. Assoc. Off. Anal. Chem.* **53**, 889–893.
Ikekawa, N., and Koizumi, N. (1976). *J. Chromatogr.* **119**, 227–232.
Ishii, K., Sarai, K., Sanemori, H., and Kawasaki, T. (1979a). *Anal. Biochem.* **97**, 191–195.
Ishii, K., Sarai, K., Sanemori, H., and Kawasaki, T. (1979b). *J. Nutr. Sci. Vitaminol.* **25**, 517–523.

Isler, O. (1971). "Carotenoids." Birkhaeuser, Basel.
Jansson, L., Nilsson, B., and Lindgren, R. (1980) *J. Chromatogr.* **181,** 242–247.
Johnson, R. J., and Metzlar, D. E. (1970). *In* "Methods in Enzymology" (D. B. McCormick and L. D. Wright, eds.), Vol. XVIII, Part A, pp. 433–470. Academic Press, New York.
Jones, G. (1978). *Clin. Chem. (Winston-Salem, N.C.)* **24,** 287–298.
Jones, G. (1980). *J. Chromatogr.* **221,** 27–37.
Jones, G., and DeLuca, H. F. (1975). *J. Lipid Res.* **16,** 448–453.
Kamman, J. F., Labuza, T. P., and Warthesen, J. J. (1980). *J. Food Sci.* **45,** 1497–1504.
Karrer, P., and Jucker, E. J. (1950). "Carotenoids." Elsevier, Amsterdam.
Keverling Buisman, J. A., Hanewald, K. H., Mulder, F. J., Roborgh, J. R., and Keuning, K. J. (1968). *J. Pharm. Sci.* **57,** 1326–1329.
Kissinger, P. T., Refshauge, C., Dreiling, R., and Adams, R. N. (1973). *Anal. Lett.* **6,** 465–477.
Kissinger, P. T., Felice, L. J., Riggin, R. M., Pachla, L. A., and Wenke, D. C. (1974). *Clin. Chem. (Winston-Salem, N.C.)* **20,** 992–997.
Knapstein, H., Puchel, P., and Scholz, H. (1979). *Fette, Seifen, Anstrichm.* **81,** 121–126.
Knudsen, C. G., Carey, S. C., and Okamura, W. H. (1980). *J. Am. Chem. Soc.* **102,** 6355–6356.
Kodicek, E., and Lawson, D. E. M. (1967). *In* "The Vitamins" (P. Gyorgy and W. N. Pearson, eds.), 2nd ed., Vol. VI, pp. 211–244. Academic Press, New York.
Kofler, M., Sommer, P. F., Bolliger, H. R., Schmidli, B., and Vecchi, M. (1962). *Vitam. Horm. (N.Y.)* **20,** 407–439.
Kohl, E. A., and Schaefer, P. C. (1981). *J. Liq. Chromatogr.* **4,** 2023–2037.
Koshy, K. T. (1980). *In* "Methods in Enzymology" (D. B. McCormick and L. D. Wright, eds.), Vol. 67, pp. 357–370. Academic Press, New York.
Koshy, K. T., and VanDerSlik, A. L. (1976). *Anal. Biochem.* **74,** 282–291.
Koshy, K. T., and VanDerSlik, A. L. (1977a). *Anal. Lett.* **10,** 523–537.
Koshy, K. T., and VanDerSlik, A. L. (1977b). *J. Agric. Food Chem.* **25,** 1246–1249.
Koshy, K. T., and VanDerSlik, A. L. (1978). *Anal. Biochem.* **85,** 283–286.
Koshy, K. T., and VanDerSlik, A. L. (1979). *J. Agric. Food Chem.* **27,** 180–183.
Koshy, K. T., and VanDerSlik, A. L. (1980). *J. Agric. Food Chem.* **28,** 161–162.
Krol, G. J., Mannan, C. A., Gemmill, F. Q., Hicks, G. E., and Kho, B. T. (1972). *J. Chromatogr.* **74,** 43–49.
Kummerow, F. A., Cho, B. H. S., Huang, W. Y.-T., Imai, H., Kamio, A., Deutsch, M. J., and Hooper, W. M. (1976). *Am. J. Clin. Nutr.* **29,** 579–584.
Lambert, P. W., Syverson, B. J., Arnaud, C. D., and Spelsberg, T. C. (1977). *J. Steroid Biochem.* **8,** 929–937.
Landen, W. O., and Eitenmiller, R. R. (1979). *J. Assoc. Off. Anal. Chem.* **62,** 283–289.
Lankelma, J., and Poppe, H. (1978). *J. Chromatogr.* **149,** 587–598.
Lankelma, J., Van Der Kleijn, E., and Jansen, M. J. T. (1980). *J. Chromatogr.* **182,** 35–45.
Lefevere, M. F., De Leenheer, A. P., and Claeys, A. E. (1979). *J. Chromatogr.* **186,** 749–762.
Lein, D. G., Campbell, H. M., and Cohen, H. (1980). *J. Assoc. Off. Anal. Chem.* **63,** 1149–1152.
Lim, K. L., Young, R. W., and Driskell, J. A. (1980). *J. Chromatogr.* **188,** 285–288.
Lofty, P. A., Jordi, H. C., and Bruno, J. V. (1981). *J. Liq. Chromatogr.* **4,** 155–164.
McCormick, A. M., Napoli, J. L., and DeLuca, H. F. (1978a). *Anal. Biochem.* **86,** 25–33.
McCormick, A. M., Napoli, J. L., Schnoes, H. K., and DeLuca, H. F. (1978b). *Biochemistry* **17,** 4085–4090.
McCormick, A. M., Napoli, J. L., Schnoes, H. K., and DeLuca, H. F. (1979). *Arch. Biochem.* **192,** 577–583.
McCormick, D. B., and Wright, L. D., eds. (1970–1980). "Methods in Enzymology" (S. P. Colowick and N. O. Kaplan, Ser. Eds.), Vol. 18, Part A (1970), Parts B and C (1971); Vol. 62 (1979); Vol. 66 (1980); and Vol. 67 (1980). Academic Press, New York.

MacIntyre, I. (1974). *FEBS Lett.* **48**, 122–125.
Mack, D. O. (1980). *J. Liq. Chromatogr.* **3**, 1005–1021.
McKenzie, R. M., Hellwege, D. M., McGregor, M. L., Rockley, N. L., Riquetti, P. J., and Nelson, E. C. (1978). *J. Chromatogr.* **155**, 379–387.
McMurray, C. H., and Blanchflower, W. J. (1979a). *J. Chromatogr.* **176**, 488–492.
McMurray, C. H., and Blanchflower, W. J. (1979b). *J. Chromatogr.* **178**, 525–531.
Mankel, A. (1979). *Dtsch. Lebensm. Rundsch.* **75**, 77–85.
Maruyama, T., Ushigusa, T., Kanematsu, H., Niiya, I., and Imamura, M. (1977). *Shokuhin Eiseigaku Zasshi* **18**, 487–492.
Mason, W. D., Amick, E. N., and Heft, W. (1980). *Anal. Lett.* **13**, 817–824.
Matthews, E. W., Byfield, P. G. H., Colston, K. W., Evans, I. M. A., Galante, L. S., and MacIntyre, I. (1974) *FEBS Lett.* **48**, 122–125.
Mino, M., Nishida, Y., Kijima, Y., Iwakoshi, M., and Nakagawa, S. (1979). *J. Nutr. Sci. Vitaminol.* **25**, 505–516.
Morita, E., and Mizuno, N. (1980). *J. Chromatogr.* **202**, 134–138.
Morton, R. A. (1965). "Biochemistry of Quinones." Academic Press, New York.
Mourot, D., Delepine, B., Boisseau, J., and Gayot, G. (1979). *Ann. Pharm. Fr.* **37**, 235–238.
Mulder, F. J., De Vries, E. J., and Borsje, B. (1979). *J. Assoc. Off. Anal. Chem.* **62**, 1031–1040.
Nair, P. P. (1966). *In* "Advances in Lipid Research" (R. Paoletti and D. Kirtchevsky, eds.), Vol. 4, pp. 227–256. Academic Press, New York.
Norman, A. W. (1972). *J. Nutr.* **102**, 1243–1246.
Nilsson, B., Johansson, B., Jansson, L., and Holmberg, L. (1978). *J. Chromatogr.* **145**, 169–172.
Nystrom, E., and Sjovall, J. (1975). *In* "Methods in Enzymology" (J. M. Lowenstein, ed.), Vol. 35, Part B, pp. 378–395. Academic Press, New York.
Okano, T., Mizuno, K., and Kobayashi, T. (1978). *J. Nutr. Sci. Vitaminol.* **24**, 511–518.
Okano, T., Matsuyama, N., Kobayashi, T., Kuroda, E., Kodama, S., and Matsuo, T. (1979). *J. Nutr. Sci. Vitaminol.* **25**, 479–493.
O'Reilly, W. J., Guelen, P. J. M., Hoes, M. J. A., and Van Der Kleyn, E. (1980). *J. Chromatogr.* **183**, 492–498.
Osadca, M., and Araujo, M. (1977). *J. Assoc. Off. Anal. Chem.* **60**, 993–997.
Osborne, D. R., and Voogt, P. (1978). "The Analysis of Nutrients in Foods." Academic Press, New York.
Paanakker, J. E., and Groenendijk, G. W. T. (1979). *J. Chromatogr.* **168**, 125–132.
Pachla, L. A., and Kissinger, P. T. (1976). *Anal. Chem.* **48**, 364–367.
Parris, N. A. (1978). *J. Chromatogr.* **157**, 161–170.
Parrish, D. B. (1977). *CRC Crit. Rev. Food Sci. Nutr.* **9**, 375–394.
Parrish, D. B. (1979). *CRC Crit. Rev. Food Sci. Nutr.* **12**, 29–57.
Parrish, D. B. (1980a). *CRC Crit. Rev. Food Sci. Nutr.* **13**, 161–187.
Parrish, D. B. (1980b). *CRC Crit. Rev. Food Sci. Nutr.* **13**, 337–352.
Pennock, J. F., Hemming, F. W., and Kerr, J. D. (1964). *Biochem. Biophys. Res. Commun.* **5**, 542–548.
Pickston, L. (1978). *N. Z. J. Sci.* **21**, 383–385.
Pilkiewicz, F. G., Pettei, M. J., Yudd, A. P., and Nakanishi, K. (1977). *Exp. Eye Res.* **24**, 421–423.
Pratt, J. M. (1972). "Inorganic Chemistry of Vitamin $B_{12}$." Academic Press, New York.
Puglisi, C. V., and De Silva, J. A. F. (1978). *J. Chromatogr.* **152**, 421–430.
Ranfft, K., and Ruckemann, H. (1978a). *Z. Lebens.-Unters. Forsch.* **166**, 13–14.
Ranfft, K., and Ruckemann, H. (1978b). *Z. Lebens.-Unters. Forsch.* **167**, 150–151.
Ray, A. C., Dwyer, J. N., and Reagor, J. C. (1977). *J. Assoc. Off. Anal. Chem.* **60**, 1296–1300.
Reed, L. S., and Archer, M. C. (1976). *J. Chromatogr.* **121**, 100–103.

Reeder, S. K., and Park, G. L. (1975). *J. Assoc. Off. Anal. Chem.* **58,** 595–598.
Reingold, R. N., and Picciano, M. F. (1982). *J. Chromatogr.* **234,** 171–179.
Roberts, A. B., Nichols, M. D., Frolick, C. A., Newton, D. L., and Sporn, M. B. (1978). *Cancer Res.* **38,** 3327–3332.
Roberts, A. B., Lamb, L. C., and Sporn, M. B. (1980). *Arch. Biochem. Biophys.* **199,** 374–383.
Rodriguez, M. S. (1978). *J. Nutr.* **108,** 1983–2103.
Rose, R. C., and Nahrwold, D. L. (1981). *Anal. Biochem.* **114,** 140–145.
Rotmans, J. P., and Kropf, A. (1975). *Vision Res.* **15,** 1301–1302.
Rouseff, R. (1979). *In* "Liquid Chromatographic Analysis of Food and Beverages" (G. Charalambous, ed.), Vol. 1, pp. 161–177. Academic Press, New York.
Ruckemann, H. (1980). *Z. Lebens.-Unters. Forsch.* **171,** 357–359.
Ruckemann, H. (1981). *Z. Lebens.-Unters. Forsch.* **173,** 113–116.
Ruckemann, H., and Ranfft, K. (1977). *Z. Lebens.-Unters. Forsch.* **164,** 272–273.
Ruckemann, H., and Ranfft, K. (1978). *Z. Lebens.-Unters. Forsch.* **166,** 151–152.
Sancmori, H., Ucki, H., and Kawasaki, T. (1980). *Anal. Biochem.* **107,** 451–455.
Seamark, D. A., Trafford, D. J. H., Hiscocks, P. G., and Makin, H. L. J. (1980). *J. Chromatogr.* **197,** 271–273.
Shaikh, B., Huang, H. S., and Zielinski, W. L. (1977a). *J. Assoc. Off. Anal. Chem.* **60,** 137–139.
Shaikh, B., Pontzer, N. J., Huang, S. S., and Zielinski, W. L. (1977b). *J. Chromatogr. Sci.* **15,** 215–217.
Shaw, W. H. C., Jefferies, J. P., and Holt, T. E. (1957). *Analyst (London)* **82,** 2–7.
Shearer, M. J., Allan, V., Haroon, Y., and Barkhan, P. (1980). *In* "Vitamin K Metabolism and Vitamin K-Dependent Proteins" (J. W. Suttie, ed.), pp. 317–327. Univ. Park Press, Baltimore, Maryland.
Shephard, R. M., and DeLuca, H. F. (1980). *Arch. Biochem. Biophys.* **202,** 43–53.
Shephard, R. M., Horst, R. L., Hamstra, A. J., and DeLuca, H. F. (1979). *Biochem. J.* **182,** 55–69.
Slover, H. T., Lehmann, J., and Valis, R. J. (1969). *J. Amer. Oil Chem. Soc.* **46,** 417–420.
Smith, M. D. (1980). *J. Chromatogr.* **182,** 285–291.
Soderhjelm, P., and Andersson, B. (1978). *J. Sci. Food Agric.* **29,** 697–702.
Sood, S. P., Sartori, L. E., Wittmer, D. P., and Haney, W. G. (1976). *Anal. Chem.* **48,** 796–798.
Stea, B., Halpern, R. M., and Smith, R. A. (1979). *J. Chromatogr.* **168,** 385–393.
Steuerle, H. (1975). *J. Chromatogr.* **115,** 447–453.
Steuerle, H. (1981). *J. Chromatogr.* **206,** 319–326.
Stewart, I. (1977a). *J. Assoc. Off. Anal. Chem.* **60,** 132–136.
Stewart, I. (1977b). *J. Agric. Food Chem.* **25,** 1132–1137.
Stewart, I., and Wheaton, T. A. (1971). *J. Chromatogr.* **55,** 325–336.
Stewart, J. T., Honigberg, I. L., Brant, J. P., Murray, W. A., Webb, J. L., and Smith, J. B. (1976). *J. Pharm. Sci.* **65,** 1536–1539.
Stillman, R., and Ma, T. S. (1974). *Mikrochim. Acta 1974,* 641–648.
Stout, R. W., Cashmore, A. R., Coward, J. K., Horvath, C. G., and Bertino, J. R. (1976). *Anal. Biochem.* **71,** 119–124.
Strohecker, R., and Henning, H. M. (1965). "Vitamin Assay: Tested Methods." Verlag Chemie, Weinheim.
Stulik, K., and Pacakova, V. (1980). *J. Chromatogr.* **192,** 135–141.
Takada, K., Okano, T., Tamura, Y., Matsui, S., and Kobayashi, T. (1979). *J. Nutr. Sci. Vitaminol.* **25,** 385–398.
Tangney, C. C., McNair, H. M., and Driskell, J. A. (1981). *J. Chromatogr.* **224,** 389–387.
Thompson, J. N., and Hatina, G. (1979). *J. Liq. Chromatogr.* **2,** 327–344.
Thompson, J. N., and Maxwell, W. B. (1977). *J. Assoc. Off. Anal. Chem.* **60,** 766–771.
Thompson, J. N., Erdody, P., and Maxwell, W. B. (1972). *Anal. Biochem.* **50,** 267–280.

Thompson, J. N., Maxwell, W. B., and L'Abbe, M. (1977). *J. Assoc. Off. Anal. Chem.* **60,** 998–1002.
Thompson, J. N., Hatina, G., and Maxwell, W. B. (1979). *In* "Trace Organic Analysis: A New Frontier in Analytical Chemistry" (H. S. Hertz and S. N. Chester, eds.). *NBS Spec. Publ. (U.S.)* **519,** 279–288.
Thompson, J. N., Hatina, G., and Maxwell, W. B. (1980). *J. Assoc. Off. Anal. Chem.* **63,** 894–898.
Thrivikraman, K. V., Refshauge, C., and Adams, R. N. (1975). *Life Sci.* **15,** 1335–1342.
Toma, R. B., and Tabekhia, M. M. (1979). *J. Food Sci.* **44,** 263–268.
Tomkins, D. F., and Tscherne, R. J. (1974). *Anal. Chem.* **46,** 1602–1604.
Tryfiates, G. P., and Sattsangi, S. (1982). *J. Chromatogr.* **227,** 181–186.
Tsao, C. S., and Salimi, S. L. (1981). *J. Chromatogr.* **224,** 277–480.
Tsukida, K., Kodama, M. I., Kawamoto, M., and Takahashi, K. (1977). *J. Nutr. Sci. Vitaminol.* **23,** 263–264.
Tsukida, K., Masahara, R., and Ito, M. (1980). *J. Chromatogr.* **192,** 395–401.
Tyler, T. A., and Shrago, R. S. (1980). *J. Liq. Chromatogr.* **3,** 269–277.
Udenfriend, S. (1969). "Fluorescence Assay in Biology and Medicine." Academic Press, New York.
Unger, K., and Nyamah, D. (1974). *Chromatographia* **7,** 63–68.
Vahlquist, A. (1980). *Experientia* **36,** 317–318.
Vanderslice, J. T., and Maire, C. E. (1980). *J. Chromatogr.* **196,** 176–179.
Vanderslice, J. T., Stewart, K. K., and Yarmas, M. M. (1979). *J. Chromatogr.* **176,** 280–285.
Vanderslice, J. T., Maire, C. E., Doherty, R. F., and Beecher, G. R. (1980). *J. Agric. Food Chem.* **28,** 1145–1149.
Van De Weerdhof, T., Wiersum, M. L., and Reissenweber, H. (1973). *J. Chromatogr.* **83,** 455–460.
Vanhaelen-Fastré, R., and Vanhaelen, M. (1979). *J. Chromatogr.* **179,** 131–142.
Van Niekerk, P. J. (1973). *Anal. Biochem.* **52,** 533–537.
Van Niekerk, P. J., and Smit, S. C. C. (1980). *J. Am. Oil Chem. Soc.* **57,** 417–421.
Vatassery, G. T., Maynard, V. R., and Hagen, D. F. (1978). *J. Chromatogr.* **161,** 299–302.
Vecchi, M., Vesely, J., and Oesterhelt, G. (1973). *J. Chromatogr.* **83,** 447–453.
Waddell, W. H., Crouch, R., Nakanishi, K., and Turro, N. J. (1976). *J. Am. Chem. Soc.* **98,** 4189–4192.
Westerberg, E., Friberg, M., and Akesson, B. (1981). *J. Liq. Chromatogr.* **4,** 109–121.
Widicus, W. A., and Kirk, J. R. (1979). *J. Assoc. Off. Anal. Chem.* **62,** 637–641.
Williams, A. K. (1980). *In* "Methods in Enzymology" (D. B. McCormick and L. D. Wright, eds.), Vol. 62, pp. 415–422. Academic Press, New York.
Williams, A. K., and Cole, P. D. (1975). *J. Food Chem.* **23,** 915–916.
Williams, R. C., Schmit, J. A., and Henry, A. (1972). *J. Chromatogr. Sci.* **10,** 494–501.
Wills, R. B. H., Shaw, C. G., and Day, W. R. (1977). *J. Chromatogr. Sci.* **15,** 262–268.
Wittmer, D., and Haney, W. G. (1974). *J. Pharm. Sci.* **63,** 588–590.
Wong, F. F. (1978). *J. Agric. Food Chem.* **26,** 1444–1446.
Woollard, D. C., and Woollard, G. A. (1981). *N. Z. J. Dairy Sci. Technol.* **16,** 99–112.
Yasumoto, K., Tadera, K., Tsuji, H., and Mitsuda, H. (1975). *J. Nutr. Sci. Vitaminol.* **21,** 117–127.
Yoshida, T., Yunoki, N., Nakazima, Y., Kaito, T., and Anmo, T. (1978). *Yakugaku Zasshi* **98,** 1319–1326.
Zakaria, M., Simpson, K., Brown, P. R., and Krstulovic, A. (1979). *J. Chromatogr.* **176,** 109–117.
Zechmeister, L. (1962). "Cis–Trans Isomeric Cartenoids, Vitamin A and Arylpolyenes." Springer, Berlin and New York.

# COMBINING LIQUID CHROMATOGRAPHY WITH MASS SPECTROMETRY

### R. C. Willoughby and R. F. Browner

School of Chemistry
Georgia Institute of Technology
Atlanta, Georgia

|      |                                                                                  |     |
|------|----------------------------------------------------------------------------------|-----|
| I.   | Introduction                                                                     | 69  |
| II.  | The Mass Spectrometer as a Detector for Liquid Chromatography                    | 70  |
|      | A. The Mass Spectrometer as an Ideal Detector                                    | 71  |
|      | B. The Compatibility of Mass Spectrometry with HPLC                              | 71  |
| III. | Off-Line Techniques                                                              | 74  |
| IV.  | On-Line Techniques                                                               | 77  |
|      | A. Direct-Coupling Techniques                                                    | 77  |
|      | B. Mechanical Transfer Techniques                                                | 83  |
|      | C. LC–MS Interfacing Using GC–MS Approaches                                      | 93  |
|      | D. Aerosol-Jet Techniques                                                        | 94  |
| V.   | Alternative Methods of Vaporization and Ionization with Possible Applications to LC–MS | 97  |
|      | A. Vaporization Enhancement                                                      | 98  |
|      | B. Alternative Ionization Techniques                                             | 98  |
| VI.  | Applications and Conclusions                                                     | 101 |
|      | References                                                                       | 105 |

## I. INTRODUCTION

Modern high-performance liquid chromatography (HPLC) has developed into a rapid and powerful technique for the analysis of compounds that are not readily handled by gas chromatography (GC). By the use of HPLC, compounds of poor thermal stability and low volatility may often be readily separated for either quantitative or qualitative analysis. As a general rule, the amount of information that can be derived from any chromatographic separation, however effective, is dependent upon the detector. Conventional methods of detection (UV/visible, refractive index, fluorescence, electrochemical, etc.) all suffer to a greater or lesser degree from limitations resulting from lack of selectivity, sensitivity, and/ or versatility, and none give structural information on compounds eluted from a column. As the range of applications for HPLC increases, the limitations of these "classical" detectors become increasingly restrictive to the growth of the technique, and this has led a number of researchers to look in new directions for HPLC detectors with more informing power.

One area of detection which has great growth potential is that of mass spectrometry (MS). The well-developed and analytically powerful application of various GC–MS interfaces has led many analysts to believe that a similarly effective and simple hybrid system may be developed for interfacing HPLC and mass spectrometry (LC–MS). However, fundamental problems of compatibility are encountered with this approach, and this has led to the development of LC–MS interfaces of considerable ingenuity, but unfortunately usually of considerable complexity as well. Of some help in this direction has been the recent development of ionization techniques that are better suited to thermally labile and involatile compounds than the traditional electron-impact (EI) and chemical ionization (CI) modes. These new ionization modes place less stringent requirements on the LC–MS interface.

The nature of the information obtained from using a mass spectrometer as an HPLC detector depends on how the two systems are combined. Practically, researchers have used MS detectors in one of two modes of operation: on-line or off-line. On-line operation refers to continuous transfer of all or part of the column effluent stream to the detector, whereas off-line refers to discontinuous or discrete transfer of column effluent to the detector. Respective advantages and disadvantages of each type of approach depend upon a wide variety of factors. These include the properties of the sample; whether structural elucidation is required; and whether quantitation is required or merely identification of particular peaks. It is important to stress that there are many factors involved in evaluating the applicability of LC–MS to any given analytical problem.

This article attempts to point out salient criteria with which to assess LC–MS in its present state of the art and discusses possible directions for future development. (LC–MS has recently been reviewed by Kenndler and Schmid, 1978; Dawkins and McLafferty, 1978; Arpino and Guiochon, 1979; McFadden, 1979, 1980; McLafferty, 1980; and Games, 1980a.)

## II. THE MASS SPECTROMETER AS A DETECTOR FOR LIQUID CHROMATOGRAPHY

Independently, both HPLC and MS have well-established analytical capabilities. The primary objective of combining these techniques is to add a new dimension to the analysis by the use of the LC–MS combination while preserving the capabilities of the separate techniques. Over a number of years, GC–MS has been developed to a point where it can be used to yield highly sensitive and detailed structural information on the individual components of quite complex mixtures. Whereas certain aspects of GC–MS technology, such as computer software and selected technical features, are directly transferable to LC–MS, nevertheless many unique and difficult problems are associated specifically with the technology of the LC–MS interface. It is in these areas that current research effort is focused.

## A. The Mass Spectrometer as an Ideal Detector

The main characteristics of an ideal LC detector are listed in Table I. All practical detectors, including the mass spectrometer, fall short of these ideals to some degree. Nevertheless, the mass spectrometer is capable of fulfilling many of these ideal characteristics, as indicated by the performance ratings in Table I. The appeal of the mass spectrometer as an LC detector lies in its ability to fulfill most of the ideal capabilities of conventional LC detectors, with the additional benefit of compound identification for a wide range of solutes. Unlike other detectors, the mass spectrometer has the unique ability to provide molecular weight and/or structural information from fragmentation spectra, which can contribute significantly to compound identification. Generally, the wide range of analyzable compounds is limited only by solutes that do not readily form gas-phase ions; however, newer volatility-enhancement and ionization techniques have extended the range of detectable compounds. Table I clearly shows the qualitative capabilities of off-line LC–MS as being superior to most on-line techniques; however, the convergence of off- and on-line qualitative capabilities becomes more apparent with each new development in on-line techniques.

## B. The Compatibility of Mass Spectrometry with HPLC

To date, all attempts to combine HPLC and MS have to some degree diminished the effective operation of one or more of the component parts in both systems. Ideally, it is desirable to have little or no compromise in the normal operating conditions of each system; however, there are certain fundamental incompatibilities between the two techniques that necessitate some degree of compromise, and these are listed in Table II. Problems associated with maintaining efficient mass transfer while avoiding sample decomposition provide one of the major obstacles to simple LC–MS interfacing. A better understanding of these basic differences in the requirements of each individual technique can help to clarify some of the practical difficulties that are encountered in developing LC–MS combinations.

In GC–MS, both analyte and mobile phase are gases and as such may be

## TABLE I
### Criteria for Ideal LC Detectors

| Ideal-detector criteria | Mass spectrometer performance rating[a] | | | Ideal-detector criteria | Mass spectrometer performance rating[a] | | |
|---|---|---|---|---|---|---|---|
| | Off-line | On-line | | | Off-line | On-line | |
| | | DLI | MB | | | DLI | MB |
| Aids in compound identification | +++ | +[b] | ++[c] | Large linear dynamic range | − | + | + |
| High sensitivity | +++ | ++ | ++ | Maintains chromatographic integrity | − | + | + |
| Response to all solutes | +++ | + | + | Inexpensive | − − | − − | − − |
| Lack of mobile-phase response | +++ | +[d] | ++ | Ease of operation | −(− −)[e] | − | − |
| Nondestructive nature | +++ | + | + | Simplicity of design | −(− −) | − | − |

[a] +, indicates fulfillment of ideal criteria; −, indicates lack of fulfillment of ideal criteria; DLI, direct liquid introduction; MB, moving belt.
[b] CI only.
[c] EI and CI available.
[d] Solvent used as CI reagent gas.
[e] Values in parentheses indicate range of behavior.

## TABLE II

### Incompatibilities between LC and MS

| Characteristic | LC | MS | Consequence of compromise |
|---|---|---|---|
| Flow rate | 2 µl/min–2 ml/min (liquid) | 0.1–2 ml/min (gas) | Low sample yield |
| Temperature | Ambient | 200°C+ | Sample decomposition |
| Pressure | Atmospheric | EI $\approx 10^{-4}$<br>CI $\approx$ 1 Torr | Low sample yield |
| Sample | Unstable, involatile | Gas-phase ions | Sample decomposition |
| Sample matrix | Solvent plus additives | Vacuum or reagent gas | Interferences |

directly accessed to the MS ionization chamber. When EI excitation is to be used, the carrier gas must be largely eliminated, but this is readily accomplished by membrane or jet-separator interfaces. Here, the relatively large difference in molecular weight between typical samples and carrier gases ($N_2$, He, etc.) makes the process relatively simple and efficient. On the other hand, the effluent from an LC column contains both sample and solvent in a liquid phase, which must be converted to a gaseous state before it can be passed to the MS ionization chamber. In addition, the chemical and physical properties of the mobile phase and the sample may be much closer than in GC, making effective separation by differences in physical properties such as rates of diffusion much less likely. If the introduction of solvent vapor into the ionization source is acceptable such that ionization takes place by a CI mode with ionized solvent vapor, this latter limitation may not be so severe. Nevertheless, LC mobile phases may possibly be quite hostile to the MS ionization-chamber environment, containing possibly acids, bases, ion-pairing agents, chelating agents, etc. Solvent impurities and column bleed may also introduce additional components likely to interfere with proper operation of the mass spectrometer. Cumulatively, these mobile-phase constituents may place severe restrictions on the performance of the mass spectrometer, including background interferences, clogging of the inlets, and physical deposition on the filament.

The energy needed to vaporize the thermally labile and low-volatility solutes as well as the elevated ion-source temperatures will often favor decomposition or structural alteration over the formation of intact gas-phase solute molecules. If decomposition or alteration occurs during the vaporization process, then spectra will be difficult to interpret and identification of unknown solutes may become impossible.

The solute is typically at or near trace concentration levels in the mobile phase. This creates a dichotomy of design objectives, particularly for on-line applications. On the one hand, the flow rate must be increased to maximize the amount of solute introduced into the mass spectrometer, but on the other hand, the maximum pressure that can be tolerated would be far exceeded if there were to be complete introduction at normal effluent flow rates. A normal flow rate of 2 ml of effluent per minute could produce as much as 1000 ml/min of gas, whereas the mass spectrometer can only accept gas in the range of 0.1–2 ml/min under normal pumping capacities. Resolving this fundamental conflict requires a reduction in flow rate by either (1) splitting the effluent stream to control mass spectrometer intake; (2) using some means of solvent separation to exclude everything but solute from the ionization chamber; (3) increasing the pumping capacity; (4) using lower flow micro-high-performance liquid chromatography ($\mu$-LC); or (5) using ionization techniques that allow inlet at higher pressures. Overcoming flow incompatibilities may necessitate specialized modifications of the inlets, ion sources, and vacuum systems of currently available commercial mass spectrometers.

## III. OFF-LINE TECHNIQUES

Unlike GC–MS, where on-line techniques are used almost exclusively, LC–MS has been widely used off-line. Generally, this involves fraction collection followed by solvent removal and subsequent introduction of the sample into the mass spectrometer. This procedure has several practical benefits:

(1) There is no limitation on the LC separation conditions (e.g., high flow of preparative LC).

(2) The interim between fraction collection and introduction into the mass spectrometer places no restrictions on treatment, storage, and transport of the sample to obtain optimum detection.

(3) The mass spectrometer is not limited in type or operating conditions by the requirements of the LC separation.

Off-line procedures therefore provide considerable flexibility in the analysis although they are generally more cumbersome and less convenient than on-line procedures.

Column effluent can be collected either manually or with automated batchwise collection devices. When simple manual collection is used such that conventional on-line detection is required to locate peaks, the system is generally only used for qualitative analysis. When automated batchwise collection procedures are used, however, quantitation is possible. Huber *et al.* (1973) described the theoretical and practical capabilities for automated batchwise collection. With

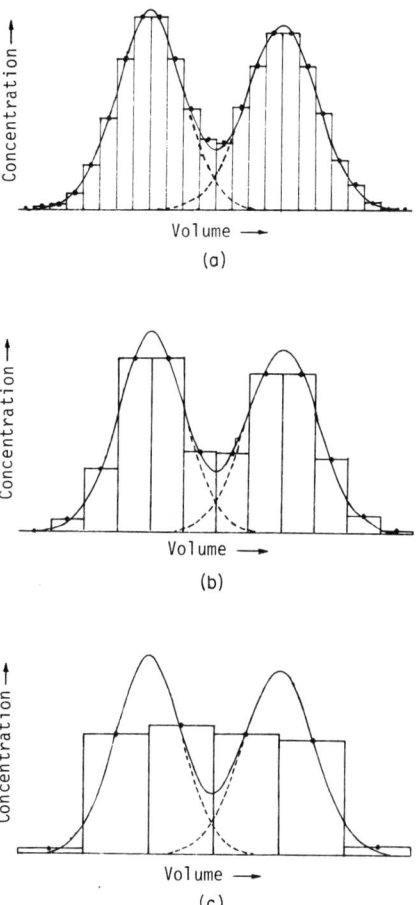

**Fig. 1.** Analog elution curve (—), histogram (-), and digital data (•) for two Gaussian peaks of equal area separated with a resolution of $R = 4$. Fraction size $\Delta V/\sigma_v = \tfrac{1}{2}$ (a), 1 (b), and 2 (c). These curves show the worst distortion of the peak, having the boundary between adjacent fractions falling at the peak maxima. [With permission, from Huber et al. (1973).]

batchwise collection, it is possible to plot histograms of fraction concentration versus elution volume. From the histograms, analog elution curves can be derived which give retention volumes and peak areas. The effects of fraction volume on these curves are shown in Fig. 1. Satisfactory accuracy can be obtained for retention volume and peak area when $\Delta V/\sigma_v = \tfrac{1}{2}$, where $\Delta V/\sigma_v$ is the relative fraction volume, $\Delta V$ is the fraction volume, and $\sigma_v$ is the volume standard deviation of the peak. The quantitative capability of batchwise collection depends on separation conditions. If eluting peaks are narrower than the fraction-volume lower limits, poor accuracy will result. Huber also demonstrated the possibilities for recycling of fractions, which allows better separations, higher purity, and potentially improved quantitative capabilities.

After collection, the sample can be treated in a variety of ways before mass analysis. Volatile solvents are easily removed by evaporation. The sample may be extracted from an involatile solvent to another, volatile solvent for subsequent ease of solvent removal. Buffers and other additives can also be removed by extraction. Chemically unstable compounds may be derivatized to allow vaporization and ionization without decomposition. The manipulation of the sample during this interim can eliminate many interferences and technical difficulties associated with on-line techniques.

Superior detection possibilities are available for unstable LC solutes using the newer ionization techniques discussed in Section V. These are generally suited to direct probe introduction of fraction solutes. Heresch and Jurenitsch (1979) and Heresch *et al.* (1979) demonstrated the quantitative use of direct-insertion MS with automated batchwise collection. Satisfactory results were obtained for natural mixtures, even though separation was incomplete, by the use of an internal standard. Since the sample is only consumed to a very small extent, a single fraction or sample can be determined repeatedly by using different ionization conditions to yield both molecular weight and structural information.

Lovins *et al.* (1973, 1974) developed an off-line interface that semi-automatically performed fraction collection, desolvation, and probe introduction. This design is shown in Fig. 2. The effluent was isolated in a capillary tube, where 10–50 μl fractions were desolvated by flash vaporization. Between 60 and 80% of the involatile solute remained on a gold gauze or adsorbant surface, whereas the solvent vapors were pumped away. The gauze served as the tip of a

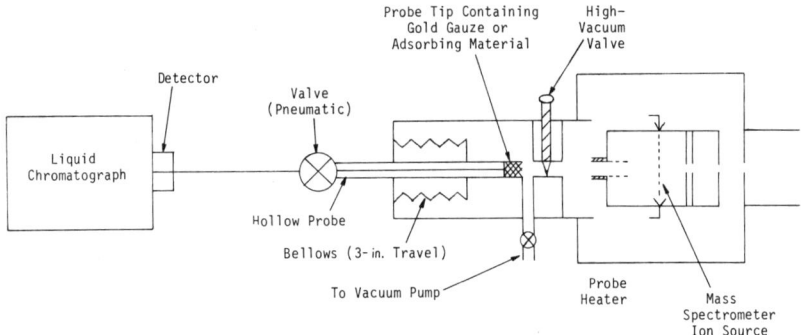

**Fig. 2.** Schematic of LC–MS interface with a motor-driven probe. [With permission, from Lovins *et al.* (1974).]

direct-inlet probe and was inserted into the EI ion source. Upon removal, the probe was baked at 280–300°C to minimize memory effects. The complete cycle took 3–5 min and in some cases required the LC to operate in the stop-flow mode (causing band spreading) to isolate closely spaced peaks. The semiautomatic operation included a motor-driven probe. However, problems were encountered with the loss of the more volatile solutes during the desolvation process.

Off-line LC–MS is clearly a less than ideal procedure, requiring considerable time and effort on the part of the analyst. Nevertheless, in the absence of totally satisfactory on-line systems, it will at least allow analysis to be performed effectively until such time as LC–MS interfaces can match the routine reliability of GC–MS interfaces.

## IV. ON-LINE TECHNIQUES

### A. Direct-Coupling Techniques

The most obvious approach to on-line LC–MS interfacing involves coupling the liquid chromatograph directly to the mass spectrometer and introducing column effluent straight into the ionization chamber. Although the potential for developing a workable interface based on this approach might be considered questionable in light of inherent incompatibilities between the two systems (as outlined in Section II,B), recent developments in both LC and MS have nevertheless given further credibility to this approach.

*1. Traditional LC Interface*

In the late 1960s Tal'roze et al. (1968, 1969a, 1972) developed a system whereby liquid solutions of volatile organic compounds were introduced directly into the ionization chamber of an EI-mode MS by immersing a tiny capillary in a test solution. The flow of liquid resulted from the pressure drop between the liquid reservoir and the vacuum of the mass spectrometer. This technique was later adapted to sampling the effluent emerging from an LC column filled with powdered-alumina gel (Tal'roze et al., 1969b, 1970). This was the first working on-line LC–MS (see Fig. 3), but it suffered from lack of suitability to samples of complex nature and thermal instability. Also, the 100–1000 Å slit, necessary to limit flow and thus maintain the source pressure, easily clogged with suspended particulates or crystals resulting from solute precipitation. It was also necessary for both solvents and solutes to be relatively volatile. This design was soon outdated due to changes in both LC and MS; specifically, HPLC replaced traditional LC for most analytical applications, and CI rather than EI ionization permitted the source to operate at higher pressures.

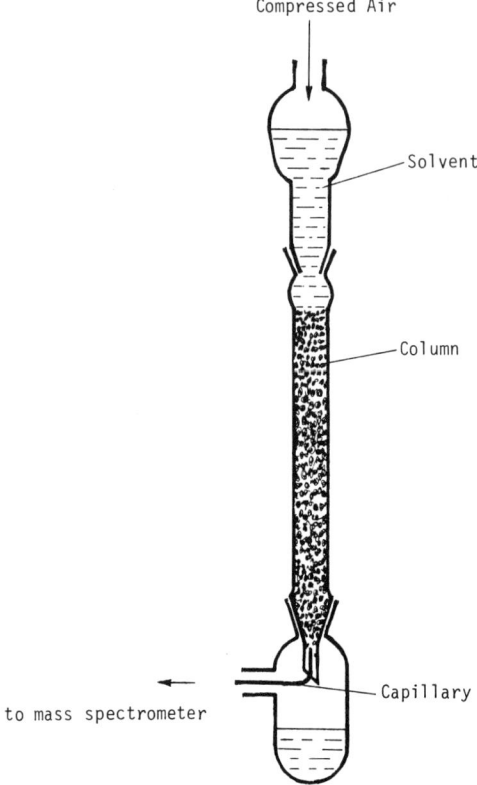

**Fig. 3.** Schematic of early LC-MS interface using small capillary to sample column effluent. [With permission, from Tal'roze *et al.* (1969b).]

## 2. Direct Liquid Introduction (DLI)

Chemical ionization mass spectrometry (CIMS) was developed for analytical applications in the late 1960s (Munson and Field, 1966; Arsenault *et al.*, 1969; Fales *et al.*, 1969). The operating conditions of CI allow several orders of magnitude higher pressure in the source when compared to EI. Thus CI is more readily adapted to direct LC effluent introduction than EI because solvent vapor itself may be used as the ionization reagent.

Baldwin and McLafferty (1973b) reported the feasibility of using "direct" CI (Baldwin and McLafferty, 1973a) to monitor HPLC effluent. In this approach, a 10-μl portion of sample solution was introduced into the mass spectrometer through a glass capillary. Using the solvent vapor as reagent gas, flow rates of 0.01 ml/min were possible while maintaining a pressure in the ionization chamber between 0.05 and 1 Torr, depending on the solvent used. This led to an interface design in which an inlet probe replaced the direct probe of a magnetic-sector mass spectrometer, allowing continuous solution introduction into the mass spectrometer (Arpino et al., 1974a). This design is shown in Fig. 4. This system can be considered a "nondestructive" detector in the sense that only approximately 1% of the LC effluent is introduced into the mass spectrometer. The remaining 99% of the effluent bypasses the mass spectrometer for either recovery or possibly further detection with auxiliary detectors. The small mass reaching the mass spectrometer unfortunately results in lower sensitivity. A major consideration with this approach is the need to find solvents that are compatible with LC separation conditions and also serve as adequate ionization reagents. However, for several solvent–solute combinations, detection limits of $10^{-8}$–$10^{-9}$ g of solute in the effluent were reported, even with the 1 : 100 split ratio.

In order to enhance the sensitivity of this system, Arpino et al. (1974b) made two modifications to the initial design. First, the pumping rate was increased by using a combination of a cryogenic pump and a 6-in. diffusion pump. The diffusion pump removed air at the source entrance while the cryogenic pump, located immediately above the source, trapped gases in the source region. This configuration allowed increased flow rates while maintaining the source pressure at $10^{-4}$ Torr. Regeneration of the cold trap was usually performed overnight. The second modification was of the ion-source block. By increasing the exit-slit size in the source it proved possible to obtain lower pressures, less high-molecular-weight clustering, and a more intense ion beam. These factors, as well as the use of an on-line minicomputer (McLafferty et al., 1975), contributed to enhanced sensitivity.

Currently, there are at least two commercial DLI interfaces that are modifications of the Baldwin–McLafferty DLI design available, from Hewlett-Packard and Ribermag. The DLI interface shown in Figs. 5a and 5b is the Hewlett-Packard DLI described by Melera (1979a,b). This interface consists of a direct probe having a stainless-steel diaphragm at the tip (Fig. 5a). The center of the diaphragm has a 5–10-μm orifice through which approximately 1–4% of the effluent stream is sampled into the desolvation chamber. A liquid jet emerges from the orifice and shatters into tiny droplets that evaporate before entering the CI source. The probe is water-cooled to avoid vapor-lock buildup in the sample stream. Figure 5b shows the large-surface-area cold fingers of the nitrogen-cooled cryogenic pump. This pump increases the pumping rate by an order of magnitude over normal diffusion-pump operation. The pumping not only enhances sensitivity by allowing increased sample introduction into the source, but also traps many impurities normally present in the mass spectrometer manifold, which reduces the background spectrum.

In order to avoid much of the instrumental complexity inherent in the Bald-

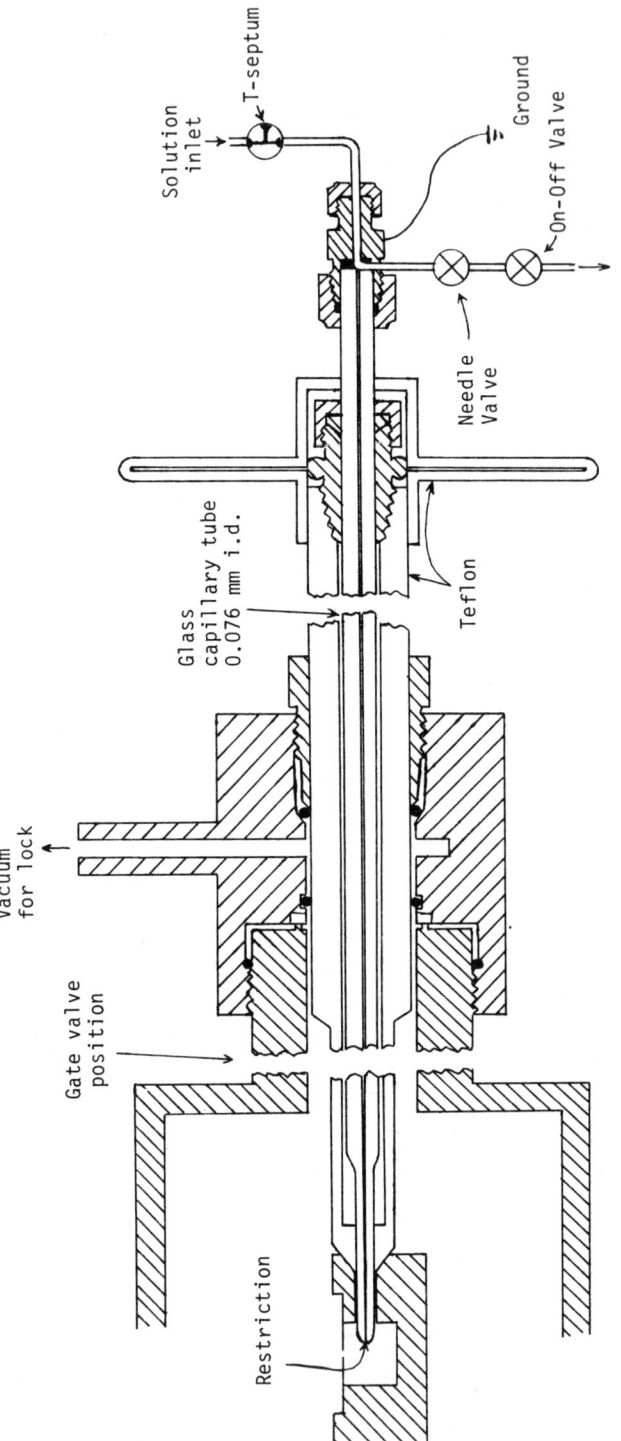

**Fig. 4.** Inlet probe for continuous introduction of solutions: replaces the direct-introduction probe of the RMH-2, utilizing the same vacuum-lock system. [With permission, from Arpino et al. (1974a).]

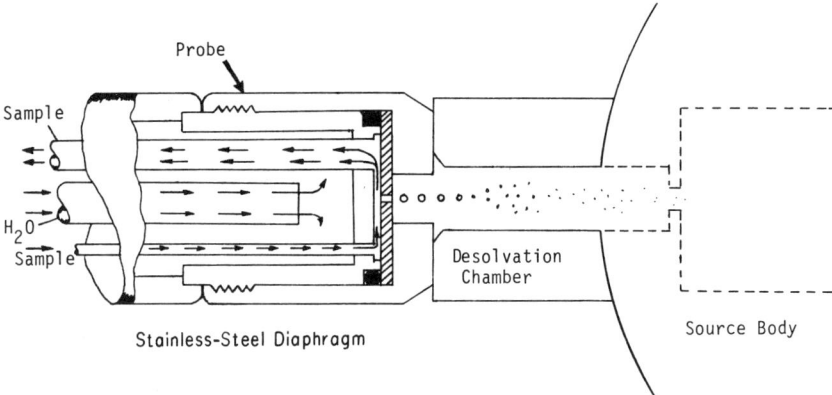

**Fig. 5a.** Probe tip of the Hewlett-Packard DLI interface showing evaporation of the droplet stream.

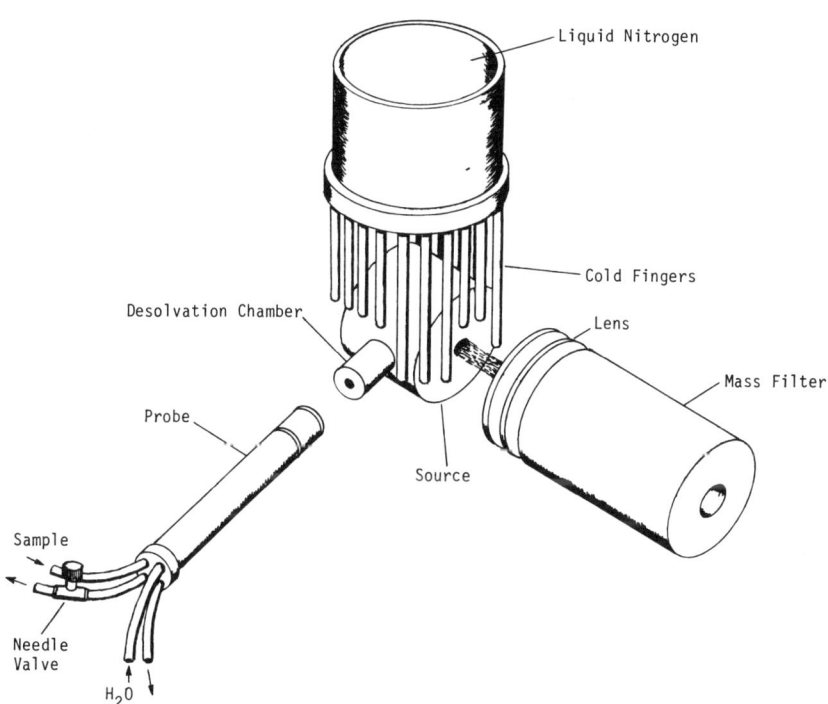

**Fig. 5b.** Diagram of the Hewlett-Packard DLI interface including the probe, source, cold fingers, and mass filter. [Figures 5a and 5b from Melera (1979a), with permission.]

win–McLafferty approach to DLI, Henion (1978a,b) chose to reduce the burden on the interface by cutting back drastically on sample flow rate. His initial approach was based on the use of a glass capillary probe which was placed in the solids inlet to the CI source of a quadrupole mass spectrometer. By using a stream splitter, only about 1% of the LC flow passed to the interface, and under these conditions no special modifications of the normal operation of the CI source were necessary. The system proved useful for drugs, metabolites, and other biologically important substances, and nanogram sensitivities were reported. As a logical step in this direction, the direct coupling of the effluent from a $\mu$-LC, operating at 2–40 $\mu$l/min solvent flow, was attempted. Henion (1979, 1981) and Henion and Maylin (1980) introduced the entire effluent from the $\mu$-LC into the ion source. This resulted in lower detection limits and better specificity than other $\mu$-LC detectors such as the micro-cell variable-UV detector. Price *et al.* (1980) and Rottschaefer *et al.* (1979) also developed probes interfaced to $\mu$-LC and achieved similar results. Schäfer and Levsen (1981) showed that flow rates up to 15 $\mu$l/min could be tolerated with their $\mu$-LC–DLI interface design over extended periods of time and that injection volumes of less than 1 $\mu$l allowed high sensitivity. Although the utility of $\mu$-LC applications to LC–MS has been demonstrated, nevertheless, limitations in the range of operating conditions may exist for $\mu$-LC when compared to normal HPLC.

In recent studies, Arpino *et al.* (1979) have systematically attempted to gain better understanding of the DLI interface. These studies have focused on developing the most effective methods of introducing the sample into the ion source while maintaining the desired pressures throughout the source and analyzer and minimizing the degradation of chromatographic peaks. Theoretical calculations of optimum pumping and slit-design parameters for the desired flow rates, solvents, and source pressures were tested on several prototype instrument designs. The results indicate that cryopumping allows safer and more reliable operating conditions. Although most CI mass spectrometers are able to tolerate flow rates of 1–10 $\mu$l/min without modification of the vacuum system or the ion source, the cryopumped system allows operation at flow rates many times this and could feasibly accept as much as a 1-ml/min solvent flow with very high active-surface-area cryopumps.

It has been determined that the formation of a stable jet is critical for proper performance of the DLI probe. A stable liquid jet tends to fragment to form a fine aerosol, which can be anticipated to result in rapid solvent evaporation due to the large surface-area/volume ratio of the droplets. The endothermic solvent vaporization tends to cool the sample droplets, preventing decomposition of thermally labile compounds at the elevated temperatures in the source region. In order to form a stable jet in split effluent streams, a small backpressure in the effluent stream, produced by a needle valve, has been found to be necessary. There are two general types of restrictions causing the formation of a stable jet in the various DLI designs: namely, the "viscous" type, consisting of capillary tubing; and the "nonviscous" type, consisting of small pinholes (as shown in Fig. 5a). Continued optimization studies by Arpino et al. (1981) conclude that the non-viscous-type restrictions outperform viscous types due to less clogging from solute precipitates and mobile-phase impurities. To obtain optimum performance, a 1–3 μm orifice should establish a high jet velocity, the diaphragm should be located close to the electron beam, and thermal energy should be used to minimize droplet lifetime. To aid in rapid evaporation, hot gas was used, resulting in a stable jet—with the added advantage of possibly using the hot gas as a CI reagent gas.

*3. Atmospheric Pressure Ionization (API)*

In an attempt to circumvent the basic pressure differential problem of LC–MS interfacing, a system using an API source has been developed (Horning et al., 1974a, 1977, 1978; Carroll et al., 1975, 1979). A similar system has also been used by Karasek and Denny (1973) and Karasek (1974) as an interface for LC and plasma chromatography. In the plasma-chromatographic design there is no pressure difference between the ionization chamber and the electrometer, whereas in the API–MS a 25-μm aperture is needed to maintain the pressure difference between the ionization region and the low-pressure mass analyzer.

The API–MS system initially used $^{63}$Ni as the primary electron source (Horning et al., 1973a,b; Carroll et al., 1974). Early work had demonstrated that the system had high sensitivity (<1 pg) and was compatible with many LC solvents, and this led them to the design of an interface specifically intended for LC–MS interfacing (see Fig. 6). The source was modified by using a larger reaction chamber and a corona discharge to replace $^{63}$Ni as the primary electron source. Investigation showed that $^{63}$Ni and the corona discharge gave identical ion spectra and limits of detection, but that the corona discharge resulted in an increase by several orders of magnitude in the dynamic response range compared to $^{63}$Ni. This is clearly a considerable advantage in quantitative studies. The major restriction with API–LC–MS is the sample decomposition that may occur at the elevated temperatures of the vaporization chamber (about 275°C). This limits API–LC–MS to more thermally stable solutes.

## B. Mechanical Transfer Techniques

With mechanical transfer processes, all or part of the chromatographic effluent is collected onto a moving wire, belt, chain, or disk. This enables the effluent to be deposited as a thin film on a surface and physically transported to the detector. The key advantages are (1) the ability to control the direction and rate of transfer of the sample to the detector and (2) the ability to desolvate the effluent between the point of collection and the detector, allowing enhanced solute-detection capabilities.

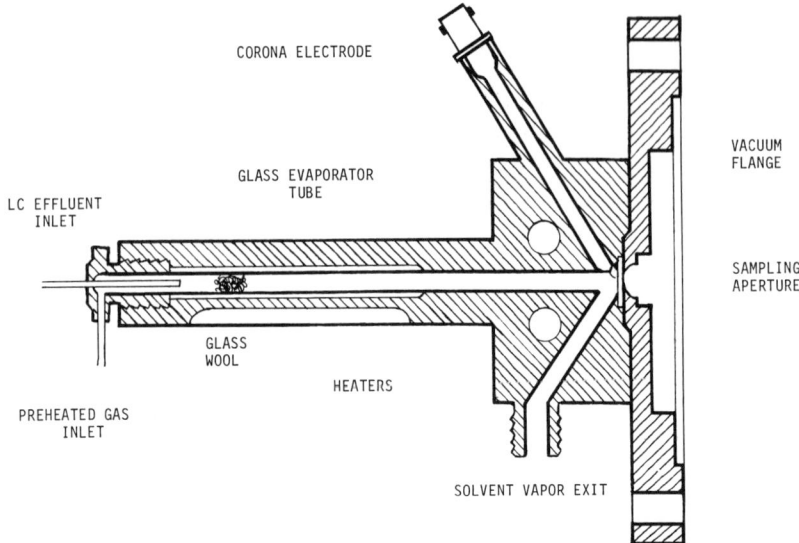

**Fig. 6.** Schematic diagram of the API interface using corona discharge as the primary electron source. [Reprinted with permission from Carroll *et al.* (1975). *Anal. Chem.* **47**, 2369. Copyright 1975, American Chemical Society.]

The mechanical transport method of chromatographic effluent collection was described by James *et al.* in 1964. With their system the sample was introduced onto a moving wire, passed through an evaporation oven to remove solvent, and then passed on to an oven where the solute was pyrolyzed. The pyrolysis products were then swept into a flame-ionization detector (FID) by a nitrogen gas stream. At the same period in time (early 1960s), Haakti and Nikkari (1963) used the same approach with a moving gold chain introduced into an FID. These early developments led to continuous on-line LC–MS using the same mechanical transport approach.

### 1. Wire Transport

The use of a wire transport as a means of interfacing a liquid chromatograph to a quadrupole mass spectrometer was first described by Scott *et al.* (1974a,b, 1979) about a decade after the wire-transport FID. This design is shown in Fig. 7. The major functions of this interface design are sample collection, desolvation, pressure reduction, and solute vaporization. Sample is applied to a stainless-steel wire (0.127 mm diameter) driven by a winding mechanism. The wire moves through three ruby-jewel vacuum locks which separate the ambient pressure from two interface vacuum chambers and the ionization chamber of the mass spectrometer. In the ionization chamber the solute is vaporized by passing

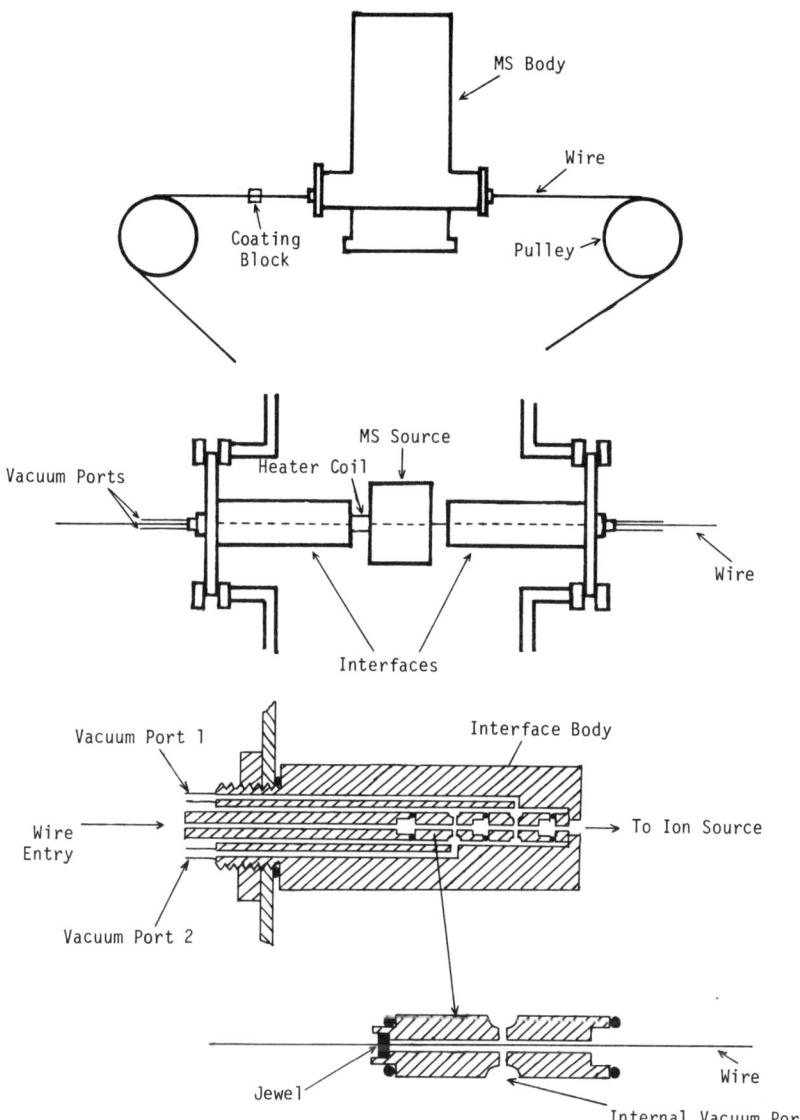

**Fig. 7.** Diagram of the moving-wire LC–MS interface. [With permission, from Scott (1979).]

current from a dc power supply through the moving wire. At higher pressures the resistive heat is dissipated by conduction, but in the low-pressure ionization chamber, the vacuum prevents conduction and the wire emits heat radiantly. A typical current range of 175–200 mA is used to vaporize most solutes at temperatures from 200 to 300°C. The wire then proceeds through three more ruby-jewel vacuum locks separating the exit-interface vacuum chambers and continues on to the collection spool.

Pressure reduction is achieved by using two interface pumping systems. The outer vacuum chambers are pumped in parallel by a 300 liters/min rotary pump, producing a pressure of 0.2 Torr. The inner interface chambers (closest to the ionization chamber) are pumped by a 2-in. oil-diffusion pump, in series with a 160 liters/min rotary pump, giving a final pressure in this region of $2 \times 10^{-3}$ Torr.

The wire entrance and exit locks have a purge T-junction which allows the use of inert gases to displace air which might otherwise leak into the system and produce background spectra. Helium has generally been used as purge gas, although CI reactant gases could also be used to obtain CI spectra.

In Scott's experiments, operating conditions were optimized using a Finnigan quadrupole mass spectrometer with its associated computer system. Under optimum conditions the sensitivity for a 0.01% solution of diazepam was found to be $4 \times 10^{-6}$ g/ml. Generally, sensitivities for most solutes ranged from $10^{-15}$ to $10^{-6}$ g/ml. Thus, sensitivity for the moving-wire LC–MS is apparently inadequate for many needed trace determinations.

The major advantages of this system (compared to, for example, the DLI systems) are its independence from solvent–ionization-chamber interactions—provided that the solvent is sufficiently volatile to be removed totally—and its compatibility with EI ionization modes. In addition, when used as an LC detector, little or no band dispersion is observed.

The moving-wire interface has not been widely adopted in its original form, but has instead acted as an impetus for and precursor to the development of current mechanical transfer systems, such as the moving-belt system of McFadden et al. (1976) (see Section IV,B,3).

Another wire interface that does not truly fall in the mechanical transport category, but consists of a resistively heated stationary wire that collects and concentrates the effluent as it runs down the wire under the influence of gravity, has been described by Christensen et al. (1981). This interface continuously preconcentrates sample (approximately 95% solvent removal is claimed) prior to its entering the mass spectrometer. The residual fluid flows into the EI ion source under needle-valve control. The performance of this interface is complicated by the varying volatility and wetting ability of solvents, deposition onto the wire by solutes, precipitation of solutes, and clogging of the needle valve. However, reasonable results were obtained for quantiative LC–MS analysis of shale-oil standard reference material, indicating potential applications.

## 2. Perforated Belt

Privett and Erdahl (1977, 1978) developed an LC–MS interface for the analysis of lipids and related compounds using the wire-transport principle. This design (see Fig. 8) allowed collection of approximately 1 ml/min of sample on a perforated stainless-steel belt. The system was initially developed for use with an FID (Privett and Erdahl, 1975). The sample was desolvated in a stream of warm nitrogen gas, and the remaining solute passed into a reaction chamber where it was evaporated into a carrier gas. The lipids were converted to hydrocarbons, with the prevalent reactions being hydrogenolysis and reductive cracking with hydrogen from the carrier gas. Approximately 15% of these products were then swept into the mass spectrometer for EI or CI analysis. The lipids could then be characterized from the hydrocarbon spectra. Detection limits in the nanogram range were reported. This approach was shown to be applicable to a wide variety of lipids and related compounds, but the lack of detailed structural information limits its use to simple or well-characterized systems.

## 3. Moving Belt

McFadden et al. (1976) constructed a mechanical transfer interface using a moving stainless-steel belt (3.2 mm wide, 0.05 mm thick), in an attempt to

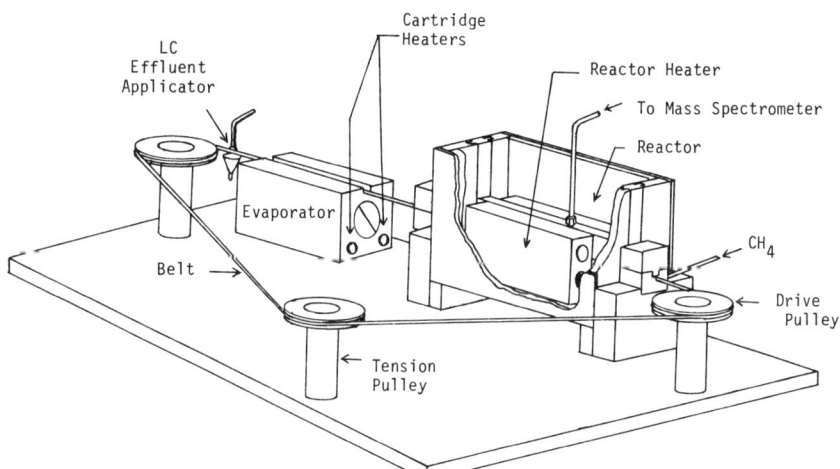

**Fig. 8.** Diagram of the perforated-belt interface coupled to a CI quadrupole mass spectrometer. [With permission, from Privett and Erdahl (1978).]

improve the efficiency of transport over the moving-wire system. With this device, sample is introduced onto a moving belt which passes through two vacuum chambers. This allows both efficient desolvation and the ability to maintain a pressure of $10^{-6}$ Torr in the ionization chamber. The vacuum chambers are separated by sapphire slits, and this allows the belt to move freely between chambers while maintaining a significant pressure differential between them. The first vacuum chamber is pumped by a 500 liters/min forepump, maintaining a pressure between 1 and 20 Torr. The second vacuum chamber is pumped by a 300 liters/min forepump, reducing the pressure to 0.1–0.5 Torr at this point. In order to achieve effective solvent vaporization in these two vacuum chambers, radiant heat or gas flow is sometimes necessary. The desolvation chambers can produce solute enrichment as high as $10^5$. The method of solute vaporization is a Nichrome heater that provides thermal and radiant heat, with the solute vaporization chamber butted directly to the ionization chamber solid-probe inlet. The effective solvent removal allows the mass spectrometer to be operated in either EI or CI modes.

This experimental design initially had a sample-transfer efficiency of 20–40% under ideal conditions. Results with samples of methyl stearate showed detection limits to be less than 1 ng, with a linear vaporization yield over the range 1–100 ng. This design showed marked improvement over the moving wire in sensitivity and potential quantitative capabilities. However, several problems associated with the early design had to be corrected. Operation was periodically interrupted by mechanical belt problems, such as slippage or stopping. Poor precision of sample yield was also observed due to uneven buildup of effluent, and peak broadening was found to occur due mostly to diffusion processes in the flash-vaporization chamber. Peak tailing was sometimes a problem and was found to be most severe when large volumes were placed on the belt or when the samples were polar in nature. It was suggested that this was due to absorption effects on the metal surfaces close to the source region. Sample carry-over presented a problem, with the belt showing ghost peaks of up to 5% of the original value on the second cycle.

In the commercial instrument marketed by Finnigan Instruments, many of these problems have been reduced or eliminated, as shown in Fig. 9 (McFadden et al., 1977). Belt-hangup problems have been essentially eliminated, and a more suitable belt material, Kapton (a DuPont polyimide), is now used. [Other materials, such as Teflon, have been suggested (Tal'roze and McLafferty, 1978).] To enhance precision, an infrared heater has been added which aids solvent evaporation and prevents spitting and spraying in the first vacuum chamber. In order to prevent contamination for recycling, a cleanup heater has been added after the flash vaporizer.

This interface can accept flow rates of up to 2.0 ml/min for volatile nonpolar solvents such as pentane and hexane. However, less volatile polar solvents

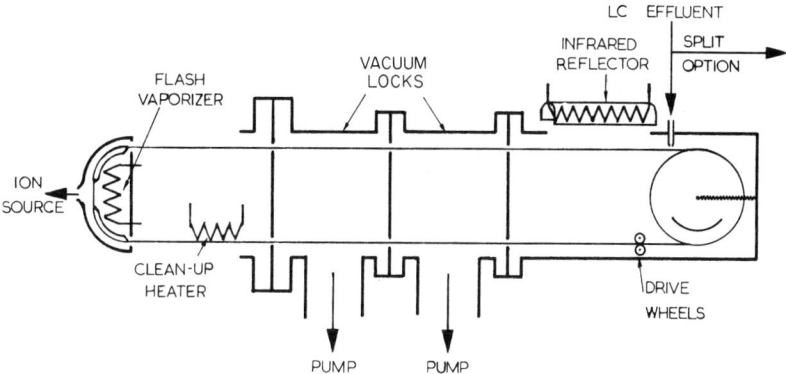

**Fig. 9.** Diagram of the Finnigan moving-belt interface. [Reprinted from McFadden *et al.* (1977). Copyright 1977 by International Scientific Communications, Inc.]

associated with reverse-phase LC are difficult to evaporate when placed on the belt at normal LC flow rates. Thus an upper limit between 0.2 and 0.3 ml/min is necessary for these solvents. To stay within the flow restriction for reverse-phase solvents without placing any flow restriction on the chromatographic system, a postcolumn zero-dead-volume splitter has been incorporated. The split option also enables collection of fractions when a significant portion of the eluent is split away from the interface. Hilker *et al.* (1979) placed an additional 40-W "stick" heater in each of the vacuum chambers. This assisted solvent vaporization and increased the acceptable flow rates for reverse-phase solvents to 1 ml/min without the need to split the eluent stream.

Millington *et al.* (1980) also reported a moving-belt LC–MS interface which has since been introduced commercially by VG-Organic. This design (see Fig. 10) operates on similar principles to the Finnigan design. The major objectives of this model are (1) to increase transfer efficiency and allowable flow rates and (2) to enhance the performance of the interface for thermally labile or involatile samples over those of previous moving-belt designs. To achieve these objectives, the belt is passed directly into the ionization chamber, with the point of solute vaporization being as close as 3 mm from the electron beam in the ion source. By vaporizing the thermally labile or involatile solutes as close to the electron beam as possible, improved EI and CI mass spectra may be obtained (Cotter, 1980). To prevent premature solute losses due to overheating of the belt and thus improve the solute yield, the shaft assembly is water-cooled. Another design characteristic of this interface is the ceramic spacer which insulates the source region from the rest of the interface. This, along with the nonconducting polyimide belt, allows the use of magnetic-sector as well as quadrupole MS.

Transfer efficiencies of this system, when compared to direct probe insertion, were reported to be from 30 to 80% for flow rates of 2 ml/min hexane, 1 ml/min

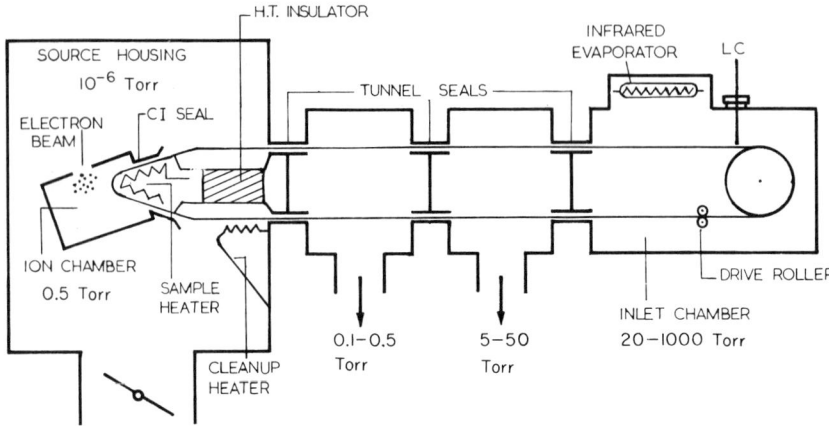

**Fig. 10.** Diagram of VG-Organic moving-belt interface. [With permission, from Millington *et al.* (1980).]

methanol, or 0.3 ml/min water. Performance was superior to that obtained with the use of a conventional direct-insertion probe. Detection limits were reported to be less than 500 pg and the dynamic range was $10^4$. The precise capabilities of this design depend upon the nature of the sample and the operating conditions. Because this version of the moving-belt design is applied to actual laboratory samples, the range and the limits of its application are demonstrated here.

Karger *et al.* (1979) incorporated an on-line segmented-flow extractor to eliminate the problem of evaporating reverse-phase solvents. Solute is continuously extracted from the reverse-phase solvent into a more volatile solvent, which is subsequently introduced onto the Finnigan moving belt. The additional benefit of the extraction process is the removal of involatile buffers or other chemical modifiers. The original design used an air-segmented stream, but it was later found that the segmented-solvent operation achieved similar results without the need for a peristaltic pump. A further modification was the replacement of mechanical with air-displacement pumping for the organic solvents (Kirby *et al.*, 1980a) (see Fig. 11). Extraction efficiency was a major concern because solute recovery depends on the nature of the organic solvent, pH, ionic strength, etc. In practice, variable solute recovery was found, and this placed serious restrictions upon solute quantitation. The use of isotopically labeled internal standards has been suggested to enable determination of extract efficiency.

As with many innovations, the segmented-flow extraction configuration has led to a variety of applications or suggested applications. For instance, its use for postcolumn derivatization has been suggested (McFadden, 1980). Kirby *et al.* (1980a,b, 1981) continued their work and found that they could enhance the extraction efficiency for ionic solutes by using ion-pairing techniques. They also

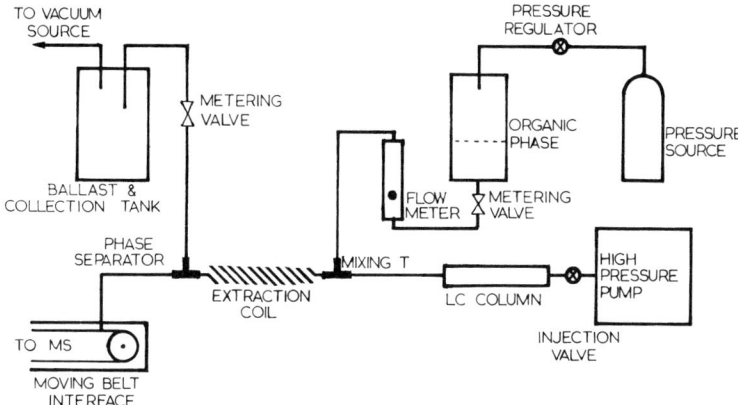

**Fig. 11.** Schematic diagram of the LC–MS system utilizing the continuous-extraction interface. [With permission, from Kirby *et al.* (1981).]

demonstrated the ability to form volatile ion-pair derivatives of ionic compounds, which allowed MS analysis.

The moving-belt interface, with the aforementioned additions or modifications, is capable of accepting a wide variety of sample and mobile-phase characteristics. However, the ability of this interface to vaporize samples with a wide range of volatilities is limited. The flash-vaporization process reportedly outperforms conventional direct-insertion problems, with the sample volatility limit falling somewhere between those for direct-probe and desorption CI (Games, 1980a). Although this range provides for a significant number of compounds beyond the current capability of GC–MS, the number is still limited by sample volatility. In an attempt to broaden this range, alternative methods for sample removal have been suggested. The use of lasers, corona discharge, and sputtering processes are several examples. The distinction between the solute-vaporization process and the ionization process becomes less apparent with the newer techniques of ionization from surfaces, and this adds an important dimension to design considerations for future moving-belt systems.

One such approach was utilized by Smith and Burger (1980), who constructed a moving-ribbon interface that can operate with secondary-ion mass spectrometry (SIMS), using an ion gun, or with laser desorption (LD) ionization. Characteristic molecular and quasi-molecular ion peaks from SIMS were complemented by collision-induced dissociation (CID), which allowed structural elucidation. Several modifications were made with their initial instrument, which resulted in the improved design seen in Fig. 12 (Smith *et al.*, 1981). A thermal desorption–EI ionization source was added to allow conventional EI spectra to compare with the SIMS spectra. The LD capability was removed. Problems with uneven

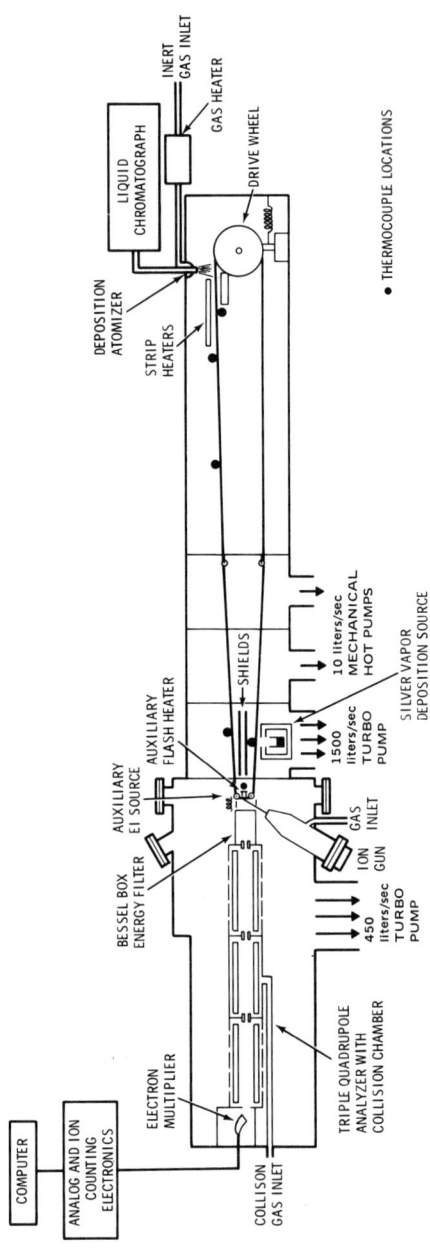

**Fig. 12.** Schematic illustration of the moving-ribbon interface. [Reprinted with permission from Smith *et al.* (1981). Copyright 1981, American Chemical Society.]

deposition of effluent and uneven vaporization when collection of effluent was performed by simple dripping have been resolved with a new aerosol-deposition method (Smith and Johnson, 1981). By depositing the effluent on the ribbon as an aerosol, and using heated nebulization gases and strip heaters, increased flow rates, improved peak shapes, and enhanced transfer efficiencies were observed.

Several modifications of the Smith *et al.* (1981) design show unique capabilities over earlier designs. The localized process associated with SIMS allows slower ribbon speeds. Storage of solute on the ribbon is possible and enables repeated analysis on a single chromatographic separation in various ionization modes. With proper adjustment of sample transfer with the SIMS mode, initial results show that quantitation is possible. Background spectra have been reduced in the SIMS mode by using high-purity silver vapor deposition on the ribbon. Preliminary analyses of amino acids have showed detection limits to be less than 10 pg.

Another novel mechanical transfer approach, by Hardin and Vestal (1980), used a pulsed laser to study desorption and ionization of low-volatility samples from a rotating stainless-steel belt with the intent of optimizing parameters for a future LC–LD–MS interface design.

## C. LC–MS Interfacing Using GC–MS Approaches

The success of GC–MS interfacing can be credited to a large degree to the development of various carrier separators. This has led many researchers to attempt similar designs for solvent separation in LC–MS interfacing. Clearly, the magnitude of the difficulties associated with solvent–solute separation in LC–MS is far greater, and so the application of GC–MS separators to LC–MS has met with only limited success.

### 1. Membrane Separation

Jones and Yang (1975) used a silicon rubber membrane separator to interface a liquid chromatograph with a mass spectrometer. A three-stage membrane separator was used, allowing flow rates greater than 2 ml/min without compromising the source pressure ($10^{-6}$ Torr). The objective was to use a membrane that would discriminate against the solvent vapors and consequently enrich the solute vapors. The effluent therefore had to be flash vaporized before entering the separator. The silicon rubber membrane used had polarity and temperature (250°C) restrictions, along with volatility requirements, that severely limited the range of application for this interface.

### 2. Jet Separation

Takeuchi *et al.* (1978) used a one-stage jet separator as the interface between μ-LC and a CIMS. The enriched sample vapor passed into the source where solvent served as the reagent gas. Constant source pressure could be obtained with varying LC flows by adjusting the pumping rate on the separator. The effluent was vaporized at 150°C before separation; therefore enrichment principles were identical to GC–MS jet separation. McFadden (1979) has pointed out that enrichments desirable for LC–MS should be on the order of $10^5$, whereas most GC–MS separators only have an enrichment of about 10. This difference and the volatility requirement have limited further development of this technique without certain modifications.

In an attempt to resolve some of the problems found with the initial design, a new vacuum-nebulizing design was developed that replaced the expanding gas jet of the earlier design with an expanding aerosol (Hirata *et al.*, 1979; Tsuge *et al.*, 1979) (see Fig. 13). Helium was used as the nebulizing gas for flow rates of effluent from 2 to 16 liters/min. Higher interface heating temperature (300°C) could be used for vaporization, because the cooling effect of droplet evaporation serves to stabilize thermally labile solutes. These modifications allowed the jet interface to function in a similar manner to the DLI interface, with the added enrichment benefit. The initial applications studies demonstrated the ability to analyze amino acids, drugs, oligomers, and other compounds of low volatility.

## D. Aerosol-Jet Techniques

The importance of establishing a liquid jet and subsequent droplet formation has been discussed as applied to DLI; however, several other LC–MS interfaces have been developed that require liquid-jet formation.

### 1. Crossed-Beam Technique

Blakley *et al.* (1978) and Vestal (1979) developed an LC–MS interface that could vaporize, ionize, and mass-analyze the sample with a minimum of contact with solid surfaces. This was performed by focusing a 50-W $CO_2$ laser across the liquid jet emerging from a capillary tube and was appropriately named ''crossed-beam'' LC–MS (see Fig. 14). The intensity of the laser allowed up to 1 ml/min flow rates of water to be vaporized. The duration of exposure to the beam was approximately $10^{-4}$ sec; consequently, this technique employed the ''rapid-heating'' volatility enhancement (see Section V,A). Once vaporized, the vapor jet was sampled through a small orifice. The vapor formed a supersonic molecular beam as it emerged from the orifice, which was directed toward a skimmer. Between the orifice and the skimmer the beam intersected an electron beam, which resulted in the primary ionization of solvent molecules leading to CI of solute molecules. A second electron beam between the skimmer and the collimator allowed EI ionization because of the lower pressures in this region. (Note the tremendous pumping capacity of this initial system.)

The initial design had problems with maintaining stable vaporization of the

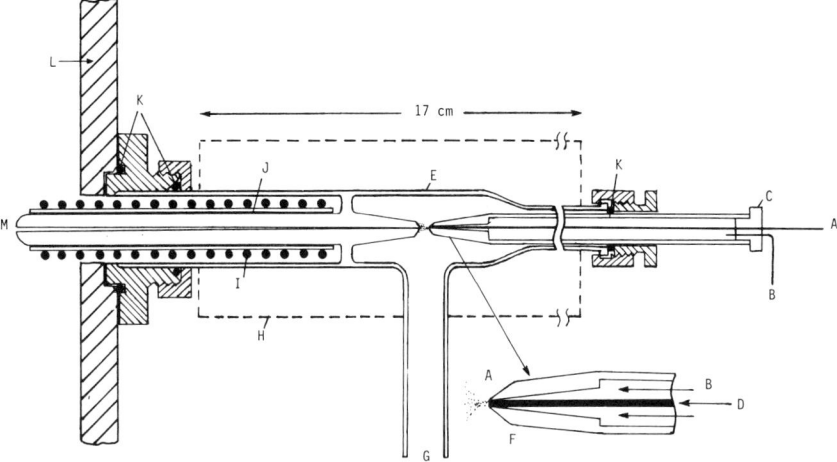

**Fig. 13.** Schematic diagram of vacuum-nebulizing interface for LC–MS coupling: (A) coaxial capillary tube from LC; (B) needle for introduction of nebulizing gas (He); (C) silicon septum; (D) effluent from LC; (E) nebulizing tube (Pyrex); (F) nebulizing tip (Pyrex); (G) to rotary pump; (H) heating oven; (I) heater for counter-capillary to MS; (J) heater sheath (Pyrex); (K) silicon O-ring; (L) body of MS; (M) to ion source of MS. [Reprinted with permission from Tsuge et al. (1979). Copyright 1979, American Chemical Society.]

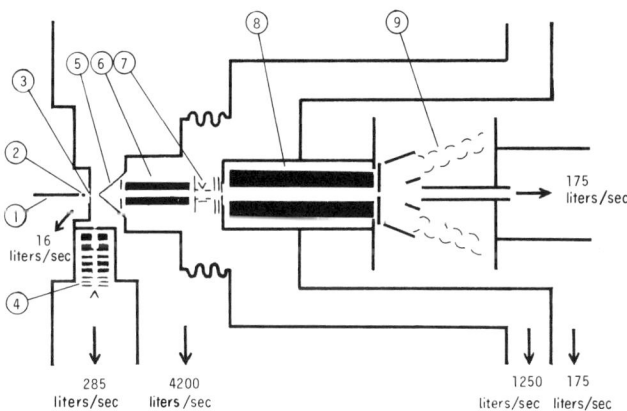

**Fig. 14.** Schematic diagram of the laser crossed-beam LC–MS interface: (1) liquid nozzle; (2) laser beam; (3) sampling aperture; (4) CI electron gun; (5) skimmer; (6) quadrupole with RF only; (7) EI ion source; (8) quadrupole mass analyzer; (9) multipliers. [With permission, from Blakley et al. (1978).]

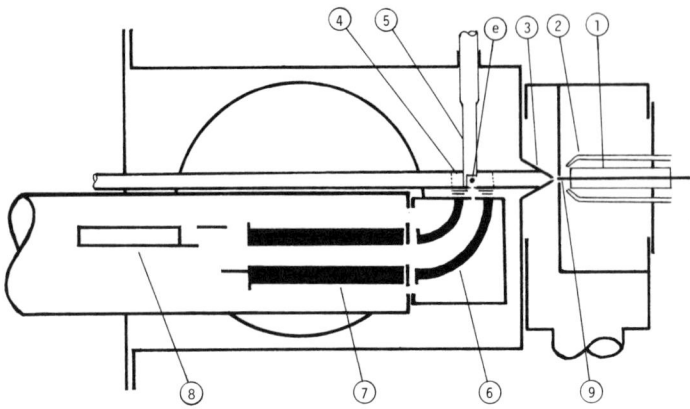

**Fig. 15.** Schematic diagram of the oxyhydrogen-torch crossed-beam LC–MS interface: (1) capillary tube carrying liquid effluent from LC; (2) oxyhydrogen torch; (3) skimmer; (4) ion source; (5) heated probe; (6) electrostatic deflector; (7) quarupole mass filter; (8) electron multiplier; (9) heated portion of stainless-steel capillary. The electron beam enters the ion source perpendicular to the plane of the figure at the point (e). [Reprinted with permission from Blakley et al. (1980a). Copyright 1980, American Chemical Society.]

liquid jet with the laser crossed-beam configuration, and these were resolved by focusing the laser directly on the capillary tip from which the liquid jet emerged. The result was stable vaporization with little indication of pyrolysis from heating the tip to red-hot, and this method was thereafter referred to as the "indirect" heating mode (McAdams et al., 1979; Blakley et al., 1979a). CI and EI spectra were obtained for a variety of involatile compounds with this modified-heating approach, but in addition, significant changes were made to simplify the design.

To lower the cost and complexity, Blakley et al. (1979b, 1980a,b) replaced the laser with oxyhydrogen flame heating (see Fig. 15) without compromising the performance of the interface. Four oxyhydrogen flames heated the capillary tip to around 1000°C, producing a high-velocity jet composed partly of vapor and partly of charged aerosol particles traveling at about $10^5$ cm/sec. These particles were solute enriched and were transmitted through the skimmer much more efficiently than the vapors. The result was that 3–5% of the solvent was transmitted compared with 50% of the solute. Once through the skimmer, the beam passed through an electron beam and impinged on a nickel-plated copper probe at 250°C, and the resulting ions were steered into a quadrupole mass analyzer. With these modifications, the cost and complexity were claimed to be comparable to that of GC–MS. Detection limits were reported to be 1–10 ng when operating at full scan and subnanogram when using selective ion monitoring.

An interesting development from this design was the discovery of a new ionization mode (Blakley et al., 1980c,d). When the charged-particle beam impinged on the heated probe without passing through the electron beam, ionization was observed. The resulting spectra were comparable to field desorption spectra, with less fragmentation than conventional CI spectra. This new ionization technique has added to the capability for conventional ionization, the capacity to accept a normal LC flow rates and the efficient transfer of solute, making this LC–MS approach more versatile than some of the other on-line approaches.

*2. Ultrasonic Nebulization Approach*

Ultrasonic nebulization has been used to produce aerosols with droplets of very small and uniform size. This facilitates rapid and uniform vaporization of the droplets. By varying the frequency of ultrasound, the droplet size can be carefully controlled. Use of this approach to produce small, rapidly vaporizing droplets from the effluent stream has been reported by Udseth *et al.* (1978) and Willoughby and Browner (1980).

Willoughby and Browner (1980) used electrostatic charging and focusing of a monodisperse droplet stream to control both direction and rate of droplet flow through their interface. Only preliminary spectra have been obtained and the results look promising, but further studies are necessary before the viability of this approach is established.

## V. ALTERNATIVE METHODS OF VAPORIZATION AND IONIZATION WITH POSSIBLE APPLICATIONS TO LC–MS

For the purpose of this discussion, it is convenient to divide compounds to be separated chromatographically into three categories on the basis of volatility, although in reality there are no clear dividing lines between categories. These categories are (1) compounds volatile and stable enough for separation with GC–MS; (2) compounds not volatile enough for GC, but with sufficient vapor pressure for normal EI or CI mass analysis; and (3) compounds that are both involatile and thermally labile. Compounds in categories (2) and (3) are typically separated with LC. The volatility restrictions of the third group make introduction, vaporization, and ionization using conventional EI or CI methods extremely difficult. The development of novel methods of vaporization and ionization may enhance the likelihood of obtaining useful mass spectra for the lower volatility compounds, which would result in a widening of the range of samples suitable for analysis by LC–MS. It is the intent of this section to describe various MS developments related to vaporization and ionization of involatile, labile species in order to indicate where major developments may be anticipated. By no means have all of the approaches described actually been incorporated into LC–MS systems.

## A. Vaporization Enhancement

In this approach, the energy applied to the sample is utilized in overcoming the interaction of the sample with its surroundings and in increasing the internal energy of the molecule, resulting in decomposition, rearrangement, and/or vaporization. Enhancement of vaporization of low-volatility samples can be obtained using several approaches, including rapid heating, applying the sample to an inert surface, and applying the sample to an activated surface.

Rapid-heating processes were found to favor desorption of intact neutral molecules if heating rates were greater than about 12°C/sec and the temperatures were above 400 K (Beuhler et al., 1974). Volatility enhancement can be achieved by rapid heating at high temperatures for short durations. These conditions have been utilized with many surface-desorption techniques. Pulsed lasers (Posthumus et al., 1978) and fission products (Macfarlane and Torgerson, 1976) have aided in the achievement of faster heating rates and higher temperatures.

Sample–surface interaction can in certain circumstances contribute to reduced volatility; conversely, if the sample is applied to a chemically inert surface, the energy needed to remove the sample from the surface will be less, thus effectively enhancing the volatility. Several inert materials have successfully been used for this purpose, namely glass (Baldwin and McLafferty, 1973a), quartz (Ohashi et al., 1978), Teflon (Hansen and Munson, 1978), and Vespel (Cotter, 1979).

Sample–sample interaction may also contribute to lower volatility, especially when polar functional groups are present and/or hydrogen bonding occurs. This type of interaction can be minimized by spreading the sample over a large surface area and has led to the development of activated emitters for field-desorption MS with extremely large surface areas.

The approach of volatility enhancement is being exploited to enhance desorption capabilities in a variety of ionization modes. Recently, there has been a surge of activity with involatile samples, resulting in a host of new or modified techniques.

## B. Alternative Ionization Techniques

Electron-impact and chemical and field ionization (FI) processes all require that the sample be introduced into the ionization chamber in the gas phase; therefore in LC–MS, where the sample is initially in solution, alternative methods of ionization using methods of sample introduction not restricted to the gas phase are worthy of examination. A comprehensive treatment of the principle behind each technique will not be attempted, but for the sake of brevity, only the utility and potential applications to LC–MS will be emphasized. Table III lists these techniques along with several nonsurface techniques that may contribute to the developing field of LC–MS.

## TABLE III

**Alternative Modes for Mass Analysis That May Contribute to LC–MS**

| MS technique | Acronym | Sample introduction[a] | Contribution to LC coupling[b] |
|---|---|---|---|
| Field desorption | FD–MS | S (emitters) | LVS |
| EI desorption | EID–MS | S (emitters) | LVS, EI plus CI spectra |
| CI desorption | CID–MS | S (emitters) | LVS |
| Plasma desorption | PD–MS | S (foil) | LVS, high molecular weights |
| Electrohydrodynamic ionization | EH–MS | S (solutions) | LVS |
| Secondary ion | SIMS | S | LVS |
| Laser ionization | LIMS | S | LVS |
| Atmospheric pressure ionization | APIMS | V | Allows high pressure, no flow restriction |
| "Mass spec" | MS–MS | V (mixtures) | Aid in separation in complex mixtures |
| Pyrolysis | Pyr–MS | V (pyrolysis products) | No volatility restriction |

[a] S, surface; V, vapor.
[b] LVS, low-volatility samples.

Field-desorption mass spectrometry (FD–MS) has developed rapidly since its introduction by Beckey (1969) for the identification of ionic, highly polar, and thermally labile organic compounds. The potential for coupling FD–MS with HPLC was realized and discussed by Schulten and Beckey (1973). The development of various activated emitters that are chemically and thermally inert has contributed to making FD comparable in sensitivity to EI methods for many compounds. The direct benefit of FD for identification of LC fractions is well established for off-line applications; however, quantitation using FD is hampered by poor reproducibility in both sample introduction and operating (desorption) conditions. Barofsky *et al.* (1978) investigated the problems associated with quantitation by studying precision, accuracy, and detection limits under various conditions of sample handling and desorption rate. Generally, variations were found to less than 20% upon normal introduction and desorption conditions; however, precalibration of the emitters and internal standards were suggested for high-accuracy work.

Improved CI spectra can result from introducing low-volatility samples directly into the electron beam on an extended probe. Baldwin and McLafferty (1973a,b) used this technique, which they called "direct CI," for studying oligopeptides and noted a high yield of $MH^+$ peaks. Similar approaches have been subsequently used by other workers and assigned a variety of names (e.g., in-beam, desorption CI, surface ionization). Hansen and Munson (1978, 1980) studied the direct CI technique using Teflon probe surfaces and found significant increases in $MH^+$ ion production. A similar approach was developed for the EI source (Dell et al., 1975) which resulted in spectra with characteristics of both EI and CI spectra. For both of these in-beam techniques, the optimum distance from the probe tip to the electron beam for $MH^+$ production was found to be 2–3 mm (Ohashi et al., 1976). The benefits to LC–MS resulting from the direct use of in-beam techniques may not be as important as the understanding of the relationship between sample and beam obtained from these techniques.

Plasma-desorption mass spectrometry (PD–MS) is a technique for producing quasi-molecular ions by bombarding the sample with high-energy heavy ions. Macfarlane et al. (1980) used $^{252}Cf$ spontaneous-fission products to bombard a sample foil. Upon impact, the temperature is elevated to about 10,000 K in approximately $10^{-12}$ sec. The rapid-heating effect restricts the deposit of energy in the unstable vibrational modes; consequently, the surface vaporization rate is faster than decomposition rates. Ionization occurs after desorption by either ion–molecule or ion-pair reactions. Of particular value is the use of this technique in the analysis of high-molecular-weight molecules (i.e., MW > 2000).

Electrohydrodynamic ionization mass spectrometry (EH–MS) produces charged droplets and/or ions by applying an electric field to a liquid meniscus at the end of a capillary tube. Quasi-molecular ions from proton or cation attachment are prevalent, with little or no observable fragmentation. Ion–molecule clusters containing both sample and solvent are also produced. The EH ion source was introduced by Colby and Evans (1973), and continued studies have demonstrated a variety of EH–MS applications for involatile organic solutes (e.g., sugars, nucleosides, polypeptides; Simons et al., 1974; Stimpson and Evans, 1978). This technique could potentially reduce the burden of solvent removal from the LC–MS interface as it operates with solutions.

Secondary-ion mass spectrometry has been used to obtain molecular weight and structural information for many low-volatility compounds. Benninghoven and Sichtermann (1977, 1978) have investigated both positive- and negative-ion spectra for peptides, drugs, and vitamins with some success, obtaining "parent-like" molecular ions for most compounds. This technique refers to the bombarding of a sample-coated surface with a primary-ion beam and sputtering the sample off the surface for subsequent mass analysis. As described in Section IV,B,3, Smith and Burger (1980) have already incorporated this technique into their LC–MS.

Laser-ionization mass spectrometry (LIMS) is capable of ionization of low-volatility compounds in addition to the rapid-heating volatility enhancement obtained using the laser. This dual function was demonstrated by Mumma and Vastola (1972) for organic salts and by Posthumus et al. (1978) for involatile organic molecules and could contribute to LC–MS in either capacity.

Atmospheric pressure ionization mass spectrometry (APIMS) permits sample ionization at atmospheric pressure. This technique was reviewed by McKeown and Siegel (1975). Sample ionization occurs by secondary ionization with reactant-gas ions. Generally, primary ionization was facilitated with either a corona-discharge source or 60-keV electrons from $^{63}$Ni foil (typically used in the electron-capture detector). The APIMS technique is highly sensitive and has less sample-flow or pressure restriction than other techniques. However, the requirement of sample volatility limits its use with many LC solutes. To resolve the volatility requirement, Tsuchiya *et al.* (1980) introduced samples into an atmospheric pressure source with a probe. The practical application of API–LC–MS has been discussed in Section IV,A,3.

"Mass spec–mass spec" (MS–MS) refers to a newer technique capable of identifying the components of a mixture by separating the primary ions of the mixture components with one mass spectrometer and subsequently using collisional activation to produce secondary-ion fragmentaton spectra with a second mass spectrometer. This is a simplified description and there are a variety of instrumental designs that are labeled by this acronym. Unfortunately, this technique requires sample vaporization, but with improved methods of introduction and vaporization this technique could become a valuable complement to LC–MS by enhancing the separation and identification capabilities of the total system (i.e., LC–MS–MS). This area is discussed further by McLafferty and Lory (1981).

Pyrolysis mass spectrometry (Pyr–MS) was handicapped for many years by the inability to reproduce fragmentation patterns, generally because of the difficulty in reproducing pyrolysis conditions. Of some help have been the effort to standardize methods, the establishment of spectral libraries, and computer search and identification techniques (e.g., factor analysis). These improvements have enabled identification and quantitation of many classes of otherwise involatile compounds (Heylin, 1980). For LC–MS analysis, using this approach in a reproducible and controlled manner would have the advantage of eliminating the sample-volatility restriction, but at the same time would involve some sacrifice of chemical information (e.g., molecular weight).

It has been the intent of this section to illustrate how dynamic the field of mass spectrometry is at the present time. Each new development may play an important role in overcoming some of the LC–MS barriers.

## VI. APPLICATIONS AND CONCLUSIONS

The number of analyses performed using the various LC–MS interfaces is rapidly growing. As more interfaces make their way from experimental research laboratories to routine analytical laboratories, the practical utility of various

## TABLE IV

### Practical Applications of LC–MS

| Sample type | LC–MS mode | Reference |
|---|---|---|
| Alkaloids | API | Horning et al. (1974a,b, 1976) |
| | DLI | Serum and Melera (1978) |
| | | Melera (1979b) |
| | MB | Games et al. (1981) |
| Aromatics: | DLI | Beggs and Melera (1979) |
| Phenols, amines, nitriles, etc. | Jet | Tsuge et al. (1979) |
| | MB | Dark and McFadden (1978) |
| | | McFadden et al. (1978) |
| | | Karger et al. (1979) |
| | | Wright and Edgerton (1979) |
| | | Martin (1980) |
| | | Thruston and McGuire (1980) |
| | API | Carroll et al. (1975) |
| Polycyclic hydrocarbons | CB[a] | Blakley et al. (1978) |
| | | McAdams et al. (1978) |
| | DLI | Arpino and Krien (1978) |
| | | Schäfer and Levsen (1981) |
| | Jet | Takeuchi et al. (1978) |
| | | Tsuge et al. (1979) |
| | MB | Dark et al. (1977) |
| | | DeRoos and Foltz (1979) |
| | | McFadden (1979) |
| | Membrane | Jones and Yang (1975) |
| | OL[b] | Lovins et al. (1973) |
| Drugs and metabolites | API | Horning et al. (1976) |
| | CB | Blakley et al. (1980e) |
| | DLI | Henion (1978a,b, 1979, 1981) |
| | | Rottschaefer et al. (1979) |
| | | Henion and Maylin (1980) |
| | | Kenyon et al. (1980) |
| | Jet | Takeuchi et al. (1978) |
| | | Hirata et al. (1979) |
| | | Tsuge et al. (1979) |
| | MB | McFadden et al. (1977) |
| | | Baty et al. (1979) |
| | | Games et al. (1979b, 1981) |
| | | Millington et al. (1980) |
| | | Roy et al. (1980) |
| | MW[c] | Scott (1979) |
| | OL | Lovins et al. (1973) |

**TABLE IV** (*continued*)

| Sample type | LC–MS mode | Reference |
|---|---|---|
| Fats, fatty acids, and waxes | DLI | Arpino et al. (1974b) |
| | | McLafferty et al. (1975) |
| | | Schmitter et al. (1979) |
| | Jet | Takeuchi et al. (1978) |
| | | Tsuge et al. (1979) |
| | MB | McFadden et al. (1977, 1978, 1979) |
| | | McFadden (1979) |
| | PB[d] | Privett and Erdahl (1977, 1978) |
| Ion pairs | MB | Kirby et al. (1980b, 1981) |
| Mold, aflatoxins, and fermentation products | MB | McFadden et al. (1977) |
| | | McFadden (1979) |
| | MW | Scott et al. (1974b) |
| Natural products | CB | McAdams et al. (1978) |
| | DLI | Serum and Melera (1978) |
| | | Melera (1979b) |
| | | Arpino and Krien (1980) |
| | MB | Evans et al. (1975) |
| | | McFadden et al. (1977, 1979) |
| | | Games et al. (1978a,b, 1979a, 1980b, 1981) |
| | | Games (1978) |
| | | Eckers et al. (1979) |
| Nucleosides, nucleotides | CB | McAdams et al. (1978) |
| | | Blakley et al. (1980a,b,d,e) |
| | MB | McFadden et al. (1978) |
| | | Quilliam and Osei-Twum (1980) |
| | | Games et al. (1981) |
| Oils | DLI | Arpino and Krien (1978) |
| Peptides, amino acids | DLI | Baldwin and McLafferty (1973b) |
| | | McLafferty and Dawkins (1975) |
| | | Dawkins et al. (1978) |
| | MB | McFadden et al. (1977) |
| | | Games et al. (1981) |
| | MW | Scott et al. (1974h) |
| | OL | Lovins et al. (1973) |
| | | Dawkins et al. (1978) |
| Pesticides and metabolites | MB | McAdams et al. (1979) |
| | | McFadden et al. (1976) |
| | | Eckers et al. (1979) |
| | | Games et al. (1980b, 1981) |
| | | Karger et al. (1979) |
| | | Dymerski et al. (1979) |
| | | Wright and Oswald (1978) |
| | | DeRoos and Foltz (1979) |

(*continued*)

**TABLE IV** (*continued*)

| Sample type | LC–MS mode | Reference |
|---|---|---|
| Saccharides | CB | Blakley et al. (1980e) |
|  | DLI | Kenyon (1980) |
|  | MB | Games (1978) |
|  |  | Games et al. (1978b) |
|  |  | DeRoos and Thompson (1979) |
| Steroids | API | Horning et al. (1976) |
|  | DLI | Arpino et al. (1974b) |
|  | MB | McFadden et al. (1976) |
|  |  | Games (1978) |

[a] CB, crossed-beam.
[b] OL, off-line.
[c] MW, moving-wire.
[d] PB, perforated-belt.

techniques or approaches will be demonstrated. Table IV summarizes the recently published applications of a wide range of samples to LC–MS. McFadden (1980) and Games (1980b) have also reviewed recent applications.

The mass spectrometer provides an exceptionally wide range of LC detection capabilities when compared to standard LC detectors. The practicality of combining a liquid chromatograph with a mass spectrometer has been demonstrated by many workers in both on- and off-line modes, and this gives the analyst a variety of viable options for sample analysis. As no individual mode satisfies completely the requirements for analysis over the entire range of LC operating conditions and eluted solutes, the analyst is left with a choice of which mode is suited to individual sample needs.

Off-line procedures offer the advantages of flexibility and fewer restrictions placed upon the analysis; unfortunately, they are also more cumbersome and lack the ease of quantitation associated with on-line analysis. However, for the ionization techniques that are applicable to low-volatility samples (e.g., field desorption), and that do not lend themselves readily to on-line interfacing, off-line approaches are quite suitable. An off-line approach is also indicated for non-routine or occasional compound identification and characterization.

On-line interfaces are commercially available in both DLI and MB designs which are directly applicable to many types of routine analysis. These systems compare favorably to GC–MS for certain classes of compounds and are capable of exploiting many of the computer data manipulation and searching potentials of GC–MS. Several other experimental on-line designs, such as API, crossed-beam, and nebulization interfaces, may offer alternative approaches to the DLI or MB systems. However, considerable commercial development is required before they will be available to operate on a simple, routine basis and at an economically acceptable level.

LC–MS has a bright future in providing enhanced detection capabilities for liquid chromatography. In the authors' opinion, the trend of future developments should go in two directions: (1) the development of simple, relatively inexpensive interfaces that are compatible with most commercial mass spectrometers as an auxiliary inlet or probe, accessible to the EI or CI sources; and (2) the development of specialized LC–MS systems that are designed specifically around one or more of the newer ionization sources intended for low-volatility-compound excitation. Developments in both areas will almost certainly expand the range of applicability of LC–MS and enhance its claim to be potentially the most universal LC detector.

## REFERENCES

Arpino, P. J., and Guiochon, G. (1979). *Anal. Chem.* **51**, 682A.
Arpino, P. J., and Krien, P. (1978). *26th Annu. Conf. Mass Spectrom. Allied Top.*, St. Louis, p. 426.
Arpino, P. J., and Krien, P. (1980). *J. Chromatogr. Sci.* **18**, 104.
Arpino, P. J., Baldwin, M. A., and McLafferty, F. W. (1974a). *Biomed. Mass Spectrom.* **1**, 80.
Arpino, P. J., Dawkins, B. G., and McLafferty, F. W. (1974b). *J. Chromatogr. Sci.* **12**, 574.
Arpino, P. J., Guiochon, G., Krien, P., and Devant, G. (1979). *J. Chromatogr.* **185**, 529.
Arpino, P. J., Krien, P., Vajta, S., and Devant, G. (1981). *J. Chromatogr.* **203**, 117.
Arsenault, G. P., Althaus, J. R., and Diveker, P. V. (1969). *Chem. Commun.*, 1414.
Baldwin, M. A., and McLafferty, F. W. (1973a). *Org. Mass Spectrom.* **7**, 1353.
Baldwin, M. A., and McLafferty, F. W. (1973b). *Org. Mass Spectrom.* **7**, 1111.
Barofsky, D. F., Barofsky, E., and Held-Aigner, R. (1978). *Adv. Mass Spectrom.* **7A**, 109.
Baty, J. D., Yorke, D. A., and Green, B. N. (1979). *27th Annu. Conf. Mass Spectrom. Allied Top.*, Seattle, p. 369.
Beckey, H. D. (1969). *Int. J. Mass Spectrom. Ion Phys.* **2**, 500.
Beggs, D. P., and Melera, A. (1979). *27th Annu. Conf. Mass Spectrom. Allied Top.*, Seattle, p. 724.
Benninghoven, A., and Sichtermann, W. (1977). *Org. Mass Spectrom.* **12**, 595.
Benninghoven, A., and Sichtermann, W. (1978). *Anal. Chem.* **50**, 1180.
Beuhler, R. J., Flanagan, E., Greene, L. J., and Friedman, L. (1974). *J. Am. Chem. Soc.* **96**, 3990.
Blakley, C. R., McAdams, M. J., and Vestal, M. L. (1978). *J. Chromatogr.* **158**, 261.
Blakley, C. R., McAdams, M. J., and Vestal, M. L. (1979a). *Adv. Mass Spectrom.* **8B**, 1616.
Blakley, C. R., McAdams, M. J., and Vestal, M. L. (1979b). *27th Annu. Conf. Mass Spectrom. Allied Top.*, Seattle, p. 622.
Blakley, C. R., Carmody, J. J., and Vestal, M. L. (1980a). *Anal. Chem.* **52**, 1636.
Blakley, C. R., Carmody, J. J., and Vestal, M. L. (1980b). *28th Annu. Conf. Mass Spectrom. Allied Top.*, New York, p. 312.
Blakley, C. R., Carmody, J. J., and Vestal, M. L. (1980c). *J. Am. Chem. Soc.* **102**, 5931.
Blakley, C. R., Carmody, J. J., and Vestal, M. L. (1980d). *28th Annu. Conf. Mass Spectrom. Allied Top.*, New York, p. 320.
Blakley, C. R., Carmody, J. J., and Vestal, M. L. (1980e). *Clin. Chem. (Winston-Salem, N.C.)* **26**, 1467.

Carroll, D. I., Dzidic, I., Stillwell, R. N., Horning, M. G., and Horning, E. C. (1974). *Anal. Chem.* **46,** 706.
Carroll, D. I., Dzidic, I., Stillwell, R. N., Haegele, K. D., and Horning, E. C. (1975). *Anal. Chem.* **47,** 2369.
Carroll, D. I., Dzidic, I., Stillwell, R. N., and Horning, E. C. (1979). *NBS Spec. Publ. (U.S.)* **519,** 655.
Christensen, R. G., Hertz, H. S., Meiselman, S., and White, E. (1981). *Anal. Chem.* **53,** 171.
Colby, B. N., and Evans, C. A. (1973). *Anal. Chem.* **45,** 1884.
Cotter, R. J. (1979). *Anal. Chem.* **51,** 317.
Cotter, R. J. (1980). *Anal. Chem.* **52,** 1589A.
Dark, W. A., and McFadden, W. H. (1978). *J. Chromatogr. Sci.* **16,** 289.
Dark, W. A., McFadden, W. H., and Bradford, D. C. (1977). *J. Chromatogr. Sci.* **15,** 454.
Dawkins, B. G., and McLafferty, F. W. (1978). *In* "GLC and HPLC Determinations of Therapeutic Agents" (K. Tsuji and W. Morozowich, eds.), p. 259. Dekker, New York.
Dawkins, B. G., Arpino, P. J., and McLafferty, F. W. (1978). *Biomed. Mass Spectrom.* **5,** 1.
Dell, A., Williams, D. H., Morris, H. R., Smith, G. A., Feeney, J., and Roberts, G. (1975). *J. Am. Chem. Soc.* **97,** 2497.
DeRoos, F. L., and Foltz, R. L. (1979). *27th Annu. Conf. Mass Spectrom. Allied Top., Seattle,* p. 358.
DeRoos, F. L., and Thompson, R. M. (1979). *Annu. Conf. Mass Spectrom. Allied Top., Seattle,* p. 358.
Dymerski, P. P., Kennedy, M., and Kaminsky, L. (1979). *NBS Spec. Publ. (U.S.)* **519,** 685.
Eckers, C., Games, D. E., Lewis, E., Rao, K. R. N., Rossiter, M., and Weerasinghe, N. C. A. (1979). *Adv. Mass Spectrom.* **8B,** 1396.
Evans, N., Games, D. E., Jackson, A. H., and Matlin, S. A. (1975). *J. Chromatogr.* **115,** 325.
Fales, H. M., Milner, G. W. A., and Vestal, M. L. (1969). *J. Am. Chem. Soc.* **91,** 3682.
Games, D. E. (1978). *Chem. Phys. Lipids* **21,** 389.
Games, D. E. (1980a). *Proc. Anal. Div. Chem. Soc.* **17,** 110.
Games, D. E. (1980b). *Proc. Anal. Div. Chem. Soc.* **17,** 322.
Games, D. E., Gower, J. L., Lee, M. G., Lewis, I. A. S., Pugh, M. E., and Rossiter, M. (1978a). *Proc. Anal. Div. Chem. Soc.* **15,** 101.
Games, D. E., Gower, J. L., Lewis, E., Pugh, M. E., and Rossiter, M. (1978b). *Methodol. Surv. Biochem.* **7,** 185.
Games, D. E., Eckers, C., Hirter, P., Lewis, E., and Rao, K. R. N. (1979a). *27th Annu. Conf. Mass Spectrom. Allied Top., Seattle,* p. 627.
Games, D. E., Lewis, E., Haskins, N. J., and Waddell, K. A. (1979b). *Adv. Mass Spectrom.* **8B,** 1233.
Games, D. E., Games, M. L., Eckers, C., Kuhnz, W., and Lewis, E. (1980a). *28th Annu. Conf. Mass Spectrom. Allied Top., New York,* p. 318.
Games, D. E., Weerasinghe, N. C. A., and Westwood, S. A. (1980b). *28th Annu. Conf. Mass Spectrom. Allied Top., New York,* p. 316.
Games, D. E., Hirter, P., Kuhnz, W., Lewis, E., Weerasinghe, N., and Westwood, S. A. (1981). *J. Chromatogr.* **203,** 131.
Haakti, E., and Nikkari, T. (1963). *Acta Chem. Scand.* **17,** 2565.
Hansen, G., and Munson, B. (1978). *Anal. Chem.* **50,** 1130.
Hansen, G., and Munson, B. (1980). *Anal. Chem.* **52,** 245.
Hardin, E. D., and Vestal, M. L. (1980). *28th Annu. Conf. Mass Spectrom. Allied Top., New York,* p. 616.
Henion, J. D. (1978a). *Anal. Chem.* **50,** 1687.
Henion, J. D. (1978b). *Adv. Mass Spectrom.* **7,** 865.

Henion, J. D. (1979). *Adv. Mass Spectrom.* **8B,** 1241.
Henion, J. D. (1981). *J. Chromatogr. Sci.* **19,** 57.
Henion, J. D., and Maylin, G. A. (1980). *Biomed. Mass Spectrom.* **7,** 115.
Heresch, F., and Jurenitsch, J. (1979). *Chromatographia* **12,** 647.
Heresch, F., Schmid, E. R., Fogy, T., and Huber, J. F. K. (1979). *Adv. Mass Spectrom.* **8B,** 1880.
Heylin, M. (ed.) (1980). *Chem. Eng. News,* **58**(36), 45.
Hilker, D., Dymerski, P. P., and Champlin, P. (1979). *27th Annu. Conf. Mass Spectrom. Allied Top., Seattle,* p. 625.
Hirata, Y., Takeuchi, T., Tsuge, S., and Yoshida, Y. (1979). *Org. Mass Spectrom.* **14,** 126.
Horning, E. C., Horning, M. G., Carroll, D. I., Dzidic, I., and Stillwell, R. N. (1973a). *Anal. Chem.* **45,** 13.
Horning, E. C., Horning, M. G., Carroll, D. I., Dzidic, I., and Stillwell, R. N. (1973b). *Life Sci.* **13,** 1331.
Horning, E. C., Carroll, D. I., Dzidic, I., Haegele, K. D., Horning, M. G., and Stillwell, R. N. (1974a). *J. Chromatogr.* **99,** 13.
Horning, E. C., Carroll, D. I., Dzidic, I., Haegele, K. D., Horning, M. G., and Stillwell, R. N. (1974b). *J. Chromatogr. Sci.* **12,** 725.
Horning, E. C., Carroll, D. I., Dzidic, I., Haegele, K. D., Horning, M. G., and Stillwell, R. N. (1976). *Kagaku no Ryoiki, Zokan* **109,** 85.
Horning, E. C., Carroll, D. I., Dzidic, I., Haegele, K. D., Lin, S.-N., Oertli, C. U., and Stillwell, R. N. (1977). *Clin. Chem. (Winston-Salem, N.C.)* **23,** 13.
Horning, E. C., Carroll, D. I., Dzidic, I., and Stillwell, R. N. (1978). *Pure Appl. Chem.* **50,** 113.
Huber, J. F. K., Van Urk-Schoen, A. M., and Sieswerda, G. B. (1973). *Z. Anal. Chem.* **264,** 257.
James, A. T., Ravenhill, J. R., and Scott, R. P. W. (1964). *In* "Gas Chromatography 1964" (A. Goldup, ed.). Institute of Petroleum, London.
Jones, P. R., and Yang, S. K. (1975). *Anal. Chem.* **47,** 1000.
Karasek, F. W. (1974). *Anal. Chem.* **46,** 710A.
Karasek, F. W., and Denney, D. W. (1973). *Anal. Lett.* **6,** 993.
Karger, B. L., Kirby, D. P., Vovros, P., Foltz, R. L., and Hidy, B. (1979). *Anal. Chem.* **51,** 2324.
Kenndler, E., and Schmid, E. R. (1978). *In* "Instrumentation for High Performance Liquid Chromatography" (J. F. K. Huber, ed.), p. 163. Elsevier, Amsterdam.
Kenyon, C. N. (1980). *28th Annu. Conf. Mass Spectrom. Allied Top., New York,* p. 608.
Kenyon, C. N., Melera, A., and Erni, F. (1980). *J. Chromatogr. Sci.* **18,** 103.
Kirby, D. P., Karger, B. L., Vouros, P., Petersen, B., and Hidy, B. (1980a). *28th Annu. Conf. Mass Spectrom. Allied Top., New York,* p. 314.
Kirby, D. P., Vouros, P., and Karger, B. L. (1980b). *Science (Washington, D.C.)* **209,** 497.
Kirby, D. P., Vouros, P., Karger, B. L., Hidy, B., and Petersen, B. (1981). *J. Chromatogr.* **203,** 139.
Lovins, R. E., Ellis, S. R., Tolbert, G. D., and McKinney, C. R. (1973). *Anal. Chem.* **45,** 1553.
Lovins, R. E., Ellis, S. R., Tolbert, G. D., and McKinney, C. R. (1974). *Adv. Mass Spectrom.* **6,** 457.
McAdams, M. J., Blakley, C. R., and Vestal, M. L. (1978). *26th Annu. Conf. Mass Spectrom. Allied Top., St. Louis.*
McAdams, M. J., Blakley, C. R., and Vestal, M. L. (1979). *27th Annu. Conf. Mass Spectrom. Allied Top., Seattle,* p. 548.
McFadden, W. H., (1979). *J. Chromatogr. Sci.* **17,** 2.
McFadden, W. H. (1980). *J. Chromatogr. Sci.* **18,** 97.
McFadden, W. H., Schwartz, H. L., and Evans, S. (1976). *J. Chromatogr.* **122,** 389.
McFadden, W. H., Bradford, D. C., Games, D. E., and Gower, J. L. (1977). *Am. Lab. (Fairfield, Conn.)* **9**(10), 55.

McFadden, W. H., Bradford, D. C., Eglinton, G., Hajlbrahim, S., Dark, W. A., and Nicolaides, N. (1978). *26th Annu. Conf. Mass Spectrom. Allied Top., St. Louis.*
McFadden, W. H., Bradford, D. C., Eglinton, G., Hajlbrahim, S., and Nicolaides, N. (1979). *J. Chromatogr. Sci.* **17,** 518.
Macfarlane, R. D., and Torgerson, D. F. (1976). *Science (Washington, D.C.)* **191,** 920.
Macfarlane, R. D., McNeal, C. J., and Hunt, J. E. (1980). *Adv. Mass Spectrom.* **817,** 349.
McKeown, M., and Siegel, M. W. (1975). *Am. Lab. (Fairfield, Conn.)* Nov., 89.
McLafferty, F. W. (1980). *In* "Biochemical Applications of Mass Spectrometry" (G. R. Waller and O. C. Dermer, eds.), Vol. I, pp. 1159–1168. Wiley (Interscience), New York.
McLafferty, F. W., and Dawkins, B. G. (1975). *Biochem. Soc. Trans.* **3,** 856.
McLafferty, F. W., and Lory, E. R. (1981). *J. Chromatogr.* **203,** 109.
McLafferty, F. W., Knutti, R., Venkataraghaven, R., Arpino, P. J., and Dawkins, B. G. (1975). *Anal. Chem.* **47,** 1503.
Martin, T. (1980). *J. Chromatogr. Sci.* **18,** 104.
Melera, A. (1979a). Hewlett-Packard Technical Paper No. MS-10. *Pittsburgh Conf. Anal. Chem. Appl. Spectrosc., Cleveland,* Paper No. 85.
Melera, A. (1979b). *Adv. Mass Spectrom.* **8B,** 1597.
Millington, D. S., Yorke, D. A., and Burns, P. (1980). *Adv. Mass Spectrom.* **8B,** 1819.
Mumma, R. O., and Vastola, F. J. (1972). *Org. Mass Spectrom.* **6,** 1373.
Munson, M. S. B., and Field, F. H. (1966). *J. Am. Chem. Soc.* **88,** 2621.
Ohashi, M., Tsujimoto, K., and Yasuda, A. (1976). *Chem. Lett.* 439.
Ohashi, M., Yamada, S., Kudo, H., and Nakayama, N. (1978). *Biomed. Mass Spectrom.* **5,** 579.
Posthumus, M. A., Kistemaker, P. G., Meuzelaar, H. L. C., and Ten Noever de Brauw, M. C. (1978). *Anal. Chem.* **50,** 985.
Price, P. C., McDonie, R. A., and Wellons, S. L. (1980). *28th Annu. Conf. Mass Spectrom. Allied Top., New York,* p. 610.
Privett, O. S., and Erdahl, W. L. (1975). *In* "Analysis of Lipids and Lipoproteins" (E. G. Perkins, ed.), p. 123. American Oil Chemists Society, Champaign, Illinois.
Privett, O. S., and Erdahl, W. L. (1977). *Lipids* **12,** 797.
Privett, O. S., and Erdahl, W. L. (1978). *Chem. Phys. Lipids* **21,** 361.
Quillium, M. A., and Osei-Twum, E. Y. (1980). *28th Annu. Conf. Mass Spectrom. Allied Top., New York,* p. 612.
Rottschaefer, S., Killmer, L. B., Roberts, G. D., Warren, R. J., and Zarembo, J. E. (1979). *Pittsburgh Conf. Anal. Chem. Appl. Spectrosc., Cleveland,* Paper No. 019.
Roy, T. A., DeRoos, F. L., Hidy, B. J., and Howard, C. C. (1980). *28th Annu. Conf. Mass Spectrom. Allied Top., New York,* p. 615.
Schäfer, K. H., and Levsen, K. (1981). *J. Chromatogr.* **206,** 245.
Schmitter, J. M., Arpino, P. J., and Guiochon, G. (1979). *J. Chromatogr.* **167,** 149.
Schulten, H. R., and Beckey, H. D. (1973). *J. Chromatogr.* **83,** 315.
Schuster, R. (1980). *Chromatographia* **13,** 379.
Scott, R. P. W. (1979). *NBS Spec. Publ. (U.S.)* **519,** 637.
Scott, R. P. W., Scott, C. G., Munroe, M., and Hess, J. (eds.) (1974a). *In* "The Poisoned Patient: The Role of the Laboratory," p. 155. Elsevier, New York.
Scott, R. P. W., Scott, C. G., Munroe, M., and Hess, J. (1974b). *J. Chromatogr.* **99,** 395.
Serum, J., and Melera, A. (1978). *26th Annu. Conf. Mass Spectrom. Allied Top., St. Louis.*
Simons, D. S., Colby, B. N., and Evans, C. A. (1974). *Int. J. Mass Spectrom. Ion Phys.* **15,** 291.
Smith, R. D., and Burger, J. E. (1980). *28th Annu. Conf. Mass Spectrom. Allied Top., New York,* p. 310.
Smith, R. D., and Johnson, A. L. (1981). *Anal. Chem.* **53,** 739.
Smith, R. D., Burger, J. E., and Johnson, A. L. (1981). *Anal. Chem.* **53,** 1615.

Stimpson, B. P., and Evans, C. A. (1978). *Biomed. Mass Spectrom.* **5,** 52.
Takeuchi, T., Hirata, Y., and Okumura, Y. (1978). *Anal. Chem.* **50,** 659.
Tal'roze, V. L., and McLafferty, F. W. (chairman) (1978). *Adv. Mass Spectrom.* **7,** 949.
Tal'roze, V. L., Karpov, G. V., Gorodetskii, I. G., and Skurat, V. E. (1968). *J. Phys. Chem. (Moscow)* **42,** 1658.
Tal'roze, V. L., Karpov, G. V., Gorodetskii, I. G., and Skurat, V. E. (1969a). *J. Phys. Chem. (Moscow)* **43,** 198.
Tal'roze, V. L., Skurat, V. E., and Karpov, G. V. (1969b). *J. Phys. Chem. (Moscow)* **43,** 241.
Tal'roze, V. L., Grishin, V. D., Skurat, V. E., and Tautsyrev, G. D. (1970). *In* "Recent Developments in Mass Spectroscopy" (K. Ogata and T. Hayakawa, eds.), p. 1218. Univ. Park Press, Baltimore, Maryland.
Tal'roze, V. L., Skurat, V. E., Gorodetskii, I. G., and Zolotoi, N. B. (1972). *J. Phys. Chem. (Moscow)* **46,** 456.
Thruston, A. D., and McGuire, J. M. (1980). *28th Annu. Conf. Mass Spectrom. Allied Top., New York,* p. 614.
Thruston, A. D., and McGuire, J. M. (1981). *Biomed. Mass Spectrom.* **8,** 47.
Tsuchiya, M., Seita, K., and Taira, T. (1980). *28th Annu. Conf. Mass Spectrom. Allied Top., New York,* p. 649.
Tsuge, S., Hirata, Y., and Takeuchi, T. (1979). *Anal. Chem.* **51,** 166.
Udseth, H. R., Orth, R. G., and Futrell, J. H. (1978). *26th Annu. Conf. Mass. Spectrom. Allied Top., St. Louis,* p. 659.
Vestal, M. L. (1979). *NBS Spec. Publ. (U.S.)* **519,** 647.
Willoughby, R. C., and Browner, R. F. (1980). *Pittsburgh Conf. Anal. Chem. Appl. Spectrosc., Atlantic City,* Paper No. 626.
Wright, L. H., and Edgerton, T. R. (1979). *27th Annu. Conf. Mass Spectrom. Allied Top., Seattle,* p. 742.
Wright, L. H., and Oswald, E. O. (1978). *26th Annu. Conf. Mass Spectrom. Allied Top., St. Louis.*

# APPLICATIONS OF STERIC EXCLUSION CHROMATOGRAPHY IN TRACE ANALYSIS

### Ronald E. Majors and Thomas V. Alfredson

Varian Associates, Inc.
Walnut Creek Instrument Division
Walnut Creek, California

|     |      |                                                                           |     |
|-----|------|---------------------------------------------------------------------------|-----|
| I.   |      | Introduction............................................................ | 112 |
| II.  |      | Basics of SEC.......................................................... | 112 |
| III. |      | Advantages of the SEC Technique .......................... | 114 |
| IV.  |      | Disadvantages of the SEC Technique ....................... | 115 |
| V.   |      | Recent Advances in SEC Columns .......................... | 116 |
| VI.  |      | Utility of SEC in Trace Analysis............................. | 117 |
| VII. |      | Approaches for Use of SEC in Trace Analysis .................. | 118 |
|      | A.   | Direct Injection..................................................... | 118 |
|      | B.   | Off-Line Collection for Prefractionation and Sample Cleanup | 127 |
| VIII.|      | Off-Line Multidimensional SEC Techniques .................... | 128 |
| IX.  |      | On-Line Multidimensional SEC Techniques..................... | 134 |
| X.   |      | Problems and Troubleshooting............................. | 143 |
|      | A.   | Problems Encountered in the Use of SEC with Direct Injection Technique ....................................................... | 143 |
|      | B.   | Problems Encountered in the Use of SEC with Off-Line Prefractionation and Sample Cleanup ......................... | 145 |
|      | C.   | Problems Encountered in the Use of SEC with On-Line Multidimensional Chromatography ........................... | 147 |
|      | D.   | Troubleshooting..................................................... | 148 |
|      |      | References............................................................ | 149 |

## I. INTRODUCTION

The technique of steric exclusion chromatography (SEC) has long been used as a method for (1) the determination of polymeric molecular weight and molecular weight distribution, (2) the analysis of polymer additives, (3) preparative fractionation, and (4) sample cleanup. There are several texts and chapters in texts which describe this technique in detail (e.g., Yau et al., 1979). Compared to high-performance liquid chromatography (HPLC) or gas chromatography (GC) techniques, SEC has not ordinarily been considered as a trace-analysis technique because the larger particle columns ($d_p \geq 40$–$70$ μm) used in the past gave somewhat broadened peaks which limited resolution and made quantitation more difficult. We shall demonstrate in the present article that SEC with modern high-efficiency columns can lend itself to trace analysis, with the added advantage that matrix interferences can often be eliminated or substantially reduced compared to other HPLC techniques.

## II. BASICS OF SEC

Briefly, SEC is a technique which uses columns filled with porous particles of different pore diameters. A solute molecule injected into such a column will diffuse into those pores that have a diameter greater than its effective diameter. The effective diameter of a solvated molecule is dependent upon its spatial configuration in that particular solvent. Molecules of the same molecular weight could have different effective diameters or, alternatively, molecules of different molecular weights could have the same effective diameter. Thus, SEC separates on the basis of molecular size and not necessarily molecular weight. Larger molecules which are excluded from small pores elute first, whereas small molecules which permeate all of the pores elute last. Intermediate-size molecules which permeate some of the pores but are excluded from others may be separated. The process of SEC is depicted pictorially in Fig. 1 and schematically in Fig. 2. Each packing used in SEC will have a defined average pore size and pore-size distribution. A calibration curve such as the one depicted in Fig. 2 generally is used to define the characteristics of a particular SEC packing (or column). This plot of log molecular weight (or more correctly, molecular size) versus elution volume has the following three important defined regions:

$V_0$ is the exclusion volume; compounds of molecular weights larger than that where the curve turns upward will elute at $V_0$; the molecular weight where this change of slope occurs is called the exclusion limit.

$V_t$ is the permeation volume; compounds of molecular weight lower than that where the curve turns downward will elute at $V_t$.

## Steric Exclusion Chromatography

DIRECTION OF SOLVENT FLOW

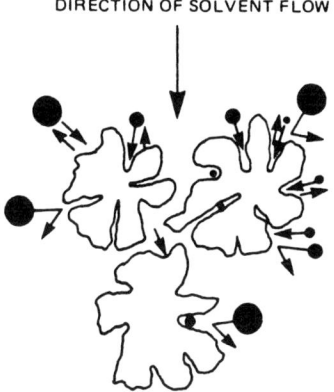

**Fig. 1.** Pictorial representation of exclusion chromatography (Johnson and Stevenson, 1978). Large solutes cannot penetrate into the pores of the packing, hence they will move with the solvent. Small molecules can diffuse into "stagnant" pools and are retained.

$V_p$ is the internal pore volume; compounds with molecular weights in this region will be separated on the basis of their relative sizes.

Equation (1) relates the elution volume $V_R$ to the other variables:

$$V_R = V_0 + KV_p \qquad (1)$$

Thus, the distribution coefficient $K$ can range from 0 to 1. If a solute is totally excluded, then $K = 0$, whereas if a solute totally permeates the pores, $K = 1$.

Note that the exclusion limit in the molecular weight operating range is not sharply defined in Fig. 2 since the pore diameters of the column packing do not have a narrow distribution. The narrower the pore distribution, the flatter the calibration curve and the better the resolving power of the column. However, the molecular weight operating range will be smaller. If the slope of the calibration curve is great, the wider will be its operating range and the lower the column's resolution.

To cover a wide range of molecular weights, several columns, each packed with a different pore-size particle, will often be used. The entire set of columns will be calibrated with standards of known molecular weight.

In "pure" SEC, unlike the other modes of HPLC, interaction between the solute and the surface of the column packing does not occur. Rarely is a surface completely inert, but specific interactions can be minimized by choice of the proper chromatographic conditions. For separation purposes, mixed mechanisms (e.g., exclusion plus partition) may not necessarily be detrimental, but interpretation of the chromatogram is sometimes rendered more difficult.

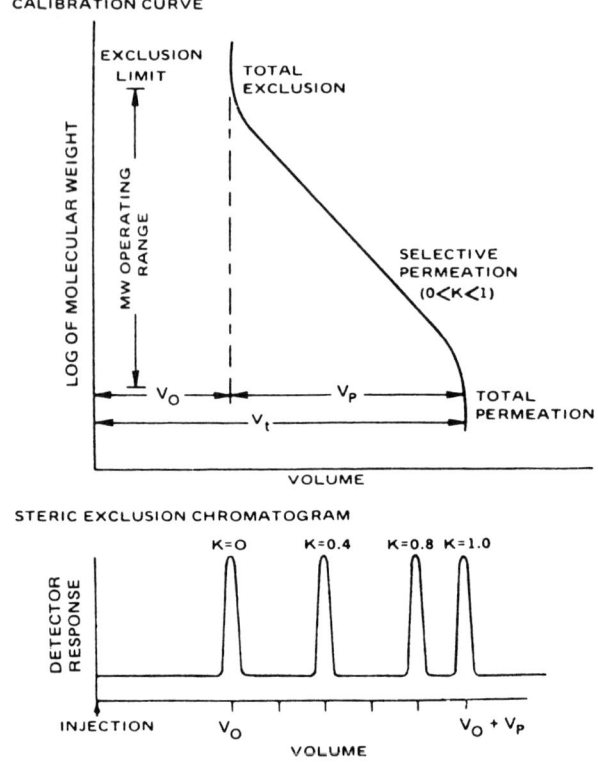

**Fig. 2.** Typical calibration curve and corresponding chromatogram for SEC (Johnson and Stevenson, 1978).

## III. ADVANTAGES OF THE SEC TECHNIQUE

The SEC technique is unique in HPLC in that in the ideal case all compounds elute between $K = 0$ and $K = 1$. Thus, the run time is predetermined. Also, the elution order is predictable since increasing elution volumes imply decreasing molecular sizes. For a homologous series, the elution order is inversely proportional to log molecular weight. Since the entire experiment is carried out isocratically, the technique requires little expertise in solvent selection. The only mobile-phase selection criterion is that the sample must be soluble in the solvent. Thus, method development time is short; the only real decision is the selection of the optimum pore size of the packing. Due to the simplicity of the technique, interpretation of the chromatogram is generally easy. In addition, no gradient instrumentation is needed. Because of the mild interactions between the solute and solid support, compounds of widely differing polarity can be handled in the same run—rarely achieved without gradient elution in the other liquid chromatography (LC) modes.

Both macroparticles (40 μm $\leq d_p \leq$ 70 μm) and microparticles ($d_p \sim$ 10 μm) have been used as SEC column packings. Compared to the larger particle columns, for microparticles using short (30 cm) columns, the pore volumes are quite small; thus solvent economy is a factor. Likewise, the lower dilution of peaks leads to greater efficiency and hence narrower peak widths and greater peak heights, which is important for trace analysis.

Furthermore, the sample capacity of SEC is generally higher than the other LC modes, which are limited by the capacity of the solid support. The sample capacity in SEC is in part dictated by the sample solubility in the mobile phase. The practical advantage of the sample capacity of SEC in trace analysis is that large amounts can be injected without column overload.

## IV. DISADVANTAGES OF THE SEC TECHNIQUE

The fact that all compounds elute in a finite volume tends to limit the resolution of the technique. The peak capacity is the lowest of all LC techniques. For a fixed-pore-size column, resolution can be increased by (1) increasing stationary-phase volume (i.e., column length and/or internal diameter); (2) decreasing particle size, usually at the expense of pressure and column fragility (for semirigid packings); and (3) decreasing flow rate at the expense of separation time. Resolution in SEC can also be increased by obtaining a column with a "flatter" calibration curve, but due to the limited number of high-performance packings on the market such columns are not always available. Indeed, one disadvantage of SEC is that the columns are more expensive, usually double those of other LC modes, and require more care in their use.

Since a size difference is required for separation, some molecules such as isomers are not usually separable by SEC. Such behavior is not always the case since interactions between solute and stationary phase do occur and mixed mechanisms are occasionally observed.

Finally, since the internal pore volume $V_p$ (the exclusion volume minus the permeation volume) of the modern SEC column is small, precise control of flow rate is required to achieve good values of molecular weight because all components must elute in a small volume. Since calibration curves are expressed in terms of log molecular weight as a function of elution volume, one can qualitatively deduce that small errors in flow rate can give rise to large errors in the estimated molecular weight. A definitive study of the magnitude of these errors in modern exclusion chromatography has been published (Yau *et al.*, 1979; Bly *et al.*, 1975). The practical implications are that highly reproducible flow rates are required for the successful operation of microparticulate SEC columns, thereby placing stringent requirements on LC pumping systems.

## V. RECENT ADVANCES IN SEC COLUMNS

The development of microparticulate silica-based packings in the early 1970s permitted the decrease of analysis times from hours to minutes for the normal- and reverse-phase modes of HPLC. The improvement in the synthesis of different pore-size, narrow particle-size-range, cross-linked semirigid polystyrene beads and rigid porous silica or glass microspheres led to similar improvements in SEC packings in the mid-1970s. Average particle sizes ranging from 5 to 15 μm became available from several suppliers, usually in packed columns. The increase in column efficiency, primarily due to increased rates of solute mass transfer, resulted in the application of shorter columns (25 or 30 cm) compared to the longer columns, usually 4 ft (~122 cm), used formerly. Polymer characterizations could then be done in less than 1 hr rather than the 4–6 hr required with the longer columns.

Commercially available microparticle SEC packings and packed columns have been reviewed (Majors, 1980; Barth, 1980). Generally, modern SEC columns range in guaranteed efficiency from 20,000 to 30,000 plates/meter when measured with a totally permeating species. They are usually available in 25-, 30-, 50-, or 60-cm lengths with internal diameters of analytical columns generally in the 4–8-mm range. Columns are available for both aqueous (i.e., gel-filtration chromatography, GFC) and organic (i.e., gel-permeation chromatography, GPC) work. Preparative columns containing microparticulates are available in diameters up to 1 in.

Packings are available to cover varying molecular weights, from less than one hundred to tens of millions. For the measurement of smaller molecules, which is by far the most useful area of the application of SEC to trace analysis, several packings are available in the 100–2000 MW range (pore diameter < 500 Å). It is in this area that, for simple mixtures, SEC may become competitive with other LC modes.

Most high-efficiency SEC supports can be divided into the broad categories of silica-based packings and cross-linked organic-gel packings. The attributes of these categories of supports can be found in Table I. In general, cross-linked organic gels offer advantages of a wide range of pore sizes and greater chemical stability (less tendency for solute adsorption) compared to silica-based packings, which tend to offer superior mechanical strength and, in some instances, improved efficiency. Due to the availability of very small pore size columns (40–100 Å) cross-linked organic-gel columns (e.g., polystyrene–divinylbenzene) have found greatest utility for the analysis of small molecules.

**TABLE I**

**Attributes of High-Efficiency SEC Supports**

| Attribute | Silica-based SEC packings | Organic-gel SEC packings |
|---|---|---|
| Mechanical strength | ++ (Rigid) | -- (Semirigid) |
| Chemical stability | -- | ++ |
| Range of available pore sizes | -- | ++ |
| Efficiency | + | - |

## VI. UTILITY OF SEC IN TRACE ANALYSIS

Generally, SEC itself is not considered to be a mode of HPLC that is useful for trace analysis. However, there are several ways in which the technique can be used in the determination of trace substances.

(1) *Direct injection.* With the highly efficient microparticles, peaks are quite sharp, permitting the measurement of small amounts of substances. For example, for a 30 cm × 7.5 mm column of MicroPak TSK 2000H, peak widths are generally 200–400 μl, allowing the ppb analysis of a compound exhibiting an extinction coefficient of $10^5$. Such a technique is most useful for the elimination of matrix interferences of higher molecular weight than the compounds of interest. Both analytical and preparative procedures can be used.

(2) *Prefractionation/concentration/measurement.* This is a technique most often used where one collects an SEC fraction containing the trace compound of interest, then removes the solvent (evaporation) to concentrate the trace substance. The trace compound is then measured directly after reconstitution into a suitable solvent. The entire technique is carried out off-line. The measurement may be electrophoretic, fluorometric, colorimetric, spectroscopic, etc.

(3) *Off-line multidimensional chromatography.* This is similar to technique (2) where a fraction (or fractions) of one chromatographic separation is collected and further separated by a secondary chromatographic step (e.g., LC, GC, TLC); SEC can be used as the primary step or as the secondary step. Of course, the coupled chromatographic technique can be carried out any number of times for further resolution.

(4) *On-line multidimensional chromatography.* Technique (3) can be automated by diverting the fraction(s) from the primary SEC column to the secondary column by means of high-speed, low-volume switching valves. The entire procedure can be automated for convenience, decreased analysis time, increased sample throughput, and reproducibility.

Techniques (2)–(4) can all be regarded as methods of prefractionation or sample cleanup using SEC as one of the separation techniques. Due to its unique ability to separate on the basis of molecular size, SEC has often been used as a cleanup technique where materials with a higher molecular weight than the compound of interest can be removed from the sample prior to GC or LC analysis. If these undesired compounds remained in the sample, they would impede the measurement of the lower molecular weight components, either by masking their detection or by contaminating the LC or GC column used. (Several examples are illustrated later.)

## VII. APPROACHES FOR USE OF SEC IN TRACE ANALYSIS

### A. Direct Injection

*1. Introduction*

The development of high-efficiency microparticulate supports for SEC dramatically increased the utility of the technique for the analysis of small molecules (MW <1000). When low-molecular-weight solutes are present in a sample at trace levels, the isolation or elimination of interference from the matrix can be a formidable analytical problem, often requiring several procedural steps. Matrices such as polysaccharides (carbohydrates), synthetic polymers, lipids, and oils are commonly encountered in many food, industrial, and biological samples. Trace-level components (drugs, pesticides, metabolites, pollutants) found in these matrices are often much smaller in molecular size than the matrix itself. As a consequence, SEC has become a viable technique for the trace analysis of many nonvolatile compounds found in a wide variety of sample matrices.

*2. Analysis by Direct Injection onto an SEC Column*

Optimum use of SEC as an analytical tool for trace analysis occurs when (1) the sample matrix is of larger molecular size than the component of interest (rule of thumb, greater than twice the solute size); and (2) adequate resolution of the solute under investigation is obtained from other components of low molecular weight. These criteria allow the analysis of trace solutes by direct injection onto an SEC analytical column.

The first criterion for solute analysis via direct injection is a general requirement for use of SEC as an analytical tool for sample cleanup. The second condition for analysis by direct injection, however, is much more complex. Although molecular weight often correlates with molecular size, molecular conformation is an important factor in SEC where separation is based upon effective solute molecular size in solution. Solvent–solute interactions can affect the separation of trace solutes, especially those with a strong hydrogen-bonding tendency (e.g., alcohols) when solvents capable of hydrogen bonding (e.g., tetrahydrofuran, THF) are used as mobile phases. Nonexclusion effects such as partitioning and adsorption can also affect trace-solute analysis using organic solvent-compatible SEC packings (e.g., polystyrene–divinylbenzene) (Yau *et al.*, 1979). Such effects seem more predominant with polystyrene gels when very nonpolar mobile phases are employed (e.g., hexane). Most compounds show little or no interaction with the gel when polar solvents such as THF are employed as the mobile phase. Hydrophobic interactions, hydrophilic interactions, and ion-exchange mechanisms are all nonexclusion effects that can occur with the use of aqueous-compatible SEC packings (e.g., silica-based supports with hydrophilic bonded phases; Barth, 1980). Most nonexclusion effects can be eliminated or controlled through the judicious use of mobile-phase modifiers. However, prediction of resolution between small molecules and matrix components that could lead to interferences with trace solutes of interest is not always straightforward.

### 3. Applications of Direct-Injection SEC Technique to Trace Analysis

High-performance exclusion chromatography using small-pore-size gels has been applied to the direct analysis of trace components in a diverse variety of sample types. In most cases, the samples need only be diluted, filtered, and injected onto the analytical SEC column.

Classical SEC separations involve the analysis of polymers and polymer formulations by gel-permeation chromatography (GPC). Polymer additives such as lubricants and stabilizers are often found at trace levels in a polymer formulation. Many such additives have dramatic effects on compound physical properties due primarily to localized activity or synergistic effects. In most GPC analyses, the pore sizes of the columns are selected to provide information about high-molecular-weight polymers. However, when column pore size is optimized for analysis of small molecules, information on trace additive levels as well as polymer content can be obtained, often in less than 20 min. Figure 3 shows an example of the analysis of trace additives in a vinyl polymer formulation (plastisol compound). Epoxidized soybean oil (ESO) is used as stabilizer and secondary plasticizer in such formulations. Fatty acid amides (erucyl and oleyl amide) are used as surface lubricants. Due to their incompatibility in the polymer matrix, fatty acid amides bloom to the surface with age, thus providing surface lubricity. Due to their highly localized action, these additives are present in very low concentrations ($<0.1\%$ by weight) in some vinyl compounds. Sample preparation for this analysis required only centrifugation to remove inorganic fillers prior to analysis by direct injection onto the SEC column.

Low levels of additives are also added to various types of oils to preserve their lifetimes or give them certain desired physical properties for specific applica-

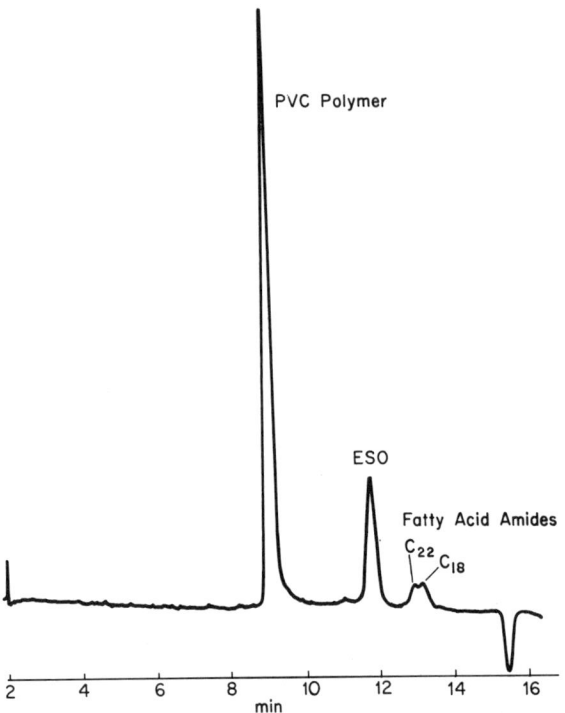

**Fig. 3.** Analysis of polymer additives using small-pore SEC columns; Column, μStyragel 500 Å × 2 plus 100 Å × 2; mobile phase, THF; flow rate, 3.0 ml/min; detector, RI (4×).

tions. Oxidative inhibitors, for example, are added to petroleum products to prevent deterioration at high temperatures and pressures encountered in automobile engines. Most oils are of higher molecular weights than the additives, and thus the exclusion technique is ideally suited for the analysis of low-molecular-weight additives without prior separation. Figure 4 exemplifies the application of small-pore SEC columns to direct analysis of additives in a lubricating oil used in aerospace and marine systems (Majors and Johnson, 1978). The additives [Antioxidant 702 (AO-702), Ethyl Corp., a hindered phenolic, and tricresyl phosphate (TCP)] were present in the oil at levels of 0.5 and 1% by weight, respectively.

Polychlorinated biphenyls (PCBs) have received much interest and notoriety in recent years due to both their ubiquity and their persistence in the environment. Although the PCB *p,p'*-DDT (a chlorinated pesticide) was outlawed several years ago in the United States, because of its environmental persistence it still occasionally appears as a contaminant in unsuspected places. The applicability of the direct-injection technique using exclusion chromatography to the analysis of

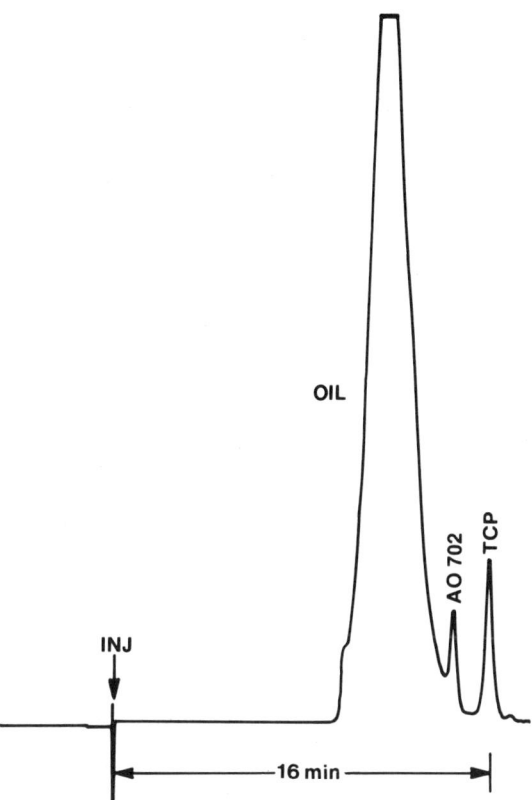

**Fig. 4.** SEC of lubricating oil with additives (Majors and Johnson, 1978): Column, MicroPak TSK 2000H, 50 cm × 7.5 mm; mobile phase, THF; flow rate, 0.5 ml/min; additives, AO-702 (MW 436) and TCP (MW 368).

a PCB pesticide in a dairy product is illustrated in Fig. 5 (Majors and Johnson, 1978). A sample of pure butter was deliberately "contaminated" with 100 ppm DDT. First, an exclusion chromatogram of a standard DDT gave an elution volume of 15.2 ml when run on a 250-Å SEC column. The unspiked butter sample, dissolved in THF and previously filtered, was chromatographed under identical conditions (Fig. 5A). (Note the presence of a small peak eluting after the higher molecular weight lipids and other components of the butter matrix.) Next, a butter sample spiked with 100 ppm DDT was treated similarly. The DDT peak is well resolved from the higher molecular weight matrix components, as seen in Fig. 5B. This experiment illustrates the ease with which direct injection using high-performance SEC columns can be used for both cleanup and analysis of trace solutes.

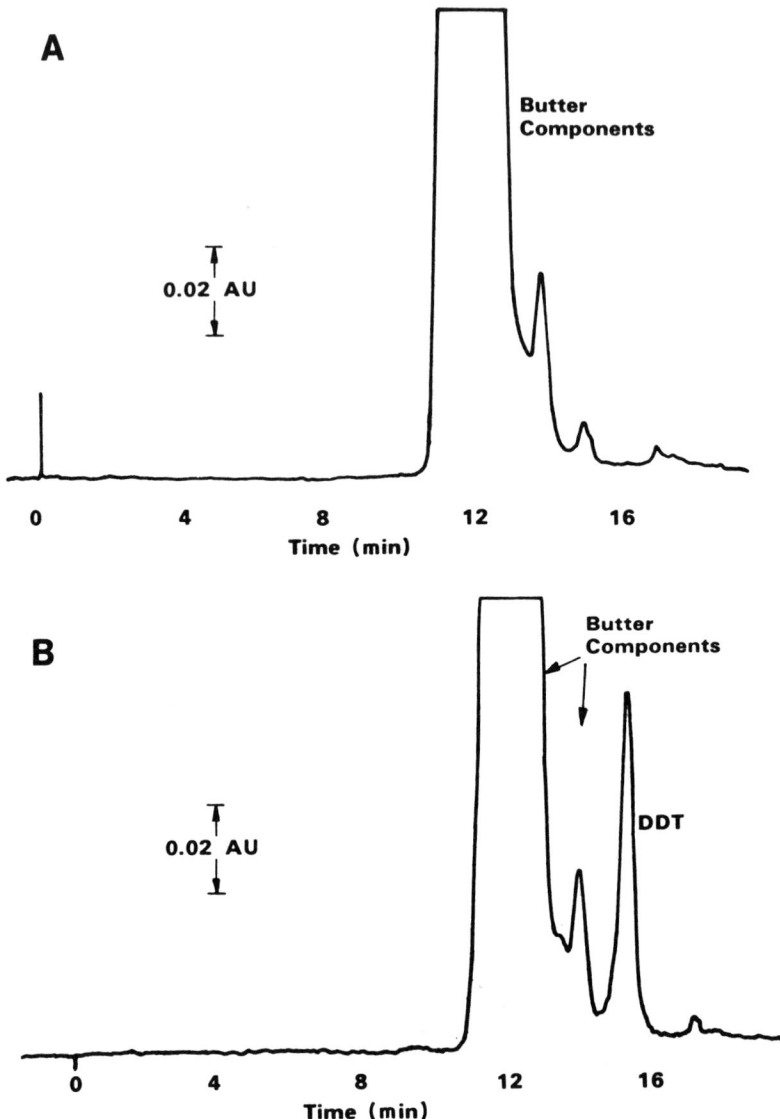

**Fig. 5.** SEC of (A) butter extract and (B) DDT-contaminated butter (Majors and Johnson, 1978): Column, MicroPak TSK 2000H, 60 cm × 7.5 mm; mobile phase, THF; flow rate, 1 ml/min; detector, UV at 215 nm, 0.2 AUFS.

Chlorinated pesticides commercially sold to agriculture are routinely monitored by many food processors. These compounds are usually present in trace levels since, in many instances, they represent unwanted contamination or unacceptable adulteration above certain concentrations often set by governmental regulations. Pentac (Hooker, 1979), a highly chlorinated cyclopentadiene, is a common pesticide used as a specific miticide in the treatment of corn and other plants. (Pentac is a trade name of the Hooker Chemical Co.) Upon processing treated corn into food-grade oil, in certain instances the pesticide may be carried along during oil manufacture.

SEC has been found to provide a rapid and reliable analytical technique for analysis of Pentac in corn oil by HPLC. Other HPLC modes such as reverse-phase offer high selectivity for vegetable oils, providing information on triglyceride content based upon separation by carbon number (Parris, 1978). However, the complex elution profiles obtained in this manner often introduce interferences in trace pesticide and additive analyses. In contrast, direct injection onto a small-pore SEC column provides adequate resolution between the oil matrix (usually MW >700) and trace components such as pesticides. Figure 6A displays the

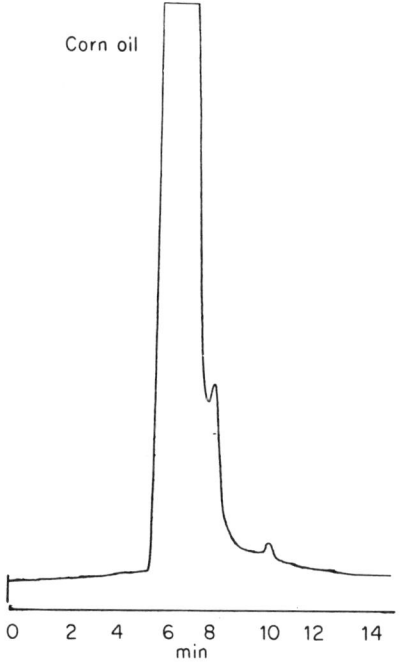

**Fig. 6A.** SEC analysis of corn oil (pure): Column, MicroPak TSK 1000H × 2, 30 cm × 7.5 mm; mobile phase, THF; flow rate, 1.5 ml/min; detector, UV at 254 nm, 0.08 AUFS.

analysis of a food-grade corn oil using a MicroPak TSK 1000H SEC column of pore size 40 Å. Figure 6B shows the trace analysis of 1 ppm Pentac in corn oil by direct injection onto the SEC column.

The antioxidants BHT and BHA are approved for food use and are frequently added to vegetable oils and other food products to preserve freshness and increase shelf life. BHT, MW 220, and BHA, MW 180, would be expected to be separated from vegetable oils which consist mainly of triglycerides with MW 600 (i.e., fatty acid composition $C_{10}$ and above). Soybean oil, used as an all-purpose salad oil, has been successfully analyzed for total antioxidant content by direct injection onto a 250-Å-pore-size SEC column (Majors and Johnson, 1978). A commercial soybean oil was diluted 1 : 1 with THF and injected directly onto a MicroPak TSK 2000H column. Figure 7 shows that the antioxidants BHT and BHA, although themselves unseparated on the exclusion column, were well resolved from the soybean oil incipients. The amount of total antioxidants found in the sample was 10 ppm.

SEC analysis by direct injection has also been shown to be useful for the determination of trace levels of vitamin A acetate in margarine (Wattenhofer,

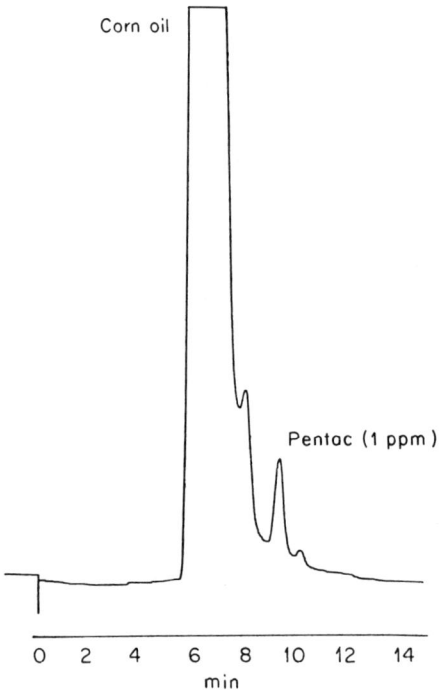

**Fig. 6B.** SEC analysis of pesticide in corn oil. Conditions same as in Fig. 6A.

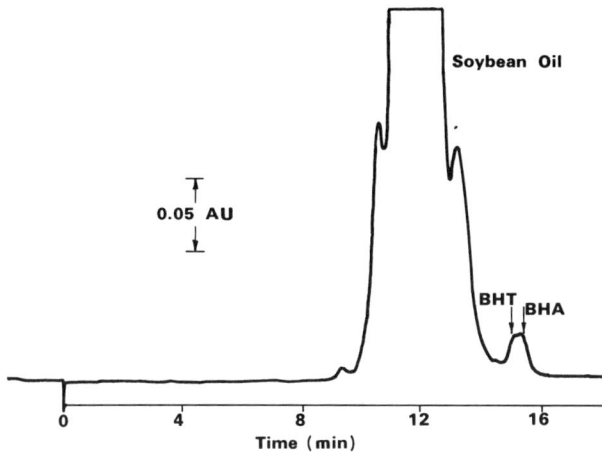

**Fig. 7.** SEC of soybean oil (Majors and Johnson, 1978): Column, MicroPak TSK 2000H, 60 cm × 7.5 mm; mobile phase, THF; flow rate, 1 ml/min; detector, UV at 215 nm; sample, soybean oil diluted 1:1 with THF prior to injection.

1981). Spectroscopic methods for quantitation of vitamin A, such as the United States Pharmacopeia method of direct measurement of three different wavelengths in the UV (United States Pharmacopeia, 1970) or the Carr–Price (1976) method with antimony trichloride followed by spectroscopic measurement, are quite complicated and involve time-consuming extraction with subsequent saponification of the lipids prior to measurement. Such sample pretreatment can lead to damage or loss of the vitamin A, which is sensitive to oxidation. Figure 8 shows the SEC analysis of vitamin A acetate in a commercial margarine sample. The only pretreatment required was centrifugation of the sample solution. Detection was at the absorbance maximum of vitamin A acetate (335 nm). As can be seen in the chromatogram, approximately 35 ppm vitamin A acetate was found in the sample using an external-standard method of quantitation. Note that refractive index (RI) detection was also used to show the fat components, which are well separated from the vitamin A.

High-speed SEC has also been applied to the trace analysis of polyoxyethylene surfactants and their decomposition products in industrial process waters (Cassidy and Niro, 1976). Small-pore-size columns (100 Å) were employed using chloroform as a mobile phase to achieve detection of trace concentrations (detection limit 0.05–0.1 ppm) of these surfactants.

Mori (1976) has employed small-pore-size SEC columns with an eluent of chloroform for determination of phthalate esters in river water in conjunction with normal- and reverse-phase adsorption LC modes. Detection at 254 nm with the SEC columns provided analysis of sub-ppm levels of these esters.

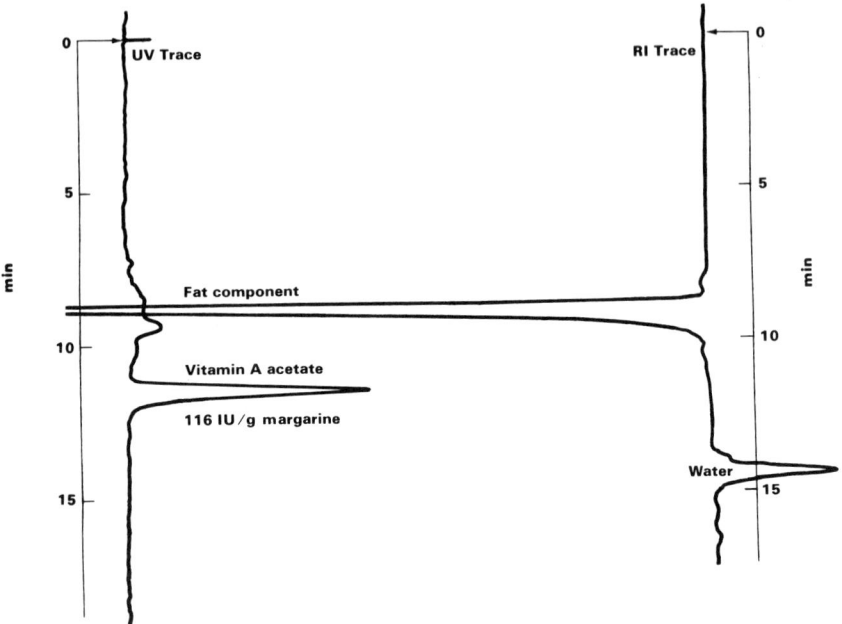

**Fig. 8.** SEC of vitamin A acetate in margarine (Wattenhofer, 1981): Column, MicroPak TSK 2000H, 50 cm × 7.5 mm; mobile phase, THF; flow rate, 1 ml/min; detectors, RI (8×) and UV at 335 nm, 0.1 AUFS.

In the past few years, several high-efficiency aqueous SEC packings have become commercially available (Majors, 1980). Although most applications of these supports involve analyses of biopolymers such as proteins, aqueous SEC packings can be applied to a variety of water-soluble compounds such as sugars and polysaccharides, vitamins, and drugs. The utility of the small-pore-size aqueous SEC columns for trace analysis should equal that of their organic-solvent-compatible counterparts.

One example that illustrates the utility of small-pore, aqueous SEC columns to the trace analysis of water-soluble compounds can be seen in Fig. 9. A candy-bar formulation was dissolved in water, filtered, and analyzed using a MicroPak TSK 2000PW column with a pore size of 250 Å (Apffel et al., 1981). Although sucrose and glucose coelute, such a simple analysis with minimal sample preparation provided quantitation of total sugar content (0.1% by weight). HPLC analysis of sugars by ion-exchange or normal-phase chromatography would require lengthy sample preparation and/or longer analysis time due to gradient reequilibration. Such an example serves to illustrate the potential of these columns yet to be exploited for trace analysis of water-soluble compounds in many food and biological matrices.

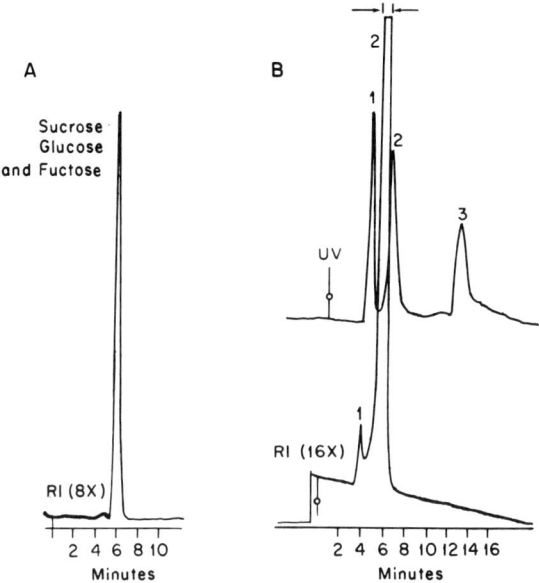

**Fig. 9.** SEC of sugars in candy-bar formulation (Apffel et al., 1981): Column, MicroPak TSK 2000PW, 30 cm × 7.5 mm; mobile phase, $H_2O$; flow rate, 1.5 ml/min. (A) Standards: Detector, RI. (B) Candy-bar extract: Detectors, RI and UV at 192 nm, 0.5 AUFS. Peaks: 1, large-molecular-size compounds; 2, sugars; 3, unknown.

## B. Off-Line Collection for Prefractionation and Sample Cleanup

Off-line SEC techniques for the cleanup of samples prior to LC or GC analysis have been used for many years, mostly with the large porous exclusion packings. The ease of collection and the handling of liquids make the off-line techniques particularly attractive. The advantages and disadvantages of off-line techniques are summarized in Table II.

Sample collection in SEC is carried out by sampling the effluent from the column, usually after a detector, either manually or by means of a fraction collector. Note that fraction collectors are now available which can be synchronized with and/or controlled by a microprocessor-based liquid chromatograph. With such an instrument, the injection/separation/fraction collection process can be fully automated. Once the fraction containing the solvent(s) of interest is obtained, further isolation of the trace substance may be carried out. Depending on the matrix, the trace substance may be extracted, dialyzed, etc., or the solvent may be removed by evaporation. Some care must be exercised during evaporation to avoid sample loss due to (1) adsorption on container walls, (2)

**TABLE II**

**Off-Line SEC Techniques**

| Advantages | Disadvantages |
|---|---|
| Easy to carry out for collection of column effluent | Difficult to automate; more cumbersome and inconvenient |
| Can concentrate trace solutes from large volumes | Greater chance of sample loss (e.g., adsorption, evaporation, oxidation) |
| Can work with second LC mode which uses incompatible solvents (e.g., SEC using water → LSC using hexane) | More time-consuming |
| | More difficult to quantitate and reproduce |

bumping of solvent, or (3) chemical transformation such as oxidation. Removal of trace organic compounds from inorganic salt buffers represents a difficult step. The SEC mobile-phase purity is of utmost concern since solvent impurities are also concentrated during evaporation. It is a gross misconception that solvent quality is of lesser importance for preparative separations in HPLC than for analytical work. Indeed, it is of *greater* importance for subsequent concentration of separated solute.

An example of how low-efficiency SEC can be used for off-line sample cleanup in trace analysis was given by Kawaguchi and Auld (1978), who measured picogram quantities of metals in metalloenzymes and demonstrated that metal quenching agents and low-molecular-weight protein contaminants could be easily removed from the enzyme of interest. Figure 10 shows that a microbore column of Sephadex G-100 was sufficient for cleanup of zinc-containing enzymes to allow the measurement of zinc in droplet fractions in the range of $10^{-10}$–$10^{-13}$ g in only micrograms of enzyme when using microwave-induced emission spectrometry for detection.

The SEC technique has been used as a cleanup or separation method and has been combined with electrophoresis (Ohtake and Koga, 1979; Lazarovits *et al.*, 1979), atomic absorption spectroscopy (Bartak, 1980), or electrochemistry (Bartak, 1980) for the measurement step.

## VIII. OFF-LINE MULTIDIMENSIONAL SEC TECHNIQUES

The most widespread applications of off-line SEC as a cleanup or prefractionation technique are in the area of multidimensional chromatography, where the secondary step is GC, LC, or sometimes TLC. Additional chromatographic steps may be required for complete analysis since the collected solute may still be contaminated with interfering compounds (i.e., overlapping peaks) because of

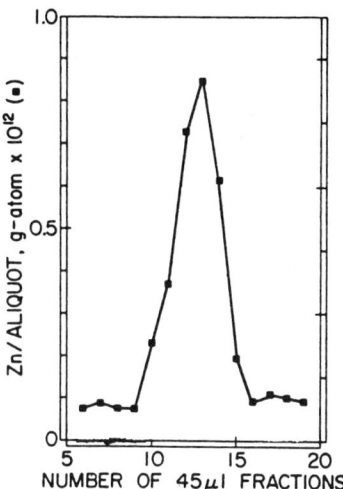

**Fig. 10.** SEC chromatogram of zinc-containing enzyme (Kawaguchi and Auld, 1975): Column, Sephadex G-100, 25 × 0.3 cm; mobile phase, 0.01 $M$ KCl, 0.01 $M$ Tris, pH 7.8; detection, microwave-induced emission spectrometer (sensitive to zinc).

the use of low-resolution SEC columns. Also, with the concentration of large volumes of mobile phase from the SEC technique, solvent impurities or additives (e.g., salt buffers) may also be concentrated and combined with the solute of interest.

For either LC or GC as the secondary step, the collected fraction may be suitable for direct injection if it is concentrated enough or if sensitive enough detection techniques are available. More often some type of concentration is required. For volatile solvents evaporation is used, often to dryness. Resolution may then be in a suitable solvent which is more compatible with the secondary chromatographic technique. For example, a large injection of a ''strong'' solvent such as methylene chloride into some nonpolar GC open tubular capillary column is not advisable as it may strip the coated phase. Reconstitution in a more compatible solvent is generally preferred. Likewise, in HPLC large injections of THF from an exclusion column onto a reverse-phase column may cause a partial migration of solute down the column, causing band spreading. In such cases, the solute will seldom be reconcentrated into a small volume after such spreading occurs, and the resolution of the overall multidimensional system is impaired.

Table III illustrates a number of applications of off-line multidimensional techniques where SEC was used as the primary separation or cleanup technique and GC was used as the secondary analytical technique. Some of these examples represent very complex matrices where a single chromatographic step would not have the resolution required to determine the solute of interest because of inter-

## TABLE III

### Off-Line SEC Applications in Trace Analysis (SEC–GC)

| Sample type | Matrix | Column | Reference |
|---|---|---|---|
| Pesticide(s) | Fish lipids | BioBeads SX | Stalling et al. (1972), Tindle and Stalling (1972) |
| | Fats, feeds, vegetables | BioBeads SX-3 | Fuchsbichler (1979) |
| | Animal and plant extracts | BioBeads SX-3 | Johnson et al. (1976) |
| | Oils and foods | BioBeads SX-2 | Griffitt and Craun (1974) |
| PCB | Fish lipids | BioBeads SX | Stalling et al. (1972), Tindle and Stalling (1972) |
| Carotenoids | Mandarin oil | Sephadex LH-20 | Deuber et al. (1979) |
| Hypoglycin | Amino acids | Sephadex G-10 | Manchester and Manchester (1980) |
| Steroidal spirolacetones | Urine | DEAE-Sephadex A-25 | Boreham et al. (1978) |
| Organophosphorous pesticide residues | Vegetables, fruits, crops | BioBeads SX-3 | Ault et al. (1979) |
| | Vegetable extract | Sephadex LH-20 | Pflugmacher and Ebing (1974) |
| Estradiol | Animal chow | Sephadex LH-20 | Bowman and Nony (1977) |
| Polycyclic aromatic hydrocarbons | Foods, oils, and fats | Sephadex LH-20 | Grimmer and Blohmke (1975) |

fering compounds. Other examples represent types of sample matrices where the repeated injection onto a GC column would contaminate the column, thereby ruining it or decreasing its lifetime, or would result in strongly retained compounds, some of which might gradually elute, causing extraneous peaks in subsequent chromatograms.

Natural products, biological extracts, and foods all represent complex sample matrices which, with SEC cleanup, result in simpler chromatograms or in longer lived columns or both. The cleanup of pesticides in a variety of matrices (e.g., vegetables, oils, fish, animal extracts), followed by GC with selective detection, is a very popular technique made easier by the availability of commercial instrumentation. Most published applications have used the "soft" gels which are of low efficiency, resulting in very broad fractions usually requiring concentration in order to determine the trace component of interest. Figure 11 depicts such an exclusion chromatogram of several vegetable-oil samples (Johnson et al., 1976) on a 20 × 270 mm column packed with BioBeads SX-3 (BioRad Laboratories, Richmond, CA) of particle-size 200–400 mesh. The mobile phase was 33%

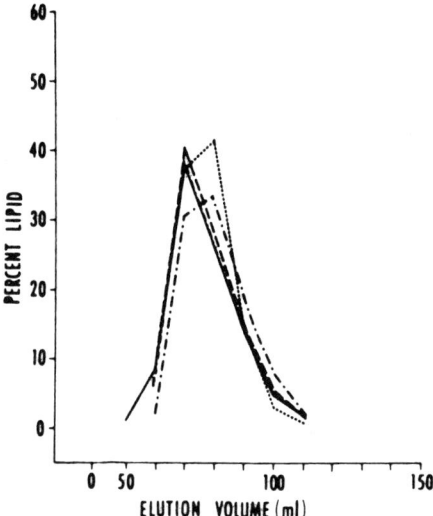

**Fig. 11.** Low-pressure exclusion chromatogram of vegetable oils (Johnson et al., 1976): Column, BioBeads SX-3, 270 mm × 30 cm; mobile phase, 33% toluene in ethyl acetate (v/v). Vegetable oil samples (500 mg of each oil): safflower (—), coconut (- · -), cottonseed (- - -), corn (···).

toluene in ethyl acetate (v/v). The method was used for sample cleanup so that chlorinated pesticides could be determined by GC. A 150-ml volume eluting after 95 ml was collected and was concentrated for injection into the GC. Note the comparison of the low-pressure, large-particle column in Fig. 11 to the SEC separation of corn oil (Fig. 6A) and soybean oil (Fig. 7) on the modern microparticulate SEC column.

The advantages of the soft gels are the following: (1) they are much less expensive than those used for high-performance exclusion chromatography; (2) they have low adsorptive properties; (3) they can readily be packed into open glass columns; and (4) by swelling or shrinking them in suitable solvents, they can yield different fractionation ranges. The disadvantages are (1) they are much slower, (2) they generally display lower resolution, (3) they give larger peak volumes, and (4) they cannot be used with high pressures since they will collapse. Large-particle-size porous glasses (e.g., BioGlas, CPG) or silicas (e.g., Porasil, LiChrosorb) are available which will take higher pressures, but they usually show adsorptive properties.

High-performance organic- and aqueous-compatible SEC packings and packed columns are available (Majors, 1980; Barth, 1980) which (1) give narrower peaks (hence smaller volume fractions) and faster separations, (2) show lower adsorption properties, and (3) take higher pressures; however, they cost much more.

## TABLE IV

### Off-Line SEC Applications to Trace Analysis (SEC–LC)

| Sample type | Matrix | Column | Reference |
|---|---|---|---|
| Retinyl palmitate | Breakfast cereal | µStyragel | Landen (1980) |
| β-Carotene/retinyl palmitate | Oil, margarine | µStyragel | Landen and Eitenmiller (1979) |
| Rodenticide | Rat tissue | BioBeads SX-3 | Koubek et al. (1979) |
| Polymer additives | Polystyrene | Styragel | Little (1971) |
| Carboxylic acids | Mixed ester (pine oil + glycol) | Styragel | Little (1971) |
| Catechol o-methyltransferase | Rat liver | DEAE-Sephadex A-25 | Pennings and Van Kempen (1979) |
| Sterigmatocystin | Corn and oats | BioBeads SX-3 | Stack et al. (1976) |
| Bilirubin photoxidation products | Serum | Shimadzu S.G.1.5 | Onishi et al. (1976) |
| Cytokinin | Plant extracts | Sephadex LH-20 | Pool and Powell (1974) |
| Polynuclear aromatic hydrocarbons | Smoke particulates | µStyragel | Liao and Browner (1978) |
| Copper-binding proteins | Liver and kidney tissue | BioGel P-10 | Port and Hunt (1979) |
| Etrimfos pesticide | Corn and alfalfa | Sephadex LH-20 | Bowman et al. (1978) |
| Diethylstilbestrol | Animal chow | Sephadex LH-20 | King et al. (1977) |
| Ferritin | Human serum | Sephadex G-200 | Worwood et al. (1976) |
| Difenacoum rodenticide | Biological materials | BioGlas 200 | Mundy and Machin (1977) |

Table IV summarizes typical applications of off-line multidimensional techniques where SEC was used as the primary separation or cleanup technique, and HPLC was used as the secondary analytical technique. Aqueous SEC (often known as gel filtration) is favored for biological extracts, whereas for organic soluble matrices (e.g., oil, polystyrene, hydrocarbons), GPC on cross-linked polystyrene is preferred.

To illustrate the developments which have been achieved with the microparticulate exclusion packings, Figs. 12A and 12B compare results obtained on the SEC separation of vitamin A and related compounds. Figure 12A shows the separation on a low-pressure column packed with Sephadex LH-20 (Pharmacia, Piscataway, NJ) using a chloroform eluent (Landen and Eitenmiller, 1979). The three compounds retinyl palmitate (1), retinyl acetate (2), and retinol (3) were separated in about 160 min at a flow rate of 1 ml/min. The column dimensions were 60 cm × 2.5 cm i.d. Compare the chromatogram of Fig. 12B where the

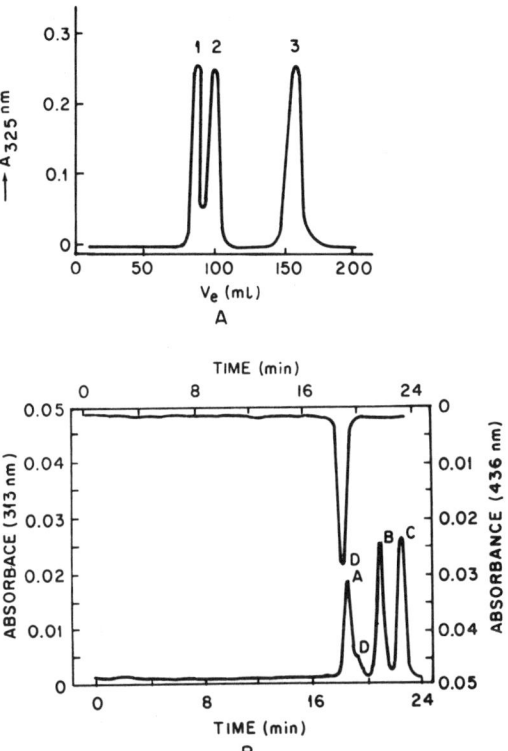

**Fig. 12.** Comparison of vitamin A and related compound separation on large-particle and microparticulate SEC packings. (A) Column, Sephadex LH-20 (Pharmacia) 60 × 2.5 cm; mobile phase, CHCl$_3$; flow rate, 1 ml/min (Holasová and Blattná, 1976). (B) Column, μStyragel, 100 Å (Waters Assoc.), 60 cm × 7.8 mm; mobile phase, CH$_2$Cl$_2$; flow rate, 0.7 ml/min (Landen and Eitenmiller, 1979).

same separation was achieved on two μStyragel 100-Å columns (Waters Assoc., Milford, MA) of 30 cm × 7.8 mm each at a methylene chloride flow rate of 0.7 ml/min. The peaks are retinyl palmitate (A), retinyl acetate (B), retinol (C), and β-carotene (D). This separation occurred in 24 min with better overall resolution. Peaks were narrower, and thus the volume of collected fractions would be smaller. Note that in both cases the separation was based on size, with the highest molecular weight compound, retinyl palmitate (MW 540.86), eluting first and retinol (MW 286.44) eluting last.

In some cases, for subsequent analysis of the SEC fraction by GC or LC, it may be of interest to use a larger pore size gel so that all of the lower molecular weight compounds elute together and are well separated from the higher molecular weight compounds. In this case, there is less interest in separating the individual lower molecular weight peaks by SEC alone.

## IX. ON-LINE MULTIDIMENSIONAL SEC TECHNIQUES

On-line multidimensional SEC is achieved through the coupling to a second column by means of a high-pressure switching valve which either traps a defined volume of collected sample, usually in a loop, and directs it to the second column (heart cutting), or diverts the mobile phase containing the desired solute(s) from the SEC column to the second column for a defined period of time (on-column concentration). From a convenience and automation viewpoint, compared to off-line techniques the on-line coupling of the two chromatographic techniques is preferred but not always feasible, as will be pointed out later. The advantages and disadvantages of on-line techniques in general are summarized in Table V.

In SEC–LC, through the use of pneumatically operated automated switching valves actuated by timers or by time-programmable events of a microprocessor-based chromatograph, the multidimensional experiments can be completely automated. Two such instrumental configurations are shown in Figs. 13 and 14. In Fig. 13, two separate pumping systems were used to perform heart-cutting experiments. In this approach, the effluent from the primary exclusion column is passed through the sample loop of a standard six-port loop injector to waste. At the appropriate time, when the desired component(s) of the sample enters the loop, its contents are injected onto the second column. The volume injected is governed by the loop volume, which is fixed. In this configuration, a single isocratic pumping system was used for the SEC column, and a dual-pump gradient system was used for the second column, a reverse-phase column (RPC). This setup allowed the flow on the primary column to be stopped and further components to be handled later. The gradient chromatograph could then handle

**TABLE V**

**On-Line Multidimensional Techniques**

| Advantages | Disadvantages |
|---|---|
| Easy to automate, especially with modern chromatographs | Requires more complex hydraulics (or pneumatics), switching valve(s), more expense |
| Less chance of sample loss since experiment is carried out in closed system | Difficult to handle trace compounds since very dilute and in large volume (can compensate for this by on-column concentration method, however) |
| Can configure switching system which best suits needs (e.g., backflushing, heart cutting, on-column concentration) | |
| Decreased total analysis time | Solvents from primary and secondary modes must be compatible, from both miscibility and strength requirements; LC solvents may affect GC stationary phase |
| More reproducible | |
| Can increase sample throughput | |

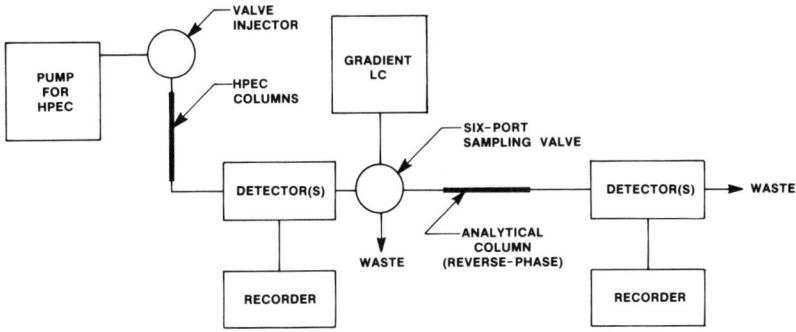

**Fig. 13.** Schematic of coupled column multidimensional system employing two pumps (Johnson et al., 1978).

the fraction(s) diverted to the RPC column. A disadvantage of column switching in the configuration depicted in Fig. 13 is that it requires two separate chromatographs.

With the advent of single-pump ternary chromatographs, a lower cost alternative can be used (Apffel et al., 1981). Figure 14 illustrates a configuration which uses a single pump with three-solvent (A, B, and C) capability. Using such a configuration, both heart cutting and on-column concentration can be carried out with no plumbing changes. By using time-programmable external events from the microprocessor-controlled LC to actuate pneumatic valves A and B at the

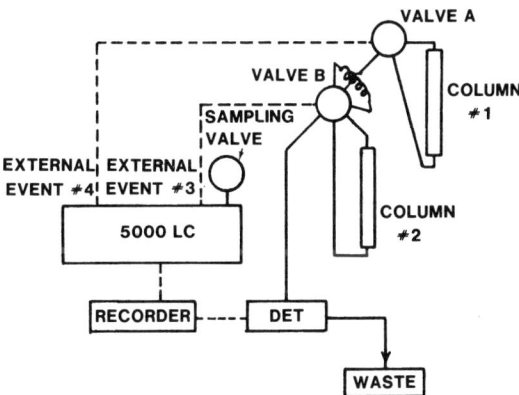

**Fig. 14.** Schematic of coupled column multidimensional system employing one pump (Apffel et al., 1981): Instrument, Varian Model 5000 LC with external events (time-programmable contact closure with switched ac); valves A and B are six-port, air-actuated switching valves; dotted lines are electrical connections; solid lines are hydraulic lines. To simplify the drawing, the hydraulic line from the sampling valve on the 5000 LC to valve A is not shown.

same or at different times, different flow paths can be achieved. In addition, by simultaneously programming the solvent composition of the three-solvent LC, the single pump can do the job of the two-pump system as is represented by Fig. 15. Thus, solvent A (SEC solvent) can be pumped through column 1 while the exclusion separation is occurring. Where the desired solute is directed to column 2, solvents B and C can be programmed to begin flow, and solvent A can be stopped and column 1 bypassed. By such synchronization of valve-switching times with pump solvent changes, the multidimensional experiments can be completely automated.

An important experimental criterion for on-line coupling is the exact timing requirements for valve switching. Since microparticulate columns produce very sharp peaks, it is of utmost importance to control the switching time since errors will ultimately affect quantitation. For this reason, it is preferable, especially in the heart-cutting schemes, to sample at the apex (rather than the shoulder) of a chromatographic peak where small variations in valve timing or solvent flow have a minimum effect on the amount of component sample. On the leading or

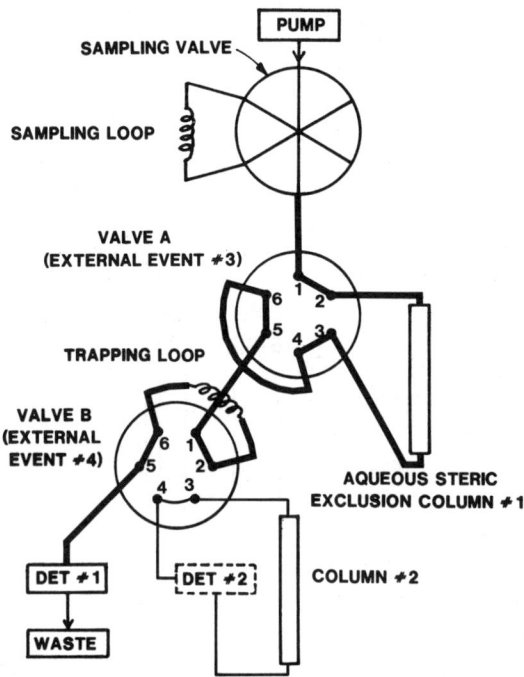

**Fig. 15.** Flow diagram for automated on-line two-dimensional chromatography (Apffel *et al.*, 1981). Heavy lines represent flow path for normal configuration.

trailing edge of a peak where the slope is the greatest, small variations in timing of flow will have a marked effect on reproducibility. Modern microprocessor-controlled liquid chromatographs are helpful in this respect. A detailed discussion of the effects on quantitation in multidimensional LC is beyond the scope of this article.

The technique of multidimensional LC–LC has proved to be highly effective when steric exclusion separation (GPC or GFC) on high-performance packings is the primary mode (Johnson et al., 1978; Apffel et al., 1981; Erni and Frei, 1978). The exclusion technique provides information concerning the molecular-weight range of the samples, usually without regard to chemical functionality. In addition, all sample components are eluted in one column volume. With microparticulate columns, a sufficient degree of separation may occur on the exclusion column alone to permit quantitation. Also, the peaks are often fairly sharp, making direct on-line coupling more feasible. Only microparticulate exclusion columns are recommended since the broad elution profiles from low-efficiency, large-particle exclusion columns limit the ultimate resolution and sensitivity of the multidimensional technique.

The RPC technique is well suited as the secondary mode. The solvents used are compatible with both GPC (organic) and GFC (aqueous) solvents. Particularly suitable is the combination of GFC, where the mobile phases are predominantly aqueous based, and RPC or ion-exchange chromatography (IEC), where the weakness of water can be used to advantage in the on-column concentration method. Thus, relatively large amounts of this "weak" solvent from the primary column that may contain only small amounts of components of interest can be injected onto the secondary column with only minimal band broadening.

A wide range of applications in both biological and industrial fields can be found for the multidimensional approach of exclusion and RPC or IEC. For example, polymer additives can be directly analyzed in a high-molecular-weight polymer without prior extraction since the pore size of the exclusion packing can be selected where the polymer is excluded, whereas the low-molecular-weight additives permeate the pores and elute as a single peak (Majors and Johnson, 1978) which can, in turn, be injected onto an RPC column. If the polymer is directly injected onto the RPC column, the column becomes contaminated by the presence of precipitated or strongly held polymer.

An even more exciting possibility is the direct injection of untreated biological fluids onto an aqueous exclusion column (Apffel et al., 1981). The newer microparticulate exclusion columns show little interaction with the sample matrix and separate by molecular size. Both drugs and drug metabolites and endogenous compounds present in serum or urine may be handled by the GPC–RPC or GFC–IEC techniques. The only sample treatment is filtration to remove particulate matter.

To illustrate the multidimensional SEC–LC technique, the on-line analysis of malathion in tomato plants was performed using GPC–RPC (Johnson et al., 1978). The exclusion-chromatographic cleanup technique has been applied to the removal of chlorinated pesticides and other contaminants from such diverse samples as grains (Stack et al., 1976), fish (Stalling et al., 1972), and animal and plant extracts (Johnson et al., 1976) using large-particle exclusion columns of 200–400 mesh. The fraction containing the pesticides was generally well diluted and required extensive concentration. The use of microparticulate exclusion columns results in much narrower fractions, and these are more amenable to multidimensional work.

Tomato plants were sprayed with a malathion solution prepared according to the manufacturer's suggestion. The plants were allowed to dry overnight. Fronds were collected, macerated, and extracted with methylene chloride. The extract was evaporated to dryness and the residue dissolved in THF. Figure 16 shows the resultant GPC chromatogram from a 20-μl injection of the green-colored extract. A MicroPak TSK 1000H and 2000H column set was used with a THF mobile phase. Using a heart-cutting approach, a 10-μl injection of the malathion region (known from injection of a malathion standard) was injected onto a reverse-phase column. A gradient elution run gave a clean, well-resolved malathion peak (Fig. 16B) with a retention time matching that of a malathion standard. At a high level of 200 ppm in the original tomato plant, the malathion was easily detected from the reverse-phase column at a wavelength of 215 nm. Due to the degradation of malathion for a similarly handled sample collected 7 days later, no malathion could be detected using on-line coupling. The off-line collection of 1 ml and subsequent concentration, however, permitted the analysis of malathion below 1 ppm (Johnson et al., 1978). Thus, on-line techniques may not always be applicable using heart cutting, especially when the concentration of the compound of interest eluted from the SEC column becomes too diluted. However, when using a solvent for the SEC technique which is a strong solvent in the secondary technique, its volume must be kept to a minimum so as to not cause a spreading out of the injected fraction on the secondary column.

In situations where heart-cutting techniques using smaller fixed-volume loops do not provide a sufficient amount of sample, the on-column concentration methods may be used for large-volume injections. Such a procedure is mainly feasible when the solvent from the primary SEC mode is a weak solvent for the

---

**Fig. 16.** Separations of tomato plant extract (Johnson et al., 1978); sample isolated 16 hr after treatment with malathion. (A) SEC separation: Column, MicroPak TSK 1000H, 60 cm × 7.5 mm; mobile phase, THF; flow rate, 1 ml/min; detection, UV at 215 nm; sample size, 20 μl. (B) RPC separation of malathion fraction: Column, MicroPak MCH, 25 cm × 2.2 mm; mobile phase, water (solvent A), water and acetonitrile (solvent B); gradient, 10–100% acetonitrile in 30 min; flow rate, 0.5 ml/min; detector, UV at 215 nm; sample size, 10 μl.

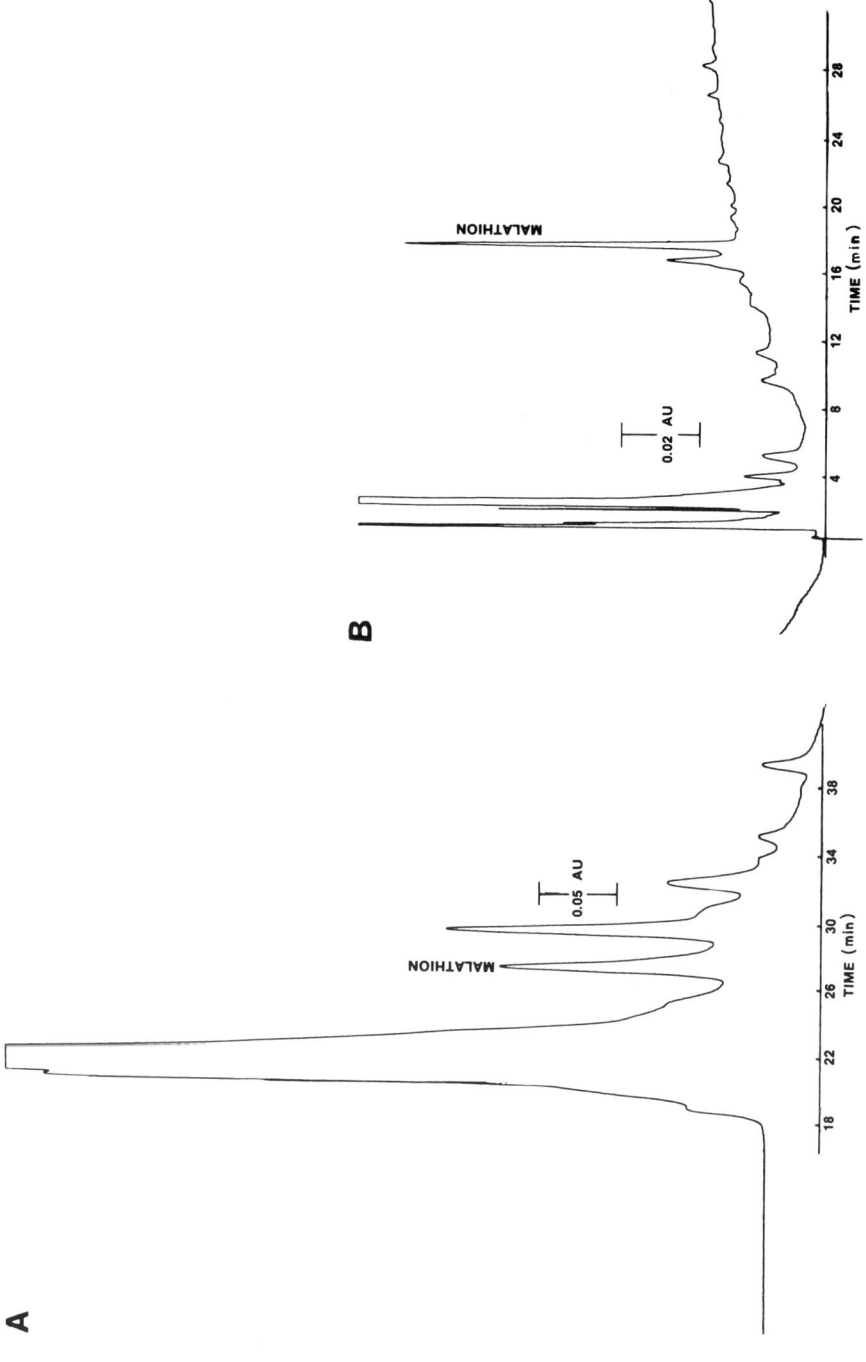

secondary mode, as is the case for aqueous SEC and RPC. The larger volume can be provided by using larger sample loops and slowing down the flow (Erni and Frei, 1978) or by using the on-column concentration method (Apffel et al., 1981) described earlier.

An example of using the two-dimensional approach was described by Erni and Frei (1978). Using a chromatographic setup as depicted in Fig. 13, a sennaglycoside extract was subjected to aqueous SEC using a column packed with 200–400-mesh CPG controlled-pore-size glass (Electronucleonics, Fairfield, NJ), and aqueous-buffer mobile phase. The resultant SEC chromatogram, shown in Fig. 17A, was sampled at seven different points by collecting 1.5-ml fractions in a 1.777-ml loop and injecting them onto a reverse-phase column. The fractions were concentrated at the head of the RPC column since the aqueous buffer was a weak solvent. Finally, the RPC column was subjected to a seven-step gradient elution using an acetonitrile–0.01 $N$ NaHCO$_3$ in water gradient. Figure 17B shows that the two-dimensional LC–LC technique gave seven simpler, better resolved chromatograms than when the extract was injected directly onto the same reverse-phase column (see Figs. 6A and 6B from Erni and Frei, 1978).

Another method of column switching which provides more versatility than that just described is to use the instrumental configuration illustrated in Figs. 14 and 15 and the on-column concentration technique. In this procedure, the effluent from the SEC column is diverted for a finite period of time, and this time dictates the sample size injected and concentrated on the RPC column. This procedure was used to separate fortified vitamins in a protein food supplement (Apffel et al., 1981). The vitamins were added at the 0.001–0.04% (w/w) level to enhance the nutritional value. First, the food supplement was suspended in water, vigorously shaken, and filtered. Next, the filtrate was injected onto an aqueous-compatible MicroPak TSK 2000SW column yielding the broad multiband peak

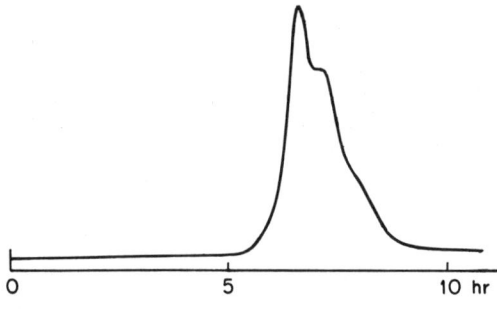

**Fig. 17A.** Separation of sennaglycoside from plant extract (Erni and Frei, 1978). SEC separation: Column, CPG, 80, 113, 170, 240 Å, 200–400 mesh, 200 × 0.4 cm; mobile phase, Titrisol (Ż. Merck, Darmstadt, West Germany) buffer, pH 6; flow rate, 1.2 ml/min; detector, UV at 254 nm.

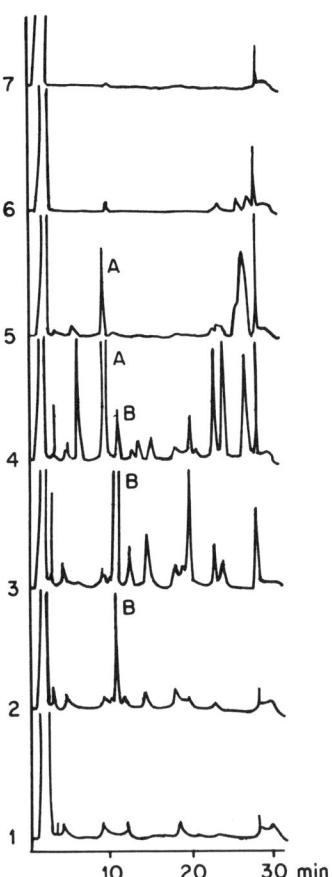

**Fig. 17B.** Separation of sennaglycoside from plant extract (Erni and Frei, 1978). RPC separation of seven fractions from SEC run: Column, 5-μm Nucleosil-$C_{18}$, 25 × 0.4 cm; mobile phase, acetonitrile–0.01 $N$ $NaHCO_3$ in water step gradient (see Erni and Frei, 1978, for details); flow rate, 2 ml/min; detector, UV at 254 nm. Peaks: A, sennoside A; B, sennoside B.

displayed in Fig. 18A. From the injection of vitamin standards, it was determined that the lower molecular weight region of the chromatogram near the peak labeled (4) contained the vitamins of interest. Since the vitamins in the food sample were very dilute, the on-column concentration method was used. Figure 18B shows the combined chromatogram where the vitamins were clearly resolved from other excipients that were coinjected onto the RPC column. Direct injection of the protein supplement onto the reverse-phase column would have resulted in irreversible adsorption of the protein, resulting in eventual column contamination and possible matrix interference.

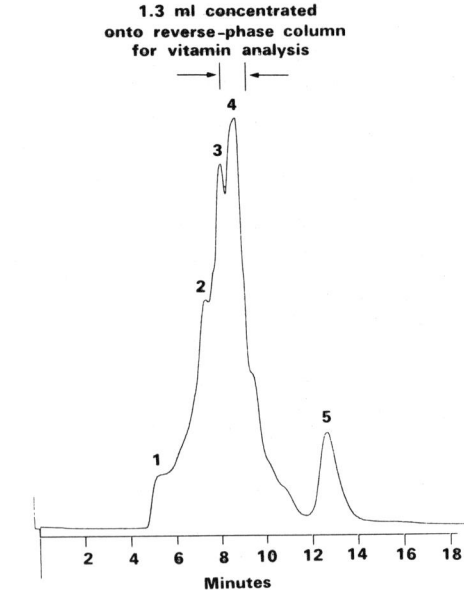

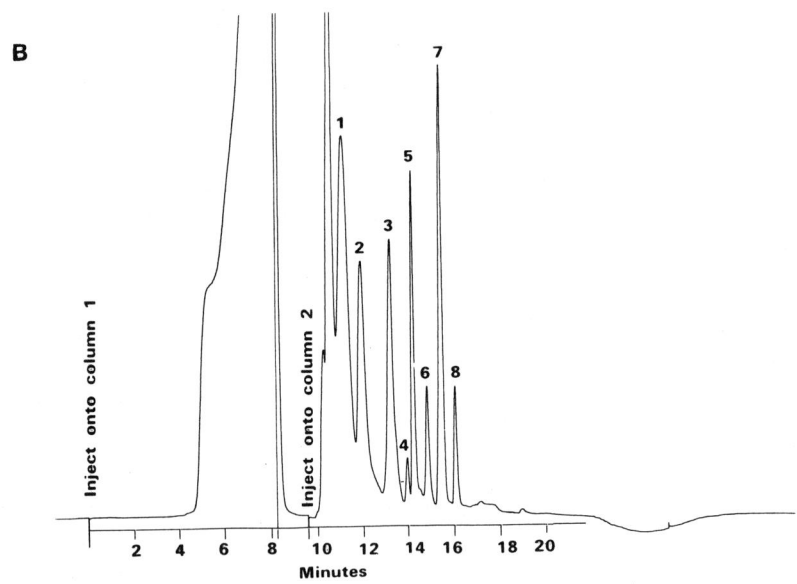

In multidimensional chromatography, the use of the SEC technique is not confined to its role as the primary chromatographic technique. Katz and Ogan (1980) have shown its application as the secondary technique in separating polycyclic aromatic hydrocarbons (PAHs) in petroleum and coal-liquid samples. The primary technique was RPC where, by proper adjustment of the mobile-phase conditions, the PAHs eluted as a single unretained band which was further resolved by an SEC column. Collected fractions from the SEC column separated on the basis of molecular size were further subjected to gradient-elution RPC to yield well-resolved, greatly simplified chromatograms. Published examples of the on-line coupling of SEC to other LC techniques using high-performance columns are listed in Table VI.

## X. PROBLEMS AND TROUBLESHOOTING

### A. Problems Encountered in the Use of SEC with Direct-Injection Technique

#### 1. Mobile-Phase Considerations

Experience has shown that a single mobile phase seems to provide the most satisfactory use of microparticulate, small-pore gel columns designed for gel filtration and gel permeation analyses. Repeated changing of solvents induces changes in gel-particle size (swelling or shrinkage) which can result in collapse of the gel particle (e.g., formation of a void in the packing bed) and failure of the column. For the trace analysis of small molecules, it is recommended that a small-pore SEC gel column be used with only one mobile phase. Several manufacturers provide small-pore SEC gel columns available in a number of solvents or provide lists of recommended solvent changes compatible with such columns. Table VII lists a number of commonly used mobile phases for small-pore SEC gels and solvents which are damaging to these columns. Due to favorable solubility properties with a wide polarity range of solutes, THF is most commonly

---

**Fig. 18.** (A) Separation of vitamins from food protein supplement (Apffel *et al.*, 1981). SEC separation of protein supplement: Column, MicroPak TSK 2000SW, 30 cm × 7.5 mm; mobile phase, 10% methanol/90% water containing 0.1 $M$ $KH_2PO_4$ and 0.01 $M$ 1-heptanesulfonic acid; detector, UV at 254 nm, 2.0 AUFS. Peaks: 1–3, protein; 4, vitamins; 5, unknown. (B) Separation of vitamins from food protein supplement (Apffel *et al.*, 1981). Reverse-phase separation after on-column concentration of 1.3-ml fraction from SEC column: Column 1, MicroPak TSK 2000SW, 30 cm × 7.5 min; column 2, MicroPak MCH-10, 30 cm × 4 mm; mobile phase, 10 to 80% methanol in water containing 0.01 $M$ $KH_2PO_4$ and 0.01 $M$ 1-heptanesulfonic acid; detector, UV at 254 nm, 0.5 AUFS. Peaks: 1, niacin; 3, pyrodoxine ($B_6$); 6, thiamine ($B_1$); 7, riboflavin ($B_2$); 2, 4, 5, 8 are unknown.

## TABLE VI
### On-Line SEC Applications to Trace Analysis

| Sample type | Primary technique | Secondary technique | Reference |
|---|---|---|---|
| Copolymers | GPC | BPC[a] | Baike and Patel (1980) |
| Malathion in tomato plants | GPC | RPC | Johnson et al. (1978) |
| Limonin in grapefruit peel | GPC | RPC | Johnson et al. (1978) |
| Catechol in urine | GFC | RPC | Apffel et al. (1981) |
| Caffeine in urine | GFC | RPC | Apffel et al. (1981) |
| Sennaglycoside extract | GFC | RPC | Erni and Frei (1978) |
| Sugars in molasses and candy | GFC | RPC | Apffel et al. (1981) |
| PAHs in petroleum and coal liquid | RPC | GPC | Katz and Ogan (1980) |

[a] BPC, bonded-phase chromatography.

preferred and employed for small-molecule analysis. If the solute of interest is not completely soluble in THF, another column permanently equilibrated or packed in a suitable solvent should be used for optimum results. Some solvent conversions are permitted for packed columns, but the manufacturer should be consulted on both permitted changeovers and the recommended experimental method.

Small-pore SEC gel columns are inherently fragile (poor mechanical strength) compared to silica-based packings. Mobile-phase velocity should be kept low (usually <2–3 ml/min flow rate, and lower if high-viscosity mobile phases are employed). Additionally, if more than one SEC column is used in series, the smallest pore column (most fragile) should be placed last to experience minimum pressure drop. Most small-pore SEC gel columns have operating-pressure limits between 100 and 150 atm.

Solvent compatibility data are limited for many of the recently developed small-pore SEC gel columns designed for gel filtration. In general, most of these small-pore gels can be used only with water, aqueous buffers, or aqueous mobile

## TABLE VII
### Solvent Guide for Small-Pore Polystyrene–Divinylbenzene Gels Used in SEC

| Compatible solvents | | Damaging solvents | |
|---|---|---|---|
| Benzene | Tetrahydrofuran | Acetone | Dimethylformamide |
| Toluene | Carbon tetrachloride | Alcohols | Acetic acid |
| Chloroform | Methylene chloride | Water | Hexafluoroisopropanol |
| | | Acetonitrile | Dimethyl sulfoxide |
| | | Butyl acetate | |

phases containing small amounts (≤10%) of an organic modifier such as methanol or acetonitrile. In many instances, the type and amount of mobile-phase modifier is dictated by specific gel–solute interactions (nonexclusion effects) often observed with gel-filtration packings. For example, the use of aqueous acetic acid (1–5%) has been found helpful for elimination of adsorption with many solutes using MicroPak TSK 1000PW columns (Alfredson *et al.*, 1980). Use of mobile-phase modifiers with microparticulate gel-filtration columns has been covered in detail by Barth (1980).

Several small-pore SEC columns of rigid silica or silica-based packings have been developed for both gel permeation and gel filtration. Although columns of siliceous particles can exhibit undesired retention of solutes by adsorption or other nonexclusion mechanisms, they offer distinct advantages for the trace analysis of small molecules. The ability to use a wide variety of solvents offers advantages of convenience and versatility with such columns. Changeover from one solvent to another is fast due to rapid equilibration. Also, the inherent mechanical strength of these columns makes them less subject to damage and much more rugged compared to small-pore SEC gels.

*2. Samples*

In general, the capacity of an SEC column is much higher than that of columns for use in other LC modes. In practice, the capacity of a small-pore SEC column is limited by sample solubility in the mobile phase and the degree of resolution required between the solute of interest and matrix components. However, the viscosity of the sample solution injected onto the column is also an important factor. Increased solution viscosity can result in significant band broadening due to "viscous streaming" or "viscous fingering" on the trailing edge of the solute band. To avoid such problems, a rough rule of thumb is that the viscosity of the injected solution be kept less than or equal to twice the viscosity of the mobile phase (Snyder and Kirkland, 1974).

## B. Problems Encountered in the Use of SEC with Off-Line Prefractionation and Sample Cleanup

Although many problems encountered in the use of SEC are general in nature and irrespective of application, specific problems can arise for a particular application. For off-line techniques employing SEC for prefractionation and sample cleanup prior to LC or GC analysis, the most common problem is related to sample collection. Several commonly used solvents in SEC contain small amounts of additives or impurities which are usually concentrated along with the solute following sample collection. Such additives and impurities often result in contamination or interference with the subsequent analytical technique applied to the collected solute. One example of this type of interference is displayed in

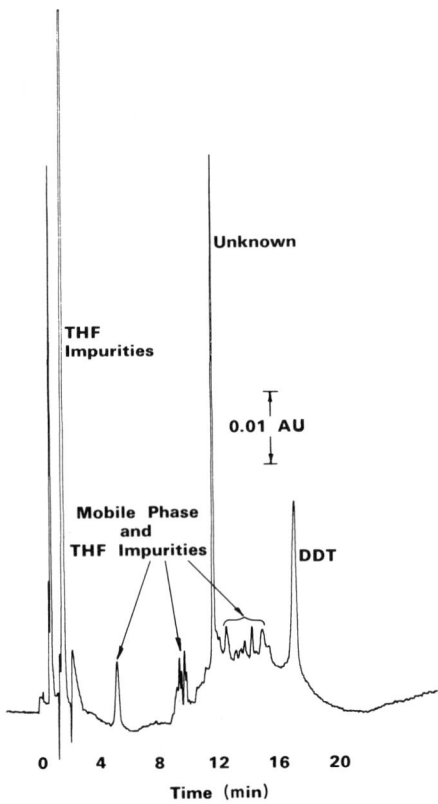

**Fig. 19.** Reverse-phase separation of DDT fraction isolated by exclusion chromatography (Majors and Johnson, 1978): Column, 25 cm × 2.2 mm MicroPak MCH; flow rate, 1 ml/min; solvent A, water; solvent B, acetonitrile; gradient, 10–60% solvent B at a rate of 5%/min; detector, UV at 210 nm.

**TABLE VIII**

**Additives and Impurities in Commonly Used SEC Mobile Phases**

| Solvent | Additive or impurity commonly present |
|---|---|
| THF[a] | BHT—peroxide inhibitor |
| Chloroform | Ethanol—photolysis inhibitor |
| 1,2,4-Trichlorobenzene | Santonox R or equivalent antioxidant |
| Carbon tetrachloride | HCl impurity |

[a] UV-grade THF contains no inhibitor but may contain low-level peroxides upon prolonged storage (>1 month) after opening.

Fig. 19. A chromatogram of reverse-phase gradient separation of DDT fraction (0.8 ml) isolated by a small-pore SEC column shows several peak impurities resulting from the THF concentrate following 20-fold concentration by evaporation. A partial list of several solvents used as SEC mobile phases and their commonly found additives or impurities is displayed in Table VIII.

Sample loss during the concentration step can occur. Adsorption on the walls of flasks or beakers during evaporation can limit recovery of sample. Oxidation of sensitive samples can occur during the handling steps. Volatile samples can be lost during the removal of solvent.

## C. Problems Encountered in the Use of SEC with On-Line Multidimensional Chromatography

On-line multidimensional chromatographic techniques (LC–LC and LC–GC) produce additional constraints that must be considered when using a small-pore SEC column for trace analysis. Most of these constraints involve compatibility of the SEC mobile phase with other chromatographic modes of analysis.

When THF is employed as the SEC mobile phase, care must be taken that it is compatible with the GC stationary phase when employing LC–GC techniques. Figure 20 displays the results of a GC stationary-phase-lifetime study with THF employed as the LC solvent injected from an automated LC–GC system (Apffel et al., 1980). THF is a powerful solvent for many GC stationary phases.

On-line multidimensional techniques employing LC–LC also produce SEC mobile-phase compatibility problems with the subsequent LC mode of separation. When two pumps are employed for LC–LC, the problem that arises is one of volume of injected SEC solvent onto the second column in the system, especially if a gel-permeation SEC column is used in conjunction with a reverse-phase column (Snyder and Kirkland, 1974). For example, THF is a very strong solvent for a reverse-phase separation using a high-aqueous-content mobile phase. Injection of a large volume of THF onto the head of the reverse-phase column could result in significant solute travel down the column with resultant problems of band broadening or nonreproducible retention times. In these cases, it has been suggested (Johnson et al., 1978) that the volume of "strong" solvent be kept to less than 5% of the column void volume. When a single pump is employed in the LC–LC technique, the SEC mobile-phase compatibility problem is further complicated. In addition to the aforementioned problem, solvent delivery system dead volumes must be taken into account for solvent changeover from the mobile phase of the SEC column to the mobile phase of the second LC mode of analysis. To remove this volume of solvent (one to several milliliters in most high-performance pumping systems), either elimination to waste through proper plumbing and activation of switching valves or automatic purge of the pump is required (Apffel et al., 1981). In some instances, incorporation of a

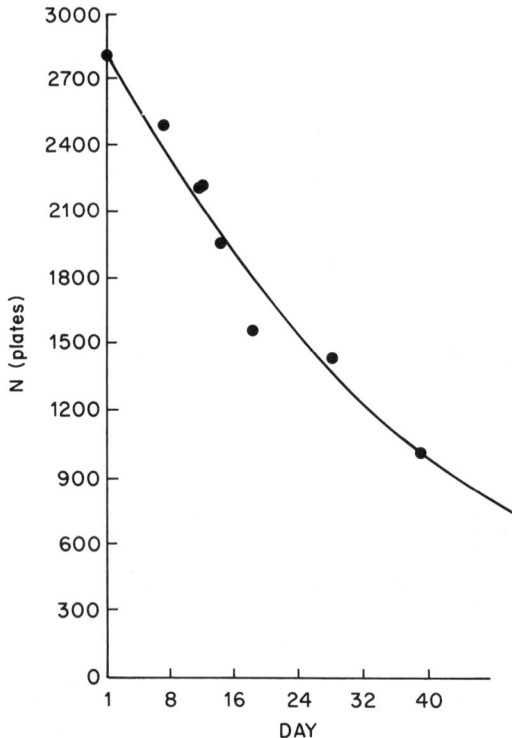

**Fig. 20.** GC column lifetime in an automated LC–GC system. (Apffel *et al.*, 1980): Stationary phase, 1% SE-30 on Chrom. W-HP; LC mobile phase, THF; injection volume onto GC column, 10 μl; 50 injections per day.

solute trapping loop into the switching scheme allows solvent to be displaced through the secondary LC column. However, additional reequilibration time required prior to analysis on the second LC column adds to the time per analysis. Switching-valve schemes have been developed for techniques involving SEC and secondary LC modes where incompatibility problems have arisen (see Apffel *et al.*, 1981) and been overcome.

## D. Troubleshooting

Troubleshooting of small-pore SEC columns employed for trace analysis is identical to that for use of SEC columns in general. Several books (e.g., Yau *et al.*, 1979) as well as the instruction manuals from most column manufacturers that usually accompany new columns treat the subject in detail. A troubleshoot-

**TABLE IX**

**Troubleshooting**

| Symptom | Cause | Solution |
|---|---|---|
| Decrease in flow or increased pressure drop | Inlet fitting partially restricted | Clean the fitting by reversing flow through column |
| Tailing and decrease in efficiency | Gap at inlet end of column | Replace missing gel |
| | Gel packing damaged | Replace column |
| | Oxygen reaction with gel | Replace column |
| | Retained reactive material in column | 1. Remove first few millimeters (which may be contaminated); replace with fresh gel or inert glass beads |
| | | 2. If above procedure does not work, replace column |
| | Sample decomposition | Prepare fresh sample |

ing guide from a MicroPak TSK SEC-column instruction manual is given here as an example (Table IX).

# REFERENCES

Alfredson, T. V., Wehr, C. T., and Tallman, L. (1980). *In* "Polymeric Separation Media" (A. R. Cooper, ed.), pp. 123–144. Plenum, New York.

Apffel, A., Majors, R., and McNair, H. (1980). Quantitation in an automated LC/GC system. Paper presented at Pittsburgh Conf. on Anal. Chem., March, 1980.

Apffel, A., Alfredson, T. V., and Majors, R. E. (1981). *J. Chromatogr.* **206**, 43–57.

Ault, J. A., Schofield, C. M., Johnson, L. D., and Waltz, R. H. (1979). *J. Agric. Food Chem.* **27**, 825–828.

Baike, S. T., and Patel, R. D. (1980). *J. Polym. Sci., Polym. Lett. Ed.* **18**, 453–456.

Bartak, D. E. (1980). *Toxocol. Res. Proj. Dir.* **5**(12).

Barth, H. G. (1980). *J. Chromatogr. Sci.* **18**, 409–429.

Bly, D. D., Stoklosa, H. J., Kirkland, J. J., and Yau, W. W. (1975). *Anal. Chem.* **47**, 1810–1813.

Boreham, D. R., Vose, C. W., Palmer, R. F., Brooks, C. T. W., and Balasubramanian, V. (1978). *J. Chromatogr.* **153**, 63–75.

Bowman, M. C., and Nony, C. R. (1977). *J. Chromatogr. Sci.* **15**, 160–163.

Bowman, M. C., Holder, C. L., and Rushing, L. G. (1978). *J. Agric. Food Chem.* **26**, 35–42.

Carr, F. H., and Price, E. A. (1976). *Biochem. J.* **20**, 497.

Cassidy, R. M., and Niro, C. M. (1976). *J. Chromatogr.* **126**, 787–794.

Deuber, A., Brandauer, H., and Ziegier, E. (1979). *Chromatographia* **12**, 737–739.

Erni, F., and Frei, R. W. (1978). *J. Chromatogr.* **149**, 561–569.

Fuchsbichler, G. (1979). *Landwirtsch. Forsch.* **32**, 341–354.

Griffitt, K. R., and Craun, J. C. (1974). *J. Assoc. Off. Anal. Chem.* **57**, 168–172.

Grimmer, G., and Blohmke, H. (1975). *J. Assoc. Off. Anal. Chem.* **58**, 725-733.
Holasová, M., and Blattná, J. (1976). *J. Chromatogr.* **123**, 225-230.
Hooker Chemical Co. (1979). Assay for Pentac®. Technical Rep., Hooker Chemical Co., Niagara Falls, New York.
Johnson, E. L., and Stevenson, R. (1978). "Basic Liquid Chromatography," pp. 149 and 152. Varian Associates, Palo Alto, California.
Johnson, L. D., Waltz, R. H., Ussary, J. P., and Kaiser, F. E. (1976). *J. Assoc. Off. Anal. Chem.* **59**, 174-187.
Johnson, E. L., Gloor, R., and Majors, R. E. (1978). *J. Chromatogr.* **149**, 571-585.
Katz, E., and Ogan, K. (1980). Multidimensional chromatography: Partition chromatography coupled with size exclusion chromatography. Presented at 5th Intern. Symp. on Polycyclic Aromatic Hydrocarbons, Battelle, Columbus, Ohio, 1980.
Kawaguchi, H., and Auld, D. S. (1975). *Clin. Chem. (Winston-Salem, N.C.)* **21**, 591-594.
King, J. R., Nony, C. R., and Bowman, M. C. (1977). *J. Chromatogr. Sci.* **15**, 14-21.
Koubek, K. G., Ussary, J. P., and Haulsee, R. E. (1979). *J. Assoc. Off. Anal. Chem.* **62**, 1297-1301.
Landen, W. O., Jr. (1980). *J. Assoc. Off. Anal. Chem.* **63**, 131-136.
Landen, W. O., Jr., and Eitenmiller, R. R. (1979). *J. Assoc. Off. Anal. Chem.* **62**, 283-289.
Lazarovits, O., Bhullar, B. S., Sugiyama, H. J., and Higgins, V. T. (1979). *Phytopathology* **69**, 1062-1068.
Liao, J. C., and Browner, R. R. (1978). *Anal. Chem.* **50**, 1683-1686.
Little, J. N. (1971). *Am. Lab. (Fairfield, Conn.)* **3**, 59-63.
Majors, R. E. (1980). *J. Chromatogr. Sci.* **18**, 488-511.
Majors, R. E., and Johnson, E. L. (1978). *J. Chromatogr.* **167**, 17-30.
Manchester, J. E., and Manchester, K. L. (1980). *J. Chromatogr.* **193**, 148-152.
Mori, S. (1976). *J. Chromatogr.* **129**, 53-60.
Mundy, D. E., and Machin, A. F. (1977). *J. Chromatogr.* **139**, 321-329.
Ohtake, H., and Koga, M. (1979). *Biochem. J.* **183**, 683-690.
Onishi, S., Fujikake, M., Ogawa, Y., and Ogawa, J. (1976). *Birth Defects, Orig. Artic. Ser.* **12**, 41-52.
Parris, N. A. (1978). *J. Chromatogr.* **157**, 161-170.
Pennings, J. M., and Van Kempen, G. M. J. (1979). *Anal. Biochem.* **98**, 452-454.
Pflugmacher, J., and Ebing, W. (1974). *J. Chromatogr.* **26**, 457-463.
Pool, R. M., and Powell, L. E. (1974). *Plant Growth Subst., Proc. Int. Conf. 8th, 1973*, 93-98.
Port, A. E., and Hunt, D. M. (1979). *Biochem. J.* **183**, 721-730.
Snyder, L. R., and Kirkland, J. J. (1974). "Introduction to Modern Liquid Chromatography," Chapter 10. Wiley, New York.
Stack, M. E., Nesheim, S., Brown, N. L., and Pohland, A. E. (1976). *J. Assoc. Off. Anal. Chem.* **59**, 966-970.
Stalling, D. L., Tindle, R. C., and Johnson, J. L. (1972). *J. Assoc. Off. Anal. Chem.* **55**, 32-38.
Tindle, R. C., and Stalling, D. L. (1972). *Anal. Chem.* **44**, 1768-1773.
"United States Pharmacopeia" (1970). 18th ed., pp. 775 and 914. Mack Publishing Co., Easton, Pennsylvania.
Wattenhofer, C. (1981). Varian Liquid Chromatography at Work Applications Note Number LC-118. Varian Instrument Division, Walnut Creek, California.
Worwood, M., Dawkins, S., Wagstaff, M., and Jacobs, A. (1976). *Biochem. J.* **157**, 97-103.
Yau, W. W., Kirkland, J. J., and Bly, D. D. (1979). "Modern Size-Exclusion Liquid Chromatography," Chapter 8. Wiley, New York.

# TRACE-ENRICHMENT TECHNIQUES FOR ORGANIC TRACE ANALYSIS

### W. A. Saner*

U.S. Coast Guard Research and Development Center
Avery Point
Groton, Connecticut

|      |                                                                                                                                           |     |
|------|-------------------------------------------------------------------------------------------------------------------------------------------|-----|
| I.   | Trace Enrichment: Definition.                                                                                                             | 152 |
|      | A. Frontal Elution                                                                                                                         | 152 |
|      | B. Analysis of Trace-Level Components Both above and below Detection Limits                                                               | 153 |
|      | C. Trace Enrichment as a Specialized Form of Liquid Chromatography                                                                         | 153 |
|      | D. Detectors                                                                                                                               | 160 |
|      | E. Techniques to Increase Extractability ($k'$) of Solutes from Aqueous Samples during the Concentration Step in Trace Enrichment         | 160 |
|      | F. Mutual-Zone Solubility                                                                                                                  | 161 |
|      | G. Forms of Trace Enrichment                                                                                                               | 163 |
| II.  | Environmental Applications                                                                                                                 | 164 |
|      | A. Direct Injection                                                                                                                        | 164 |
|      | B. Column Switching, Off-Line                                                                                                              | 175 |
|      | C. Column Switching, On-Line                                                                                                               | 191 |
| III. | Clinical Applications                                                                                                                      | 202 |
|      | A. Direct Injection                                                                                                                        | 202 |
|      | B. Column Switching, Off-Line                                                                                                              | 207 |
|      | C. Column Switching, On-Line                                                                                                               | 208 |
| IV.  | Pharmacological Applications                                                                                                               | 214 |
|      | A. Direct Injection                                                                                                                        | 214 |
|      | B. Column Switching, Off-Line                                                                                                              | 217 |
|      | C. Column Switching, On-Line                                                                                                               | 219 |
|      | References                                                                                                                                 | 220 |

*Present address: U.S. Army Corps of Engineers, Ohio River Division Laboratory, Mariemont, Ohio 45227.

## I. TRACE ENRICHMENT: DEFINITION

The technique of trace enrichment is a means of sample concentration by the injection of large sampling volumes onto a liquid-chromatographic (LC) column packing. A mobile phase of sufficiently weak elution strength is used so that the compounds of interest are concentrated on the head of the column. Sample extraction and concentration are effected on the chromatographic stationary phase, which also serves as the medium to provide the actual physical (chromatographic) separation (preparative or analytical) of the enriched compounds.

### A. Frontal Elution

During the loading (enrichment) stage, the sample is introduced to the chromatographic bed in a continuous manner rather than as a discrete plug (conventional sample injection). This process represents a form of frontal development whereby the sample-introduction process is complete prior to breakthrough of the compounds of interest from the end of the column. Continued sample introduction beyond breakthrough results in a partial chromatographic separation in the formation of fronts rather than bands as in conventional elution chromatography. An example of frontal elution is shown in Fig. 1 (Karger et al., 1973). The least retained component (A) is first to emerge from the column, followed by fronts B and C in order of increasing capacity factor ($k'$ values). Although frontal analysis cannot achieve actual physical separation of peaks, it defines the process during the enrichment step. As such it is useful for concentrating trace-level contaminants and, conversely, for purifying large volumes of liquids (and gases). During the trace-enrichment process, the loading volume would not ordinarily be of such an excessive amount that breakthrough would be allowed to occur. However, if the analysis were specific for compound C, then the intentional elution of fronts A and B from the column could be permitted without disturbing the quantitative enrichment of C.

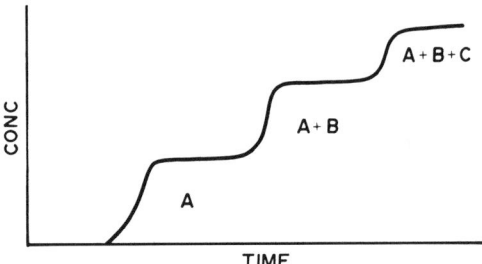

**Fig. 1.** Frontal analysis chromatogram (Karger, 1973).

After the sample has been enriched onto the column (with or without the formation of fronts from the end of the column or within the bed of the column itself), the adsorbed components are desorbed and migrate through the column by means of elution chromatography. (This is the chromatographic mode used exclusively in analytical separations under either isocratic- or gradient-elution conditions.) Sample-component linear velocities are less than the mobile-phase linear velocity. This ultimately results in a physical separation (partial or complete, depending on chromatographic resolution) of the components whose elution profiles are the typical chromatographic peak (ideally Gaussian).

## B. Analysis of Trace-Level Components Both above and below Detection Limits

The problem of the analysis of trace-level components is a two-part one. When the components of interest are below the detection limits of the detection devices, there must be a preconcentration (enrichment) step prior to the actual chromatographic separation. In some instances, however, the components of interest may actually be above the detection limits but may be present with a large relative amount of background interferences, so that neither detection nor accurate quantification is possible. In order to simplify the starting mixture, various chromatographic "cuts" can be made prior to the actual analytical chromatographic step. This procedure fractionates the sample and reduces the level of excipients relative to the desired components. The net result is an increase in the concentration of the components of interest relative to background. As such, this fractionation procedure is described in more detail later in Section III,B.

## C. Trace Enrichment as a Specialized Form of Liquid Chromatography

Particularly in light of the advent of new and improved reverse-phase stationary packings (including octadecyl, phenyl, octyl, propyl, and amino), the potential usefulness of trace enrichment for environmental, industrial (waste and waste-treatment monitoring), and biomedical applications becomes readily apparent; all involve an aqueous sample matrix which is compatible with the reverse-phase material so that adsorption can occur. These phases exert strongest retention of nonpolar and moderately polar compounds by using an aqueous mobile phase. The adsorbed components are easily recoverable in essentially pure form simply by desorption with an increased percentage of organic modifier added into the aqueous mobile phase (water miscible, e.g., methanol, acetonitrile, tetrahydrofuran).

Displacement or elution (isocratic or gradient) chromatography processes can be used to recover adsorbate. Both forms of chromatography have been useful, but under different circumstances. In the displacement mode, the mobile phase is of such high elution strength that the sample is "pushed" through the chromatographic bed by the advancing mobile-phase front. Little, if any, chromatographic separation ensues. However, displacement is useful for fast and total recovery of sample in very little solvent volume. [This can then be concentrated by evaporation if needed prior to an elution (analytical) chromatographic separation.] When interferences are more severe, isocratic elution (instead of displacement) chromatography may be necessary to provide increased resolution. When background interferences are substantial, the resolving power of gradient elution may be necessary because isocratic elution may not suffice.

Trace enrichment can be either intentional or unintentional on the chromatographer's part. Column packings (especially reverse phases) can continuously adsorb trace-level contaminants from the mobile phase as long as their $k'$ values are sufficiently great that migration down the columns is negligible. These trace contaminants will continue to build up on the head of the column until the mobile-phase elution strength has increased sufficiently to cause the adsorbed fraction to migrate down the column. Any further introduction of trace-level contaminants into the mobile phase will not result in enrichment but rather in continuous bleed in the mobile phase. (This bleed is first preceded by an increased concentraton, i.e., peak, of the contaminant as the enriched portion elutes.)

In addressing the problem of unintentional trace enrichment occurring in reverse-phase systems, the source(s) of the contamination can be attributable to the aqueous portion and/or the organic portion, which together comprise the mobile phase. Bristol (1980) studied this problem and developed a convenient solvent-purity test that distinguishes among trace organic impurities present either in organic solvents miscible with water or in water itself.

In this work, gradient elution was carried out on an ODS column ($\mu$Bondapak $C_{18}$, 4.0-mm i.d. × 30 cm). The standard operating procedure included a 4.0-ml/min flow rate and a linear, 10-min gradient from 0 to 100% in either forward or reverse direction.

The solvent-purity test for a binary solvent system consisted of four to five gradient elutions. First, 40 ml of water (A) was enriched, followed by a gradient from 0 to 100% organic modifier (B); then B was enriched (at least 10 ml), followed by a reverse gradient from 100 to 0% B. The third gradient consisted of enriching A (like the first), but changing the volume. Finally, after B had flushed the column (40 ml minimum), the cumulative volume of B pumped across the column during a reverse gradient (to 100% A) was reduced to one-fourth by lowering the flow rate from 4.0 to 1.0 ml/min. The flow rate was reset to 4.0 ml/min and a forward gradient to 100% B was initiated quickly to keep the volume of A pumped to a minimum. The differences among the gradient profiles were

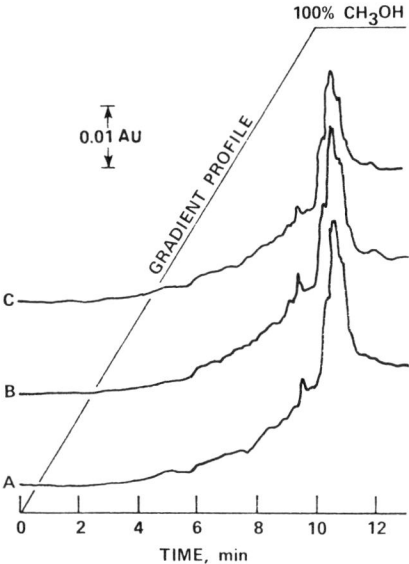

**Fig. 2.** Reverse-phase LC gradient-elution profiles of: (A) 0 ml; (B) 160 ml; (C) 40 ml of Millipore-Q treated water monitored at 254 nm, 0.1 AUFS. Mobile-phase composition was programmed from 0 to 100% methanol at 10%/min. For (C) only 5 ml rather than the usual 20 ml of methanol was used in returning the gradient from 100% methanol to initial conditions (Bristol, 1980).

interpreted to reveal sources and relative levels of trace organic impurities in the individual solvents.

Figure 2 shows that when various amounts (0, 40, 160 ml) of Millipore-Q water was enriched (first gradient mentioned above), peaks did not increase in size with enrichment volumes as would be expected if the water was their means of introduction into the chromatographic system. Furthermore, when distilled and deionized water samples were similarly enriched, they produced identical profiles indistinguishable from those of the Millipore-Q water. This behavior was also contrary to the expected, since these samples should have contained various types and amounts of organic impurities. Bristol concluded that the different water profiles were masked by impurities present in the organic portion of the mobile phase (methanol).

The substitution of acetonitrile as organic modifier in place of methanol provided additional evidence that the observed chromatographic peaks were methanol derived. Figure 3 shows that the enrichment of 40-ml volumes of house-deionized water and Millipore-Q water, both with and without carbon filtering, did indeed produce different LC profiles to reveal differences between water sources. Furthermore, the observed baseline drift was only 0.003 AU for the

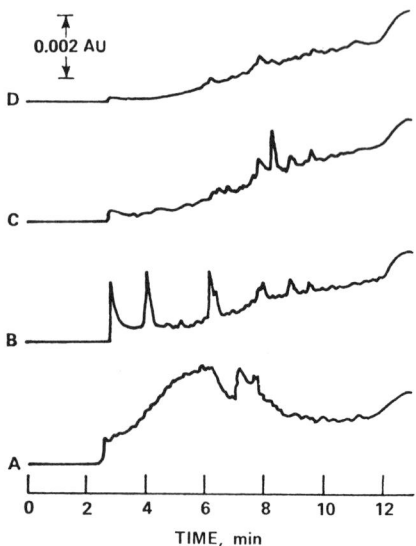

**Fig. 3.** Reverse-phase LC gradient-elution profiles of: (A) 40 ml of house-deionized water; (B) 40 ml of Millipore-Q system water, with manufacturer's recommended cartridge configuration; (C) 40 ml of Millipore-Q system water, with carbon cartridge at end of train; (D) 0 ml water. Monitored at 254 nm, 0.02 AUFS. Mobile-phase composition was programmed from 0 to 100% acetonitrile at 10%/min (Bristol, 1980).

acetonitrile/water gradients, whereas a much greater drift (0.02 AU) was observed for the methanol/water gradients (Fig. 2). This suggests that the methanol contained seven times the UV-absorbing equivalents, due to trace impurities, as did the acetonitrile. Last, because peak heights for chromatograms C and D (Fig. 3) are directly proportional to volume of water enriched, almost all the observed impurities can be attributed to the water and not the acetonitrile.

Nevertheless, the acetonitrile was also found to contain contaminants. Figure 4 shows the results of enriching different volumes (10 and 40 ml) of acetonitrile before programming a reverse gradient from 100 to 0% acetonitrile. Because the peaks are proportional to the amounts of acetonitrile pumped, they can be attributed to this solvent. Bristol surmised that these peaks were probably very polar or ionogenic solutes commonly present in "pure" acetonitrile that became adsorbed or underwent ion-exchange reactions with underivatized silanols on the silica support.

In addition to determining the nature and amounts of contamination present in water and water-miscible organic solvent pairs, nonaqueous reverse-phase solvents such as acetonitrile/methylene chloride and hexane/isopropanol can be similarly studied.

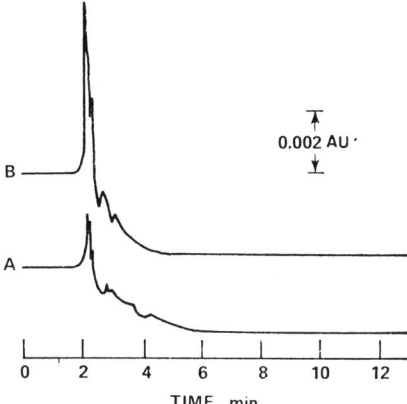

**Fig. 4.** Reverse-phase LC reverse-gradient elution profiles of: (A) 10 ml; (B) 40 ml of acetonitrile monitored at 254 nm, 0.02 AUFS. Mobile-phase composition was programmed from 100 to 0% acetonitrile at 10%/min (Bristol, 1980).

Non-water-soluble organic solvents can also be enriched on straight-phase systems in order to assess their purities. For example, trace-level contaminants in hexane can be concentrated on a silica-gel column. This enrichment results in a concentration of polar and ionogenic impurities contained in the hexane.

Huber and Becker (1977) used the displacement mode of chromatography to effect desorption of *p*-cresol after its enrichment from cyclohexane, using a straight-phase concentrator column (LiChrosorb Si-60, 5 μm). They also enriched dibenzanthracene from trimethylpentane, using the same straight-phase concentrator column. (The maximum concentration effect possible for adsorbates by enrichment is attainable in the displacement mode, although it provides generally poorer separation of multicomponent samples.)

Figure 5 shows the chromatograms resulting from the transition from isocratic elution with the loading solvent as mobile phase (frontal chromatography) to dichloromethane as the mobile phase (displacement chromatography). For this study 1,2,5,6-dibenzanthracene (DBA) was enriched from 2.5 ml of 1 ppm DBA in 2,2,4-trimethylpentane onto a column packed with LiChrosorb Si-60. An enrichment factor of 18 was possible with this loading volume when using dichloromethane as the displacer. The DBA becomes concentrated into a small volume in the area of the displacer front. DBA was also trace-enriched from 250 ml of a saturated aqueous solution onto LiChrosorb RP-18 (5-μm) packing material. Using dioxane as displacer, DBA was concentrated by a factor of 1250, quantitatively into a peak volume of 0.2 ml.

Huber and Becker also studied the effect of the total amount of adsorbate trace-enriched on peak width (final concentration) by enriching constant 50-ml

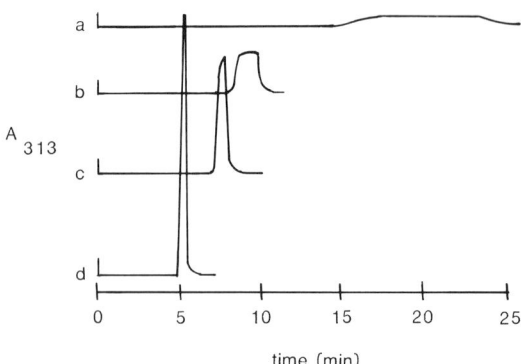

**Fig. 5.** Transition from isocratic elution to displacement chromatography. Column, LiChrosorb Si-60 (5 μm); sample, 2.5 ml of DBA solution in 2,2,4-trimethylpentane, 1 mg/liter; flow velocity, 0.8 mm/sec. Column influent: (a) 2,2,4-trimethylpentane; (b) 2,2,4-trimethylpentane/dichloromethane (9:1, v/v); (c) 2,2,4-trimethylpentane/dichloromethane (8:2, v/v); (d) dichloromethane (Huber and Becker, 1977).

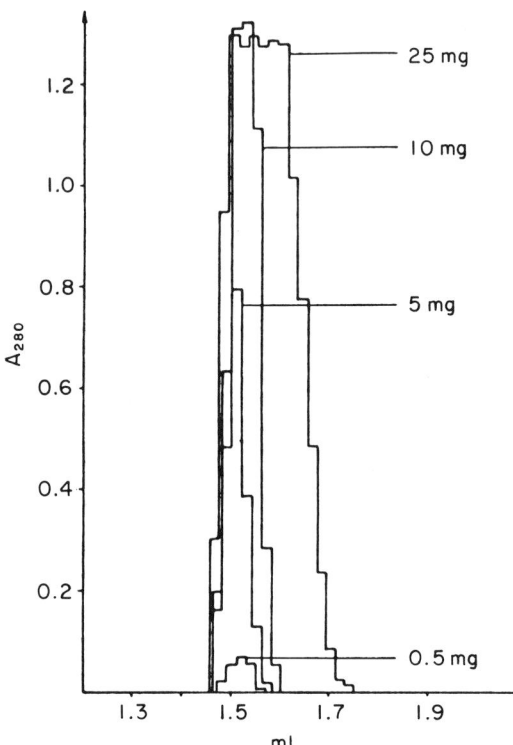

**Fig. 6.** Influence of the amount of trace species on its final concentration. Column, LiChrosorb Si-60 (5 μm); sample, 50 ml of $p$-cresol solutions in cyclohexane containing different amounts of $p$-cresol; displacer, dioxane; flow velocity, 0.6 mm/sec (Huber and Becker, 1977).

volumes of varying concentration (10–500 ppm) *p*-cresol in cyclohexane (silica concentration column). Dioxane was used as the displacer. Figure 6 demonstrates that no increase in peak width is evident up to 10 mg of *p*-cresol loading; however, the 25-mg loading displays no further concentration effect since the peak absorbance remained constant and the peak width increased instead (to 0.2 ml). This increased width is not a function of column overload, but is instead simply a reflection on the absolute limit of enrichment. This limit was defined by the solubility of the enriched compound in the mobile phase (dioxane). A larger loading on the column cannot increase the concentration (of *p*-cresol) beyond this limit—the only possible outcome is an increase in zone width. The displacement of *p*-cresol resulting from the loading of 520 ml of a solution of 10 ppm *p*-cresol in cyclohexane (5.2 mg *p*-cresol) was not solubility limited and afforded an enrichment factor of 5200 (*p*-cresol peak volume 0.1 ml).

Band broadening in trace enrichment can also occur due to column overload. If the sample (enrichment) volume becomes too large, then the fraction of the column occupied by the sample becomes too large. The remaining length of column is insufficient, and part of the adsorbed component is eluted before it can be overtaken by the solvent front. This is a combination of chromatographic

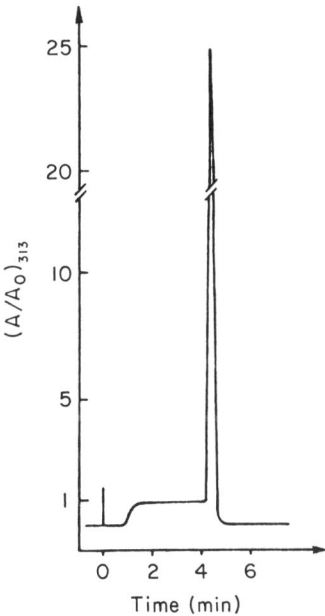

**Fig. 7.** Displacement chromatography on an overloaded column. Column, LiChrosorb Si-60 (5 μm); sample, 6.25 ml of DBA dissolved in 2,2,4-trimethylpentane, concentration 1 mg/liter; displacer, dichloromethane; flow velocity, 1 mm/sec (Huber and Becker, 1977).

modes operating sequentially; displacement chromatography is preceded by frontal elution. Figure 7 demonstrates this chromatographically with DBA in 2,2,4-trimethylpentane (1 ppm) using a column of LiChrosorb Si-60 and dichloromethane as displacer.

## D. Detectors

As LC columns have advanced technologically, so also have LC detectors. Improvements in sensitivity have resulted in much lower detection limits. However, Karger *et al.* (1974) also recognized the potential of large injection volumes as an often neglected means of further improving the detection limits.

The key factor to enhance the mass of sample components is the injection volume. Furthermore, it was realized that greater volumes can be loaded as the $k'$ value is increased; programming techniques offer a means of lower detection limits, since large injection volumes are possible as $k'$ will be very large at the start of a run.

The primary LC detection mode in trace enrichment is UV absorption. Additionally, electrochemical detection has been used in a significant portion of clinical/biomedical applications for oxidizable/reducible compounds. Fluorescence is also widely used in environmental work. Refractive index is generally unsatisfactory due to both lack of relative sensitivity and operating constraints.

## E. Techniques to Increase Extractability ($k'$) of Solutes from Aqueous Samples during the Concentration Step in Trace Enrichment

Those techniques that reduce or eliminate band tailing and increase $k'$ values for poorly retained components are not only useful in elution chromatography but are employable in trace enrichment in order to increase solute/stationary-phase interactions.

The potential extractability of a compound from its sample-loading solvent during trace enrichment primarily depends upon the mobile-phase solubility of the compound. Specifically, for reverse phases, water solubility (relative to adsorption on the stationary phase) will determine the extractability of that compound from aqueous solutions during the concentration step.

In order to increase the affinity of a solute for a reverse phase like $C_{18}$ (or to decrease water solubility), both ion-suppression or ion-pairing techniques are available. Ion suppression can be thought of as a simple form of ion pairing in that only the pH of the sample solution need be controlled, without the further addition of an ion-pairing reagent. However, when a silica-gel-based LC column is used, care should be taken not to exceed pH ~8.0 since silica dissolution can result. Primarily for this reason, basic compounds such as amines cannot be analyzed by ion suppression. Instead, these compounds are analyzed at acid pH

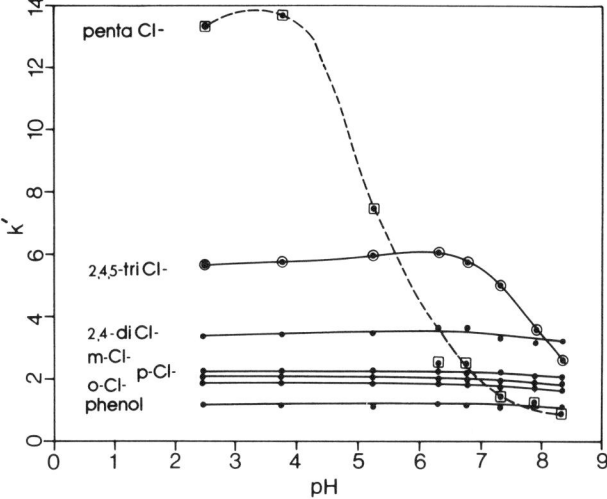

**Fig. 8.** Variation in the capacity factor $k'$ of various chlorinated phenols with mobile-phase pH (Mayer and Shoup, 1981).

conditions in order to induce protonation. These ionized species can then be effectively ion paired and analyzed.

Ion suppression (ion pairing with $H^+$ and $OH^-$ as the ion-pairing reagents) enables enrichment to occur on the concentrator column by increasing $k'$ values and can also eliminate tailing on the analytical column. Figure 8 illustrates the $k'$ dependency on pH for the acidic compound pentachlorophenol (PCP). In this application a low pH is necessary in order to increase $k'$ sufficiently for the enrichment step, but higher pH may be useful in providing an additional aspect for enhancing the analytical separation. Generally, in ion suppression, the pH is kept at least one pH unit below the $pK_a$ (for acids) or one pH unit above the $pK_b$ (for bases). In this way compounds in question exist in solution in un-ionized forms. (For cases where both acids and bases must be analyzed, the sample must then be split and half treated separately as for acids, half as for bases.) On the other hand, the ion-pairing technique operates at a pH which ensures that the compounds in question are ionized. They exist in solution, each associated with an ion of equal and opposite charge. This results in a neutral associated pair which has enhanced affinity for the hydrophobic reverse-phase packings.

## F. Mutual-Zone Solubility

As an LC column becomes loaded with one or another compound, other trace-level components present in the sample may solubilize into the adsorbed-component zone already on the column. This results in either an increase or a decrease

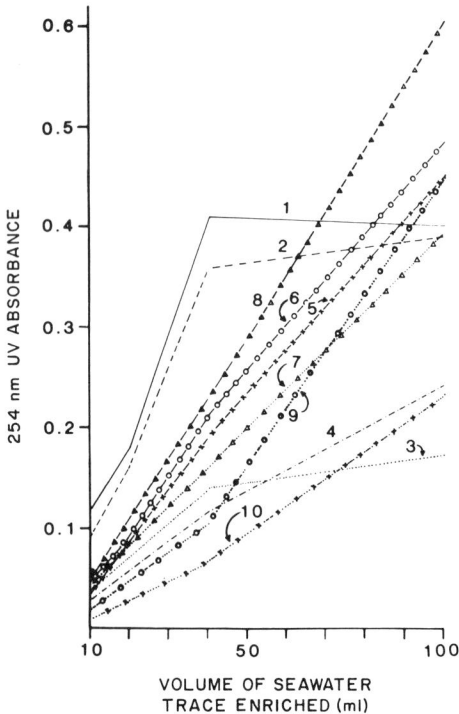

**Fig. 9.** Rates of change in UV absorption for 10 chromatographic peaks trace enriched from various volumes of Bunker-C-oil-contaminated seawater (4 weeks oil/water contact time). [Reprinted with permission from Saner *et al.* (1979). *Anal. Chem.* **51,** 2180–2188. Copyright 1979, American Chemical Society.]

in extraction efficiency for these components, depending on whether their solubilities in the adsorbed compounds have increased or decreased.

An example of a mutual-zone solubility effect was demonstrated by Saner *et al.* (1979), who studied the linearity of peak height with sampling volume for 10 chromatographic peaks. Various volumes of seawater in contact with a bunker-C oil for 4 weeks were trace-enriched onto a precolumn (2.1-mm i.d. × 7 cm) packed with CO:PELL ODS (~30-μm pellicular particles) at a loading rate of 9.9 ml/min. Desorption was carried out by a water/methanol gradient using a reverse-phase $C_{18}$ column (Chromegabond $C_{18}$, 4.6-mm i.d. × 15 cm) for the analytical (separator) column. The rate of change in UV absorption with increasing sampling volumes for each of 10 chromatographic peaks is shown in Fig. 9. Growth curves for peaks 1–6 generally show decreasing amounts of convexity, indicating breakthrough or its approach; curves 7 and 8 remain linear over the sampling volume range; growth curves for peaks 9 and 10 are concave. This increase in the accumulation rate for peaks 9 and 10 with increasing sampling volume could be a manifestation of mutual-zone solubility; the increased adsorption of these peaks for the larger sampling volumes could simply reflect a change in the stationary phase at the head of the column. The increase in growth rates for peaks 9 and 10 could result from their increased extraction efficiency due to their affinity for (solubility in) other adsorbates which are accumulating simultaneously at the head of the concentrator column. In other words, the presence of adsorbates on the column actually presented a new packing material which provided an increasingly more efficient stationary phase for the adsorption of peaks 9 and 10 than did the original ODS material.

## G. Forms of Trace Enrichment

### 1. Direct Injection

Direct injection is the introduction of sample solution directly onto an analytical LC column (a precolumn is usually present in order to preserve the expensive analytical column from particulates and nonreversible adsorption). The sampling volumes can range from microliter amounts up to liter volumes, depending primarily on the concentration(s) of the solute(s) of interest present in the sampling solutions and their $k'$ value(s).

Since the concentration effect can only take place up to breakthrough of the solute front during the loading process, only loading volumes less than the breakthrough volume will allow quantitative recoveries. For loading volumes up to about 5 ml, a conventional loop injector in conjunction with a hand-held syringe is feasible; for greater volumes pump-loading of the sample is necessary using a specific pumping rate over predetermined time intervals for specific volumes.

### 2. Off-Line Enrichment

By carrying out the concentration step off-line from the analytical column, on-site sample loading can be accomplished. This is particularly valuable when it is impossible or impracticable to bring the entire liquid chromatograph to the sampling scene. This capability is attractive for environmental analyses where sample acquisition and preparation (extraction) can be done simultaneously on-site. No large sampling volumes need be transported to the place of analysis. Also circumvented is the problem of adsorption of trace-level components on sampling-container walls during the storage interval between sampling and analysis.

For on-site applications, disposable cartridges are a recent innovation that allows simple sample extraction (loading) via a hand-held syringe. The enriched samples are recovered from the cartridges by means of solvent desorption and are then either concentrated further or injected, as is, onto the analytical column. When using a precolumn (guard) as the concentrator column, desorption is accomplished by physically connecting it on-line ahead of the analytical column and desorbing it by means of solvent programming.

The off-line enrichment technique is advantageous since quickly eluting background contamination does not elute through the analytical column and the detector to cause baseline and column perturbations. However, it is disadvantageous because it cannot be automated, and each analysis requires some degree of dexterity in either application and desorption of sample or connection of the precolumn (concentrator) on-line ahead of the analytical column.

*3. On-Line Enrichment*

This form of enrichment involves a concentrator/guard column which is automatically (or manually) positioned for the concentration step (with the column effluent going to waste) and then flushed of extraneous contaminants after the loading process. It is then placed on-line ahead of the analytical column for desorption and separation. The concentrator column is commonly located in the loading loop of a multiport injection valve. During the enrichment step the analytical column is conditioned with a mobile phase while the concentrator column effluent passes to waste. After loading is completed, the switching valve moves the concentrator column on-line to be desorbed onto the analytical column. Many applications provide for back-flushing the concentrator column onto the analytical column. On-line enrichment can provide more precise sampling volumes than off-line enrichment permits. The ability to automate the on-line enrichment/desorption sequence is of particular utility.

## II. ENVIRONMENTAL APPLICATIONS

### A. Direct Injection

Creed (1976) used direct-injection trace enrichment to study trace organic contamination in distilled water, well water, reservoir water, and river water (as manufacturing-plant-process water both before and after use). Direct injection onto an anylical μBondapak $C_{18}$ column (3.9-mm i.d. × 30 cm) was either by pumping a certain volume of water sample through the column or by manual injection (for smaller injection volumes).

In order to study the organic contamination introduced by a manufacturing process, a 50-ml volume of river water before use was enriched (Fig. 10); only 1 ml of river water from the stream after use in the manufacturing plant was enriched due to its much higher relative organic load (Fig. 11).

The leaching of additives from plastic containers was studied by enriching 15-ml volumes of distilled water, meant for human use, packed in a polyethylene container. The resulting chromatogram is shown in Fig. 12. Although phthalates

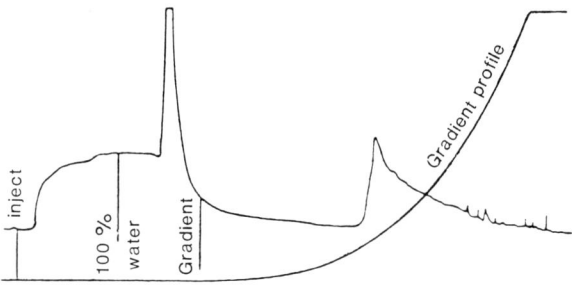

**Fig. 10.** Chromatogram provides "fingerprint" of river water before its use in manufacturing process. Gradient profile, 100% water to 100% acetonitrile over 1 hr; flow rate, 1 ml/min; UV at 0.32 AUFS (Creed, 1976).

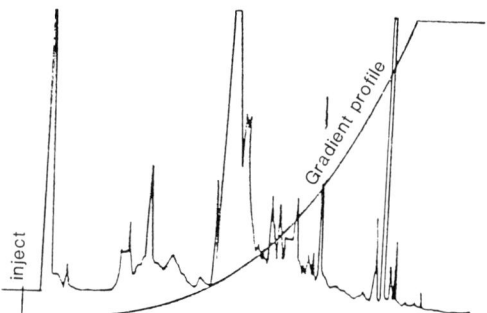

**Fig. 11.** After being used in manufacturing process, river water shows high concentration of organics before being pumped into waste-treatment plant. UV at 0.32 AUFS (Creed, 1976).

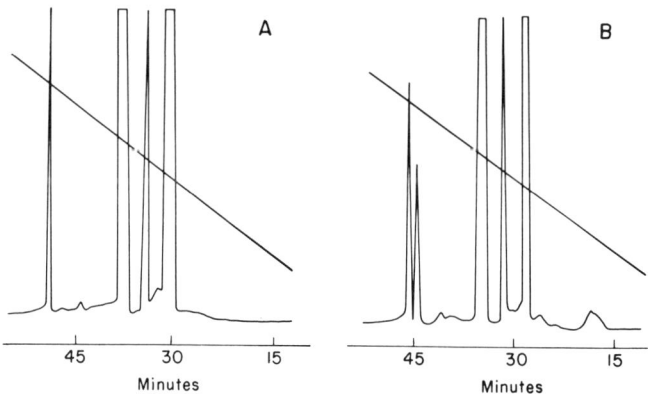

**Fig. 12.** Chromatogram of 15-ml sample of commercial distilled water (A) shows four large peaks which turn out to correspond to four of the five peaks in chromatogram of 7.5-ml sample of standard mixture of phthalates (B) (Creed, 1976).

are not normally contained in linear polyethylene, the four chromatographic peaks observed corresponded in retention time to phthalate standards. Creed attributed the probable sources of these plasticizers to either the lid of the bottle or the transport lines in the bottle-filling process.

Frei (1978) also studied trace-level organic contamination in distilled water by direct injection of 165 ml onto a Nucleosil $C_{18}$ analytical column (3.0-mm i.d. × 10 cm). Two major impurity peaks were desorbed. They were attributable to softeners from the plastic bottle in which the water sample had been stored a short time. These same peaks were absent from the chromatogram when the enriched water was stored in glass containers.

The advantages of trace enrichment in terms of sample recovery and lack of phase changes (which can contribute to rearrangement or degradation of some sample components) were pointed out by Little and Fallick (1975). These workers directly injected 200 ml of Blackstone River (Massachusetts) water onto a μBondapak $C_{18}$ column (3.9-mm i.d. × 30 cm). Column effluent monitored by UV absorption during the loading process showed an increase in absorption which was attributed to some highly polar fractions that were essentially unre-

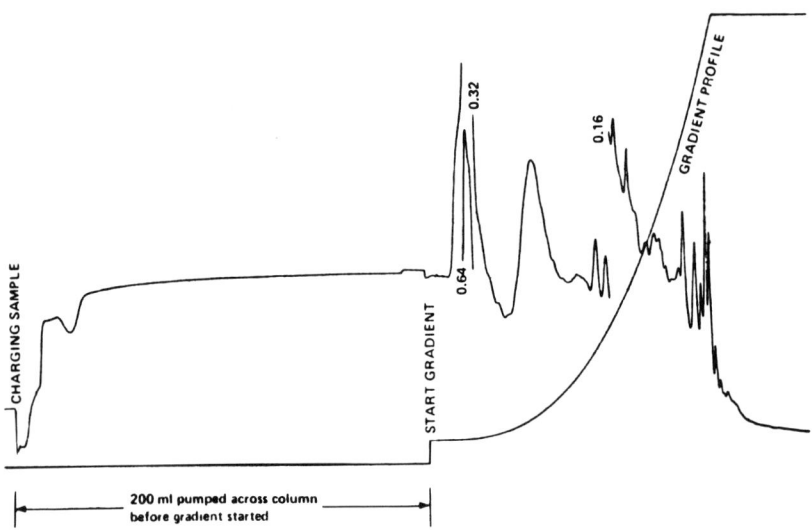

**Fig. 13.** Demonstration of trace enrichment with aromatics in river water. The compounds are very nonpolar (probably contain only carbon and hydrogen) and are aromatic as indicated by the strong UV absorption; presumably they are some form of polynuclear aromatics. The chromatograph was operated isocratically with river water for 200 ml, then in gradient mode from 5 to 95% acetonitrile in water. Curve 8 on Model 660 Programmer for 30 min. Sample, Blackstone River (MA) water. Flow rate, 2.0 ml/min. Detector, UV. Column, 4 mm × 30 cm μBondapak $C_{18}$ (Little and Fallick, 1975).

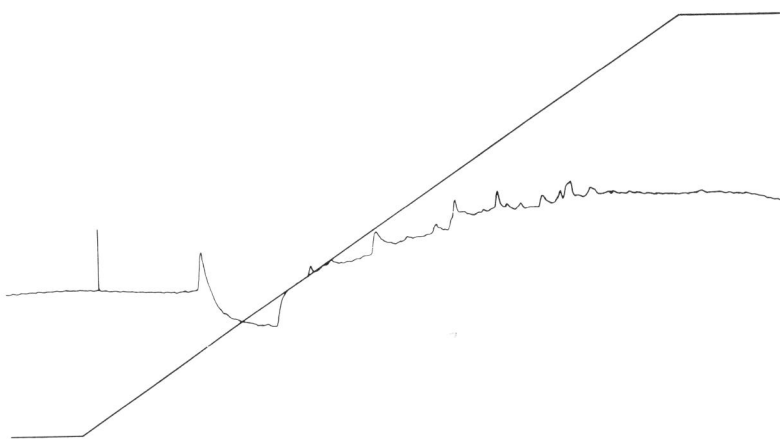

**Fig. 14.** Fresh distilled water through a 30-in. carbon column. [Reprinted from Sampson (1977). Copyright 1977 by International Scientific Communications, Inc.]

tained by the packing material (frontal elution). Adsorbed organics were eluted by means of a water/acetonitrile gradient and are shown in the chromatogram in Fig. 13. The authors mentioned that smaller injection volumes (0.5–2.0 ml) may be all that is necessary for injection in order to concentrate sufficient organic contaminants for detection from more highly contaminated water samples.

A joint study by Millipore Corporation and Waters Associates on the efficacy of various water purity techniques for removal of organic contaminants was reported by Sampson (1977). Various volumes from 4.0- to 160-ml water samples were directly loaded onto a μBondapak $C_{18}$ analytical column (3.9-mm i.d. × 30 cm).

Figures 14 and 15 are trace-enrichment chromatograms of 160-ml volumes each, of distilled water with different pretreatments; one sample (Fig. 14) was passed through a carbon adsorption column and the other (Fig. 15) was heated to boiling in order to drive off low-molecular-weight organics carried over in the condensate during the distillation process. The chromatogram in Fig. 14 shows that the carbon adsorption step removed much of the trace organic impurities from the distilled water, whereas the chromatogram shown in Fig. 15 demonstrates the dubious utility of boiling the distilled water.

Tetrachloroethylene was determined by Kummert *et al.* (1978) in natural waters by direct injection of 2-, 5-, and 10-ml volumes onto a μBondapak $C_{18}$ column (3.9-mm i.d. × 30 cm). Concentration by trace enrichment, rather than by solvent extraction followed by solvent stripping, avoided loss of this highly volatile analyte during the evaporation step. Detection was by UV absorption at

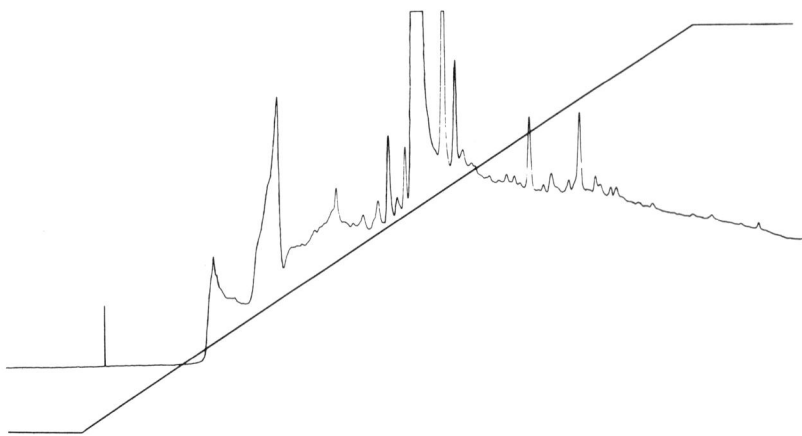

**Fig. 15.** Boiled distilled water. [Reprinted from Sampson (1977). Copyright 1977 by International Scientific Communications, Inc.]

208 nm. The water samples were prefiltered prior to injection. Isocratic operation of the column at methanol-in-water concentrations between 50 and 70% allowed most UV absorbing interferences to elute just prior to the tetrachloroethylene peak (~20-ml elution volume).

A detection limit of 0.06 μmol/liter (~10 ppb) was found for tetrachloroethylene in natural waters, whereas a limit of 0.006 μmol/liter (1.0

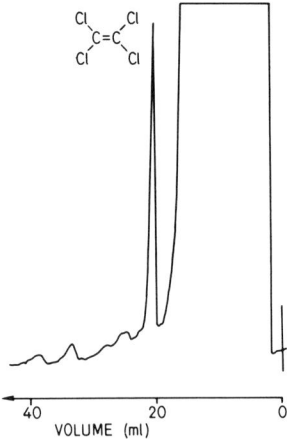

**Fig. 16.** HPLC analysis of a groundwater sample contaminated with 0.65 μmol/liter tetrachloroethylene. Conditions: μBondapak $C_{18}$, methanol/water (65:35) at 2 ml/min, 208-nm detection wavelength, 0.1 AUFS, 10 ml injected. [Reprinted with permission from Kummert et al. (1978). Anal. Chem. **50**, 1637–1639. Copyright 1978, American Chemical Society.]

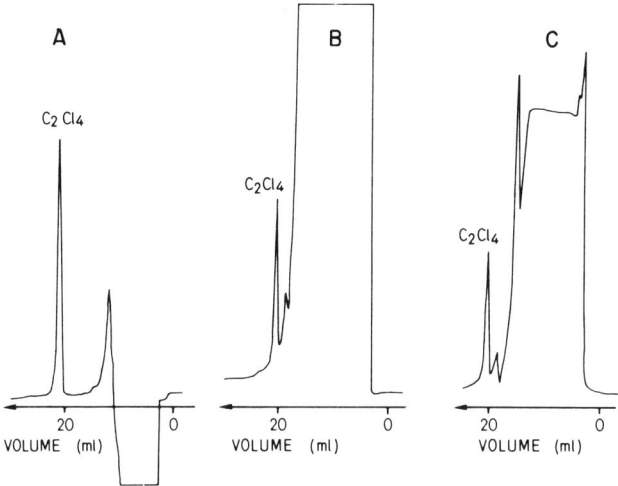

**Fig. 17.** HPLC analyses of water samples spiked with known amounts of tetrachloroethylene: (A) doubly distilled water with 0.60 μmol/liter $C_2Cl_4$; (B) lake water with 0.35 μmol/liter $C_2Cl_4$; (C) river water with 0.30 μmol/liter $C_2Cl_4$. Conditions: μBondapak $C_{18}$, methanol/water (60:40) at 1.5 ml/min, 208-nm detection wavelength, 0.1 AUFS, 10 ml injected. [Reprinted with permission from Kummert et al. (1978). Anal. Chem. **50**, 1637–1639. Copyright 1978, American Chemical Society.]

ppb) was achieved for a spiked sample of distilled water (signal-to-noise ratio, ≥3; injection volume, 10 ml). Analysis of a contaminated groundwater sample is shown in Fig. 16; Fig. 17 shows the results of a spiking experiment using distilled, lake, and river waters.

This direct-injection method was successfully used to locate the source of contamination in an actual case of aquifer pollution caused by tetrachloroethylene. The major advantage of the high-performance liquid chromatography (HPLC) method is fast turnaround and simplicity of operation. The achievable detection limit for tetrachloroethylene is an order of magnitude less than the maximum allowable concentration limit in treated effluents as set by Swiss law (0.7 μmol/liter).

In contrast to the 10-ml injection volumes described above for determination of tetrachloroethlene, Kirkland (1975) used direct injection of 100-μl volumes onto a reverse-phase psm-50/$C_{18}$ column (3.2-mm i.d. × 50 cm). The raw creek-water samples were filtered prior to injection. He reported a detection of 25 ppb phenol in surface waters with a 100-μl injection volume. An apparent concentration of 50 ppb phenol, as shown in Fig. 18, was found in the creek water.

The adsorption of phthalate esters from aqueous solutions was studied by Otsuki (1977) in order to develop a method for their determination in water by

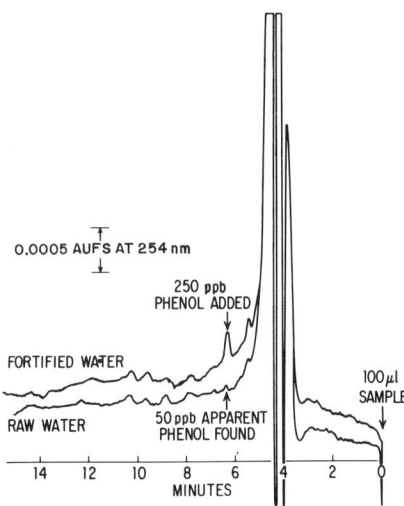

**Fig. 18.** Trace analysis of phenol in surface water. Temperature, 50°C; mobile phase, 60% methanol/water; pressure, 1180 psi; flow rate, 0.40 ml/min; UV detector; sample, 100 μl of Brandywine Creek water. [Reprinted with permission from Kirkland (1975). Copyright 1975, Pergamon Press, Ltd.]

adsorption chromatography. A μBondapak $C_{18}$ column (3.9-mm i.d. × 30 cm) was used in conjunction with a 10–15-min gradient from water to methanol—curve 2 (Waters programmer)—and 2.0-ml/min flow rate.

The trace-enrichment procedure consisted of injecting 1–5 ml of a solution of phthalate standard (0.2–0.3 μg/μl) in methanol by means of a U6-K injector as redistilled water was being pumped through the column at 3.0 ml/min. An alternatine procedure consisted of loading, also at 3.0 ml/min, sufficient volumes (unspecified) of phthalate-spiked water to provide from 0.2 to 2.0 μg phthalate. The column was then washed with 20 ml of redistilled water prior to gradient elution described above. A 10-min hold at 50% methanol was included in the gradient in order to elute most naturally occurring interferences having an absorption near 254 nm that were enriched during the loading procedure. This was done to avoid peak overlap of the interferences with those of the phthalate esters.

The ability of the column to retain adsorbed phthalates using a 50:50 methanol/water mobile phase for various holding times is shown in Fig. 19. There is minimal loss of the shortest chain alkyl (*n*-butyl) phthalate with holding times of 15 min or less. By keeping the holding time down to 10 min, there is no effective loss of adsorbed phthalates with side chains of equal or greater length than $n$-$C_4$. This procedure resulted in a detection limit of 0.1 μg phthalate at 0.05 AUFS.

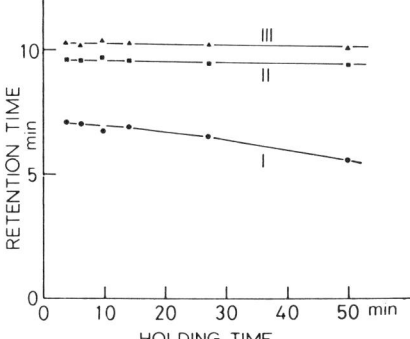

**Fig. 19.** Effect of holding time on apparent retention time for a mobile-phase composition of 50:50 water/methanol. Flow rate, 2 ml/min; gradient elution with 10-min holding process; setting time, 10 min. (I), Di-*n*-butyl; (II), di-*n*-heptyl; (III), di-2-ethylhexyl phthalate. The retention time was taken from the starting point of the further gradient (Otsuki, 1977).

The resulting chromatogram from a direct water injection of a 5-ml tap-water sample spiked with pesticides at the ppb level is shown in Fig. 20 (Waters Associates, 1976). A μBondapak $C_{18}$ column (3.9-mm i.d. × 30 cm) was used. A water/60% acetonitrile gradient in 20 min was used to separate 11 pesticides ranging in concentration (in the original water sample) from 5 to 50 ppb.

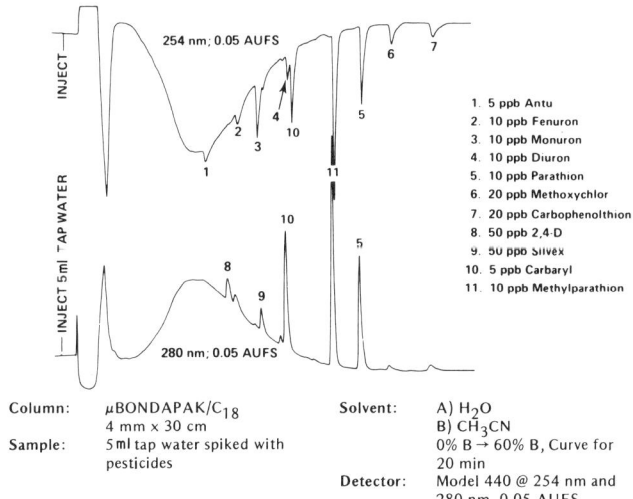

**Fig. 20.** Pesticide residues determined from the direct injection of 5 ml of a spiked tap-water sample (Waters Associates, 1976).

In another work Otsuki and Takaku (1979) used a similar direct-injection enrichment as described above (Otsuki, 1977) in developing a rapid method for the determination of the herbicide Abate in water in order to investigate its effects on small-stream ecosystems.

A Bondapak phenyl/Corasil (37–50-μm particle) column (2.0-mm i.d. × 122 cm) served as concentrator/separator column. A linear gradient from water to acetonitrile in 5 min was used to desorb the Abate at a flow rate of 2.0 ml/min.

The enrichment procedure consisted of injecting from 1 to 20 μl of a standard solution of Abate in acetonitrile (10 or 100 ng/μl) by means of a U6-K injector as purified water was pumped through the column at 2.0 ml/min. A second procedure consisted of injecting a maximum of 2.0 ml volumes of water containing 5–2000 ng Abate. The loaded column was rinsed with 10 ml water prior to gradient elution.

In preliminary recovery experiments, between 40 and 80% of the Abate was recovered from spiked samples. This low recovery was due to the tendency of Abate to concentrate on the surface of the fortified water samples due to its hydrophobicity. Aqueous sodium lauryl sulfate (SLS) (no absorption at the detection wavelengths of 254 and 280 nm) was added as an emulsifier to help prepare and preserve homogeneous fortified samples of Abate-spiked water.

Table I lists the recovery efficiencies of Abate at the 1-ppm level (1 ml injected volumes) versus concentration of SLS solution added. The addition of 2 ml of a 10% solution of the SLS to 100-ml water samples resulted in 100% recovery. Recoveries were essentially the same for Abate enriched from distilled water and from filtered or unfiltered pond water. The addition of emulsifier kept

**TABLE I**

**Effect of Addition of Sodium Lauryl Sulfate Solution on Abate Recovery[a]**

| Sample type[b] | Concentration of SLS added (w/v%) | Recovery[c] Average (%) | $SD^d$ (%) |
|---|---|---|---|
| Distilled water | 0.1 | 55.8 | 1.5 |
|  | 1 | 61.1 | 6.4 |
|  | 10 | 97.7 | 4.0 |
| Filtered pond water | 10 | 101.9 | 6.8 |
| Pond water[e] | 10 | 93.1 | 7.1 |

[a] Reprinted with permission from Otsuki and Takaku (1979). *Anal. Chem.* **51**, 833–835. Copyright 1979, American Chemical Society.
[b] 2 ml of 10% solution of SLS added to 100-ml water samples.
[c] At 1-ppm level (1-ml injected volumes).
[d] $n = 5$.
[e] Pond water was analyzed after it was fortified and filtered with Whatman GF/C glass-fiber filter.

Abate evenly in the water column. Otsuki and Takaku suggested that an emulsifier be routinely added to water samples to keep a homogeneous solution and avoid adsorption of hydrophobic compounds on suspended matter such as bacteria and algae. Using UV absorption at 254 nm and at 2.0 ml/min flow rate, it was possible to determine Abate at a concentration of 5 ppb with an injection volume of 1 ml.

In order to study the rate of dissipation of an experimental herbicide, Fluridone, West and Parka (1981) developed a direct-injection LC technique which offered decreased sample-analysis time (by eliminating the need for bromination of Fluridone for detection by electron-capture gas chromatography) and improved analytical precision.

A μBondapak $C_{18}$ column (3.9-mm i.d. × 30 cm) was used with a Whatman guard column containing a pellicular $C_{18}$ packing (CO:PELL ODS). The chromatograph was operated isocratically at 1.1 ml/min using 60:40 methanol/water. Detection was carried out at 254 nm (0.05 or 0.01 AUFS).

Fluridone was determined in pond water by filtering an aliquant; 1-ml volumes of the filtrate were injected. For comparison purposes 100-ml aliquots of the pond water were also extracted three times with 20-ml volumes of dichloromethane. The combined extracts were taken to dryness and reconstituted in 2 ml of the mobile phase; 200-μl volumes of the concentrated extracts were injected.

Table II compares the results of the direct-injection method to the liquid/liquid extraction method. The direct-injection technique was more precise; both techniques were faster and provided greater precision than the gas chromatographic

**TABLE II**

Recovery and Precision of Methods for Trace Levels of Fluridone in Pond Water[a]

| HPLC method | Fluridone added (ppm) | Number of replicates | Recovery (%) | | Coefficient of variation |
|---|---|---|---|---|---|
| | | | Range | Average | |
| Direct injection | 0.05 | 8 | 90–100 | 95 | 3.5 |
| | 0.100 | 6 | 96–100 | 97 | 2.1 |
| | 1.000 | 8 | 99–102 | 100 | 1.8 |
| Extraction | 0.001 | 9 | 105–149 | 125 | 14.7 |
| | 0.010 | 9 | 84–113 | 101 | 8.7 |
| | 0.100 | 6 | 86–100 | 92 | 6.0 |

[a]Reprinted with permission from West and Parka (1981). *J. Agric. Food Chem.* **29,** 223–226. Copyright 1981, American Chemical Society.

(GC) technique. The detection limit of Fluridone using the direct-injection method was ~5 ppb.

The priority pollutants benzidine and dichlorobenzidine are used extensively in the synthesis of dyes. They are suspected carcinogens. Their determination in the aqueous environment was undertaken by Riggin and Howard (1979) using reverse-phase LC with electrochemical detection (for oxidizable compounds). Direct-injection and solvent-extraction techniques were investigated. In addition, off-line trace enrichment was also utilized (Section II,B).

Direct injections of 50-μl volumes were applied to a LiChrosorb RP-2 column (4.6-mm i.d. × 25 cm). An electrochemical detector monitored column effluent (50:50 acetonitrile/water) at a flow rate of 0.8 ml/min. The mobile phase was buffered at pH 4.7 with 0.1 $M$ sodium acetate. One-liter waste-water samples were collected in sample bottles containing 75 g NaCl, 5 g $KHSO_4$, and 200 ml methylene chloride. (The NaCl served to increase extractability of the nonpolar organic compounds into the methylene chloride.) Addition of $KHSO_4$ served to maintain pH ~2.0. Although this pH gave best results, at this pH any diphenylhydrazine was found to degrade into benzidine, artificially inflating the benzidine concentration. This diphenylhydrazine conversion was overcome by

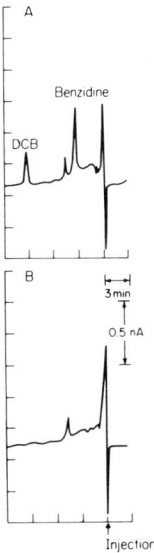

**Fig. 21.** Chromatograms for direct injection of aqueous effluent from an organic-chemical plant: (A) spiked with 10 ppb dichlorobenzidine (DCB) and benzidine; (B) unspiked. [Reprinted with permission from Riggin and Howard (1979). *Anal. Chem.* **51**, 210–214. Copyright 1979, American Chemical Society.]

employing 0.1 $M$ sodium acetate (degradation still occurred, but not to benzidine).

Chromatograms of the direct injection of 50-μl volumes of chemical-plant effluent are shown in Fig. 21. The chromatogram A was spiked to 10 ppb each of dichlorobenzidine and benzidine; chromatogram B was unspiked. Although the direct-injection technique was the most susceptible to interferences of all the sample-preparation techniques investigated, it was very rapid and provided a detection limit of ~1.0 ppb. However, when interferences were severe, or when 1-ppb sensitivity was inadequate, off-line enrichment was more appropriate.

## B. Column Switching, Off-Line

As previously mentioned, Riggin and Howard (1979) also used off-line enrichment of benzidine and dichlorobenzidine. This was done by first adjusting the aqueous sample solutions to pH 7.0 (with 0.2 $M$ phosphate buffer) and then passing 10-ml volumes through Sep-Pak $C_{18}$ cartridges using glass syringes. After washing of the cartridges with 5.0 ml of distilled water, benzidine and dichlorobenzidine were desorbed with 3.0 ml methanol. The alcohol was evaporated and the sample was reconstituted to 1.0 ml in 0.1 $M$ sodium acetate buffer, pH 4.7 ($\times 10$ concentration factor). In comparison to direct injection, off-line enrichment provided an order of magnitude lower detection limit to ~100 ppt (ng/liter). Figure 22 compares the chromatogram obtained by enriching, off-line, 10 ppb benzidine and dichlorobenzidine spiked into municipal sewage samples to that of the unspiked sewage and to a standard corresponding to 100% recovery of the spiked samples.

In a study by Saner and Gilbert (1980), Sep-Pak $C_{18}$ cartridges were similarly used to adsorb the pesticide Dursban from contaminated water after a fire in a chemical-packaging plant. A methylene chloride extraction of the water samples was also performed in order to compare results to the enrichment technique.

A Zorbax ODS analytical column (4.6-mm i.d. × 50 cm) was used with a Whatman guard column (4.6-mm i.d. × 7 cm) packed with CO:PELL ODS. A water/acetonitrile segmented gradient provided separation, and UV absorption at 280 nm was used for detection.

Aliquots (25 ml each) of the water samples were also extracted using two 20-ml volumes of methylene chloride. The combined extracts were dried and concentrated to 2.5 ml ($\times 10$ concentration factor); 100-μl injection volumes of the extracts were used. For the enrichment on Sep-Paks, 20-ml aliquots of the aqueous samples were applied using a hand-held syringe. Since a proportional amount of suspended particulates in the raw-water samples was included in the aliquots, a buildup of particulates occurred on the heads of the cartridges during the extraction step. Two milliliters of methanol served to recover adsorbed

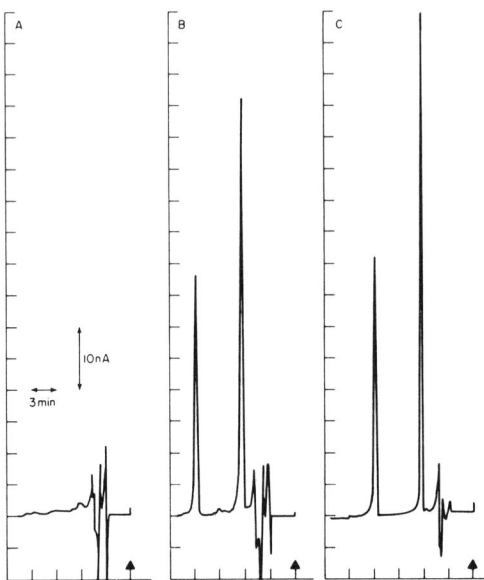

**Fig. 22.** Chromatogram for resin concentration of municipal sewage: (A) unspiked; (B) spiked with 10 ppb benzidine and dichlorobenzidine; (C) standard corresponding to 100% recovery of spike. [Reprinted with permission from Riggin and Howard (1979). *Anal. Chem.* **51,** 210–214. Copyright 1979, American Chemical Society.]

compounds from both the cartridge packing and from the particulates at the cartridge inlet; 100-μl injection volumes of this methanol rinse were used.

The extraction efficiencies for both techniques are very similar; Table III lists the results for the extraction of Dursban from spiked deionized water over the concentration range from 100 ppb to 10 ppm. On the other hand, the respective levels of contamination by Dursban in actual environmental samples are very dissimilar overall. Table IV lists the Dursban concentrations found. The differences were ultimately traced to the presence of a large complement of adsorbed Dursban present in the environmental samples. The reduced concentrations of Dursban calculated from the data provided by the methylene chloride extractions were primarily attributed to loss of some of the larger mass particulates during sedimentation before removal of the aliquots for methylene chloride extraction.

The use of Sep-Pak $C_{18}$ cartridges for extraction of the nonadsorbed Dursban (dissolved) fraction resulted in nearly complete recovery (95%). However, due to the large adsorbed fraction present in the environmental samples, a mechanical filtration of this fraction from the water and a buildup at the cartridge inlet during the "enrichment" step was the primary mode of removal of Dursban

### TABLE III

Extraction Efficiencies for Methylene Chloride Liquid/Liquid Extraction and Sep-Pak $C_{18}$ Adsorption Techniques for Dursban in Deionized Water[a]

| Dursban in water (ppm) | Percentage recovery of Dursban | |
|---|---|---|
| | $MeCl_2$ (25-ml sampling volume)[b] | Sep-Pak $C_{18}$ (20-ml sampling volume)[c] |
| 10.0 | 97.4 | 89.7 |
| 1.0 | 89.1 | 104.5 |
| 0.1 | 91.3 | 90.8 |

[a] Reprinted from Saner and Gilbert (1980), p. 1758, by courtesy of Marcel Dekker, Inc.
[b] Average, 92.6; relative standard deviation, 4.3%.
[c] Average, 95.0; relative standard deviation, 8.7%.

from the water samples, and not one of enrichment per se. It is interesting to note that the use of a direct injection of the contaminated water (after filtering) onto the column would have provided a determination only of the dissolved portion of Dursban and would have indicated a much lower concentration than the off-line enrichment provided using the Sep-Pak cartridges (where both adsorbed and dissolved species were codetermined).

Sep-Pak cartridges have also been used to enrich chlorophenoxy herbicides (2,4-di-; 2,4,5-tri-; etc.) from 100-ml tap-water samples spiked at the ppb level with each herbicide (Waters Associates, 1978a). The herbicides were desorbed from the cartridge using 1.0 ml of acetonitrile/water, and 1–200-µl volumes were then injected onto a µBondapak $C_{18}$ column (3.9-mm i.d. × 25 cm) with

### TABLE IV

Dursban Concentrations Determined in Environmental Water Samples Using Methylene Chloride Liquid/Liquid Extraction and Sep-Pak $C_{18}$ Cartridge Adsorption Techniques[a]

| Sample | Dursban concn (ppb) | |
|---|---|---|
| | $MeCl_2$ | Sep-Pak $C_{18}$ |
| 1 | 4240 | 14,100 |
| 2 | 80 | 250 |
| 3 | 40 | 1900 |
| 4 | 40 | 50 |
| 5 | 70 | 200 |

[a] Reprinted from Saner and Gilbert (1980), p. 1761, by courtesy of Marcel Dekker, Inc.

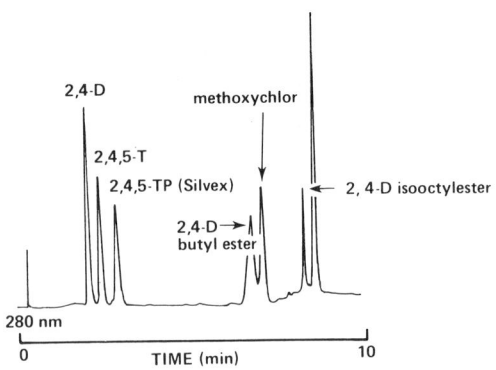

**Fig. 23.** Liquid chromatogram of aliquot from the Sep-Pak $C_{18}$ cartridge extraction of a 100-ml sample of tap water spiked at the ppb level with seven herbicides (Waters Associates, 1978a).

detection by UV absorption at 254 and 280 nm. The resultant chromatogram is shown in Fig. 23. The simplicity and rapidity of this approach are in stark contrast to a GC method requiring chemical conversion of the free acids to their methyl esters and up to 6 hr analysis time per sample.

A microprecolumn technique for the enrichment of phthalate esters, followed by separation on a microanalytical column, was described by Ishii *et al.* (1978). Since only a very small injection volume could be used with the microseparator column, much of the sample solution (if prior solvent extraction was used) was not required. In order to concentrate trace-level components in a small amount of sample without loss, this microprecolumn technique was devised.

The precolumn (0.35-mm i.d. × 18 mm), composed of stainless-steel tubing, was packed with Hitachigel 3010 (styrene/divinyl benzene copolymer, 20-μm particle); the separator column (0.5-mm i.d. × 170 mm), composed of PTFE tubing, was packed with the same kind of gel, but of 5-μm particle size (TSK gel LS111). A diagram of the precolumn and separator column is shown in Fig. 24. The columns were connected by means of a Teflon tube without any void volume. Detection was by UV absorption at 240 nm, using a flow cell between 0.05 and 0.4 μl in volume.

The precolumn was disconnected from the separator column and washed with 200 ml of methanol followed by 200 ml of distilled water in order to remove organic impurities, then the methanol, respectively, prior to enrichment. The sample loading was carried out at flow rates not exceeding 100 μl/min. The loaded precolumn was rinsed with 200 μl of distilled water followed by passage of air and was then connected to the separator column as shown in Fig. 24. The desorption and separation were done isocratically using methanol/methylene dichloride (5:1) as the mobile phase at a flow rate of between 4 and 7 μl/min.

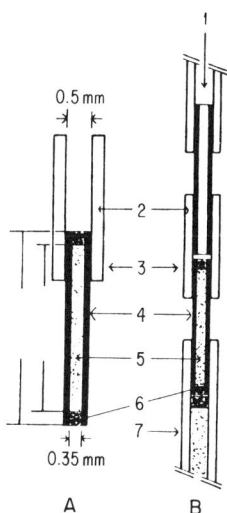

**Fig. 24.** Diagram of a microprecolumn (A) and its connection to the separation column (B): (1) carrier liquid; (2) Teflon tube for connection; (3) microprecolumn; (4) stainless-steel tubing; (5) packing material; (6) glass wool; (7) microseparation column (Ishii *et al.*, 1978).

Table V compares chromatographic results from the conventional (small volume) injection of phthalate esters (methanolic standard solutions) to results obtained using the microprecolumn enrichment technique. Capacity factors, HETP values, and resolution are compared and are shown to be essentially the same for enrichment volumes ranging from 67 to 5 ml compared to conventional 0.06-$\mu$l microinjection volumes of the standards.

The retention of diethyl phthalate (DEP) and dibutyl phthalate (DBP) versus various volumes of aqueous samples enriched on the microprecolumn are shown graphically in Fig. 25. The volumes of aqueous solutions enriched (10–100 $\mu$l) correspond to 100–1000 ng of the respective phthalate. The linearity of the data demonstrated that the esters were quantitatively retained by the microprecolumn up to 1000 ng (100-$\mu$l enrichment volumes).

The extraction efficiency of more dilute aqueous solutions of DEP was also investigated. Concentrations of 7.7, 82, and 330 ppb were enriched from 10-, 6-, and 1-ml aqueous volumes, respectively. Recovered phthalate was determined from the calibration curve (Fig. 25), and these results are compared in Table VI to the total calculated amount of DEP to which the precolumn was exposed. These results indicated that DEP was quantitatively enriched and retained by the precolumn to the 7.7-ppb level from injection volumes up to 10 ml.

Sep-Pak $C_{18}$ cartridges were shown to be useful for the enrichment of PCBs and the determination of very low concentrations of PCBs from water (Waters

## TABLE V
### Effect of Sample Enrichment with the Microprecolumn on Chromatographic Data[a]

| Test ester[b] | Conditions[c] | Sample concn in methanol (%) | Injection volume (μl) | Sample concn in water (ppm) | Volume enrichment injection (μl) | Capacity factor | HETP (mm) | Resolution[d] |
|---|---|---|---|---|---|---|---|---|
| DEP | A | 0.43 | 0.06 | — | — | 0.87 | 0.30 | 1.35 |
|     | B | —    | —    | 17 | 67 | 0.88 | 0.32 | 1.35 |
| DBP | A | 2.76 | 0.06 | — | — | 1.33 | 0.29 | — |
|     | B | —    | —    | 12 | 100 | 1.35 | 0.32 | — |
| DOP | A | 1.98 | 0.06 | — | — | 2.55 | 0.37 | 2.44 |
|     | B | —    | —    | 2 | 5000 | 2.56 | 0.34 | 2.46 |

[a] Ishii et al. (1978).
[b] DEP, diethyl phthalate; DBP, dibutyl phthalate; DOP, dioctyl phthalate.
[c] A, injection of methanol solution into the precolumn connected to the separation column; B, enrichment of the aqueous solution on the precolumn.
[d] Between DBP and the other esters.

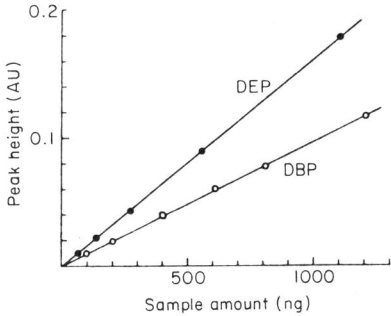

**Fig. 25.** Calibration graphs for DEP (17 ppm) and DBP (12 ppm) in water using the microprecolumn and microseparation column; mobile phase, methanol/dichloromethane (5:1); flow rate, 4.17 μl/min (Ishii et al., 1978).

Associates, 1978b). Figure 26 compares the enrichment chromatograms of 1.0-ppb aqueous solutions of Arochlor 1232 spiked into reverse-osmosis water (A), and potable water (C) after removal of interferences (B). One-liter volumes of 1.0-ppb-spiked aqueous solutions of the PCB were enriched on the Sep-Pak cartridges after first rinsing them with 2.0 ml of methanol, then 2.0 ml of organically pure water. A preliminary separation of PCB from interferences adsorbed during the enrichment step consisted of desorbing most using a 2.0-ml rinse of the loaded cartridge with 50% methanol. The PCB was retained and desorbed with 1.0 ml of a stronger mobile phase, i.e., methanol. Chromatogram B shows the interferences which would have coeluted with PCB to obscure the Arochlor pattern without a preliminary fractionation step. Since the reverse-osmosis water (A) contains very little organic interferences, the adsorbed components consisted almost exclusively of PCB. Most of the extraneous organic interferences in the potable water were removed by the elementary fractionation

**TABLE VI**

Efficiency of Retention of DEP on Microprecolumn from Dilute Aqueous Solution[a]

| DEP concn in water (ppb) | Sample volume for enrichment (ml) | Amount of DEP calculated from sample concn and volume (ng) | Amount of DEP determined from calibration graph (ng) |
|---|---|---|---|
| 330 | 1 | 330 | 330 |
| 82 | 6 | 492 | 469 |
| 7.7 | 10 | 77 | 83 |

[a] From Ishii et al. (1978).

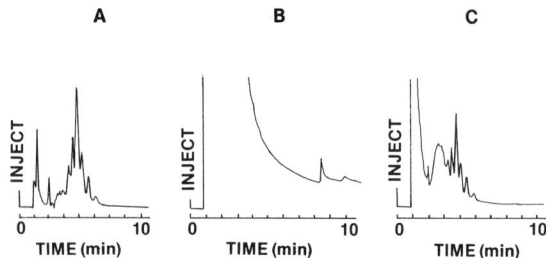

**Fig. 26.** Enrichment using Sep-Pak $C_{18}$ cartridges of 1.0-liter volumes of 1.0 ppb Arochlor 1232 in reverse-osmosis water (A) and potable water (C) after desorption from the cartridge of interferences present in the potable water (B) (Waters Associates, 1978b).

step. This allowed an easily recognizable PCB pattern in chromatogram C. The recovery of Arochlor 1232 from 1-liter water samples was 95 ± 5% for reverse-osmosis water and 105 ± 5% for potable water.

May *et al.* (1975) developed a headspace sampling procedure for volatile hydrocarbons in seawater followed by a "coupled-column" (off-line trace en-

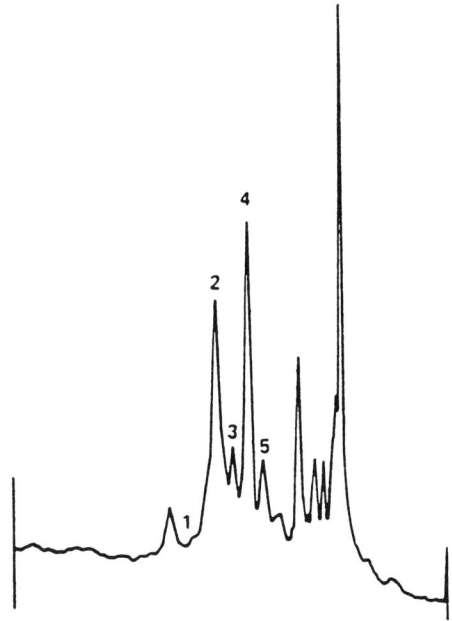

**Fig. 27.** Coupled-column liquid chromatogram from site of fuel-oil spill. Numbered peaks represent the internal standard consisting of (1) naphthalene, (2) phenanthrene, (3) pyrene, (4) benzo[*a*]pyrene, and (5) dibenzanthracene added at 20 ppb each (May *et al.*, 1975).

richment) technique for the nonvolatile hydrocarbon fraction. These trace-level techniques were developed in response to the need to analyze to the sub-ppb level for pollution baseline studies.

The concentrator column (6.0-mm i.d. × 6.5 cm) was packed with μBondapak $C_{18}$ (pellicular, 37–50 μm) and that portion of a 750-ml water sample remaining after headspace sampling was enriched at a flow rate of 10.0 ml/min. The loaded concentrator column was then connected ahead of a μBondapak $C_{18}$ column (3.9-mm i.d. × 25 cm) and desorbed by means of a gradient from (30 : 70) methanol/water to methanol in 40 min, at 3.0 ml/min flow rate, and with detection by UV absorption at 254 nm.

Figure 27 shows the liquid chromatogram resulting from the enrichment of a water sample taken from the site of a fuel-oil spill 6 weeks after cleanup. The numbered peaks are standards, each spiked at 20 ppb into the water sample.

In a similar application, Chesler *et al.* (1976) measured trace-level hydrocarbon concentrations in a baseline study of Prince William Sound and the Gulf of Alaska. Figures 28–30 show the chromatograms resulting from enrichment of Katalla River water, Old Valdez seawater, and Squirrel Bay seawater, respectively. All extraction and desorption procedures were identical to those of May *et al.* (1975). Only the Squirrel Bay sample is clean relative to the other two. This is consistent with what was known; the Katalla River is a documented oil-

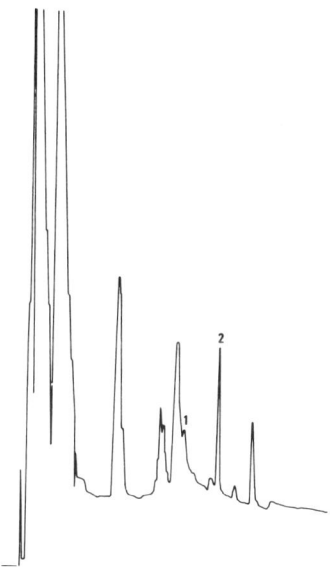

**Fig. 28.** Coupled-column LC analysis of Katalla River water. Peaks labeled (1) and (2) are the internal standard compounds phenanthrene and benzpyrene (Chesler *et al.*, 1976).

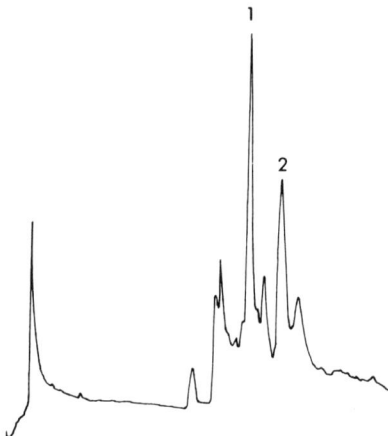

**Fig. 29.** Coupled-column LC analysis of Old Valdez seawater. Peaks labeled (1) and (2) are the internal standard compounds phenanthrene and benzpyrene (Chesler *et al.*, 1976).

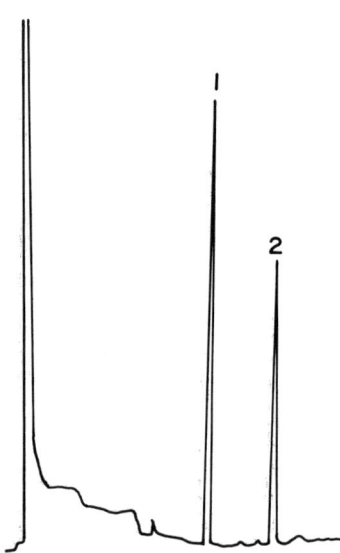

**Fig. 30.** Coupled-column LC analysis of Squirrel Bay seawater. Peaks labeled (1) and (2) are the internal standard compounds phenanthrene and benzpyrene (Chesler *et al.*, 1976).

seepage area and Old Valdez was the site of an oil spill during the 1964 earthquake. In contrast, Squirrel Bay appeared to be pristine.

The seawater-soluble fraction of a bunker-C oil was investigated by Saner *et al* (1979) using concentrator columns (2.1-mm i.d. × 7.0 cm) packed with CO:PELL ODS. The water-soluble fraction of the oil was obtained by carefully layering the oil onto the surface of the water contained in a 3-liter jug. The water was neither filtered before nor agitated during the oil/water contact time (1–35 days). A glass tube extended through the oil layer to the bottom of the jug, into which a Teflon line was inserted for withdrawal of water from beneath the oil layer for loading (9.9 ml/min flow rate) onto the concentrator columns. After loading, the concentrator columns were first rinsed with 15 ml of organically pure water and were then connected on-line ahead of the analytical column (E S Industries Chromegabond $C_{18}$, 4.6-mm i.d. × 15 cm). A two-segment gradient (both linear) from water to methanol effected separation at 1.0 ml/min flow rate. Figure 31 compares a 250-ml seawater blank and a system blank (no injection). The enrichment of 250 ml of seawater exposed to the bunker-C oil for 1 day is shown in Fig. 32. An additional enrichment chromatogram of the same oil-contaminated seawater is shown in Fig. 33; however, the loading volume was decreased to 25 ml, while the oil/water contact time increased to 22 days. The differences between chromatograms for 1- and 22-day oil/water contact times were primarily attributed to increased solubilization of oil components with time.

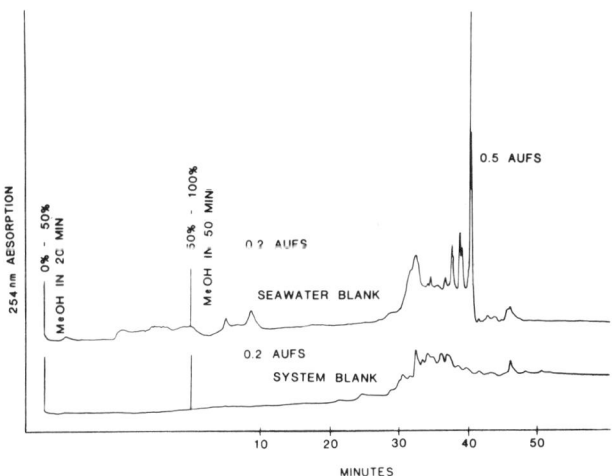

**Fig. 31.** Comparison of LC system blank and trace-enrichment chromatogram of 250 ml of uncontaminated seawater. [Reprinted with permission from Saner *et al.* (1979). *Anal. Chem.* **51**, 2180–2188. Copyright 1979, American Chemical Society.]

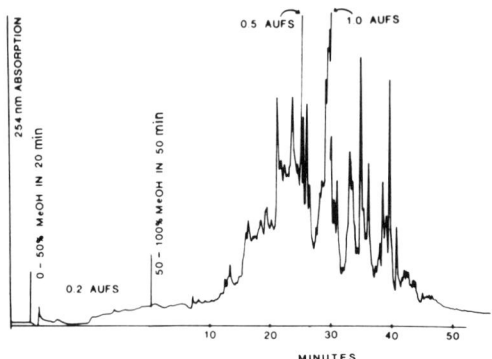

**Fig. 32.** Trace-enrichment chromatogram of 250 ml of Bunker-C-oil-contaminated seawater (24-hr oil/water contact time). [Reprinted with permission from Saner *et al.* (1979). *Anal. Chem.* **51**, 2180–2188. Copyright 1979, American Chemical Society.]

However, microbial action and extraction differences induced by sampling size differences were also recognized as possible contributory factors.

Hertz *et al.* (1976) used an essentially identical procedure as enumerated above (May *et al.*, 1975; Chesler *et al.*, 1976) for assessing oil spills quantitatively by off-line trace enrichment ("coupled-column" technique) of contaminated water. (All extraction and desorption procedures were, in fact, identical except that a water/acetonitrile gradient was used in place of a water/methanol one.)

The samples consisted of an oil/water emulsion (M-1); two sediments (S-1 and S-2); a control sediment from beyond the spill area (S-3); a homogenized tissue

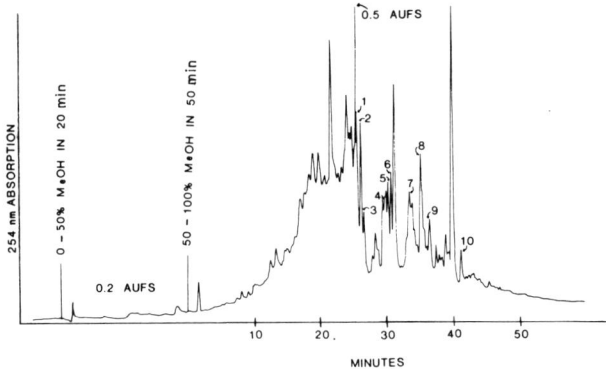

**Fig. 33.** Trace-enrichment chromatogram of 25 ml of Bunker-C-oil-contaminated seawater (3-week oil/water contact time). [Reprinted with permission from Saner *et al.* (1979). *Anal. Chem.* **51**, 2180–2188. Copyright 1979, American Chemical Society.]

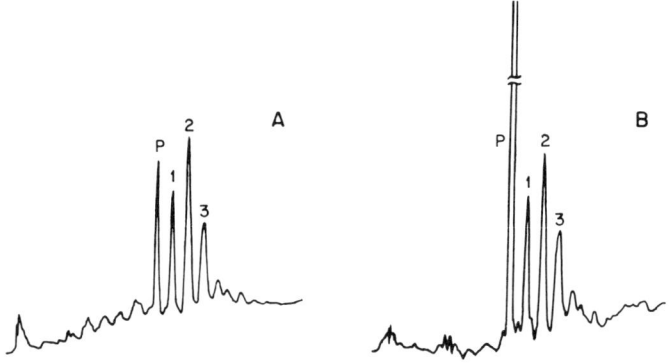

**Fig. 34.** Liquid chromatograms from coupled-column LC following headspace sampling: (A) sediment sample S-1; (B) unweathered spill oil. Peak labeled (P) is internal standard phenanthrene; (1), (2), and (3) are aromatic hydrocarbons used for quantitation (Hertz et al., 1976).

sample (T-1); and a sample of the original spill oil (a light Saudi Arabian crude). About 750 ml of the emulsion were enriched, whereas the crude-oil sample, the tissue sample, and the sediment samples were each mixed with separate volumes of 600 ml of hydrocarbon-free water and were then extracted as for water. In addition, Soxhlet extractions were performed for comparison purposes.

Figure 34 shows the liquid chromatograms resulting from the enrichment of water in contact with sediment S-1 and of water in contact with the crude oil. (During the 4-hr oil/water contact time, a purge-and-trap procedure of volatiles for analysis by GC was carried out.)

Table VII compares the hydrocarbon content of the various samples as deter-

**TABLE VII**

**Hydrocarbon Content of Samples Obtained from Catastrophic Oil Spill**[a]

| | | Percentage hydrocarbons by various methods[b] | | |
|---|---|---|---|---|
| Sample | Type | Soxhlet extraction corrected for water content (Karl Fischer) | GC | Coupled-column LC |
| M-1 | Oil/water emulsion | 67(1) | 50 ± 20(4) | 80 ± 12(3) |
| S-1 | Sediment | 0.2(1) | 0.2 ± 0.06(3) | 0.7 ± 0.07(2) |
| S-2 | Sediment | 16(1) | 9 ± 3(3) | 10 ± 1(2) |
| S-3 | Control sediment | Trace(1) | 0.0002 ± 0.00004(3) | <0.0002(2) |
| T-1 | Tissue | 7.7(1) | 0.03 ± 0.01(3) | 1.8 ± 0.2(2) |

[a] From Hertz et al. (1976).
[b] Values in parentheses indicate number of separate determinations.

mined by Soxhlet extraction and GC (purge-and-trap) and LC (off-line trace enrichment). The various techniques all produced generally comparable results (except for T-1), particularly when viewed in light of sample inhomogeneity and the small sample volumes taken (due to their high petroleum contents). The tissue sample (T-1) gave high Soxhlet results since it contained large amounts of biogenic extractable compounds of nonpetroleum origin. Since petroleum-derived aromatics are bioaccumulated by marine bivalves, the higher concentration of petroleum hydrocarbons in T-1 determined by LC relative to the comparable GC determination is not surprising. In addition, since marine organisms can depurate aliphatic compounds, this would tend to lower GC results for this sample.

May et al. (1978) determined the solubilities of various aromatic hydrocarbons in distilled water and seawater using the same off-line enrichment procedure as described before (Hertz et al., 1976). Saturated aqueous solutions of various aromatics were enriched onto the concentrator column ($\mu$Bondapak $C_{18}$, 6.0-mm i.d. $\times$ 6.0 cm) by pumping water through a "generator" column containing glass beads (60–80 mesh). These beads were coated to 1% by weight with the various aromatics of interest. The adsorbed aromatics were desorbed from the concentrator column, using a water/acetonitrile gradient, onto a $\mu$Bondapak $C_{18}$ separator column (3.9-mm i.d. $\times$ 25 cm).

The concentrator column provided better than 99% extraction efficiency for volumes of 25 ml or less of the aqueous aromatic solutions. Benzene, however, was not quantitatively recovered, so instead the saturated benzene solution was pumped into a stainless-steel loop for direct injection of discrete volumes onto the separator columns. The aqueous solubilities determined for the various aromatics investigated using the above-described generator/concentrator/separator-column system are listed in Table VIII.

The aromatic solubilities ranged from 1791 ppm (mg/kg) for benzene to 1.8 ppb ($\mu$g/kg) for chrysene. These values compared well with the consensus-literature values, except for anthracene. This, however, could stem from the fact that commercial-grade anthracene commonly contains $\sim$20% phenanthrene. This contaminant has roughly 20 times greater water solubility than does anthracene. Only a selective analytical technique which can differentiate these isomers, such as fluorescence, would come into agreement with the coupled-column technique described here.

Polycyclic aromatic hydrocarbons (PAHs) and their chlorination reaction products in water at the ppt to ppb levels were determined by Oyler et al. (1978) using an off-line enrichment technique.

The concentrator column (7.0-mm i.d. $\times$ 5.0 cm) was packed with $\mu$Bondapak $C_{18}$ Corasil or Porasil B (37–50 $\mu$m). A stainless-steel filter holder (47 mm) was used with glass-microfiber filters in-line ahead of the concentrator column in order to remove particulates from environmental samples. Since the

## TABLE VIII

### Aqueous Solubilities of Some Aromatic Hydrocarbons As Determined by Several Investigators[a,b]

| Compound | MW[c] | This work | | Davis (1942) 29°C | 25°C | | | |
| --- | --- | --- | --- | --- | --- | --- | --- | --- |
| | | 25°C | 29°C | | MacKay and Shui (1977) | Schwarz (1977) | Wauchope and Getzen (1972) | Others |
| Benzene | 78.1 | 1791 ± 10 | | | | | | 1780, 1755, 1755, 1796 |
| Naphthalene | 128.2 | 31.69 ± 0.23 | | | 31.7 ± 0.2 | 30.3 ± 0.3 | 31.2 | 34.4 |
| Fluorene | 166.2 | 1.685 ± 0.005 | | | 1.98 ± 0.04 | | 1.90 | |
| Anthracene | 178.2 | 0.0446 ± 0.0002 | 0.0570 ± 0.003 | 0.075 ± 0.005 | 0.073 ± 0.005 | 0.041 ± 0.0003 | 0.075 | 0.075 |
| Phenanthrene | 178.2 | 1.C02 ± 0.011 | 1.220 ± 0.013 | 1.600 ± 0.050 | 1.290 ± 0.070 | 1.151 ± 0.015 | 1.180 | 0.994 |
| 2-Methylanthracene | 192.3 | 0.0213 ± 0.003 | | | | | | |
| 1-Methylphenanthrene | 192.3 | 0.269 ± 0.003 | | | | | | |
| Fluoranthene | 202.3 | 0.206 ± 0.002 | 0.264 ± 0.002 | 0.240 ± 0.020 | 0.260 ± 0.020 | | 0.265 | 0.240 |
| Pyrene | 202.3 | 0.132 ± 0.001 | 0.162 ± 0.001 | 0.165 ± 0.007 | 0.135 ± 0.005 | 0.129 ± 0.002 | 0.148 | |
| 1,2-Benzanthracene | 228.3 | 0.0C94 ± 0.0001 | 0.0122 ± 0.0001 | 0.011 ± 0.001 | 0.014 ± 0.0002 | | | 0.010 |
| Chrysene | 228.3 | 0.0C18 ± 0.0001 | 0.0022 ± 0.0001 | 0.0015 ± 0.0004 | 0.002 ± 0.0002 | | | 0.006 |

[a] Reprinted with permission from May et al. (1978). Anal. Chem. **50**, 997–1000. Copyright 1978, American Chemical Society.
[b] Solubilities in mg/kg.
[c] MW, molecular weight.

## TABLE IX

**Determination of PAH Recovery from Aqueous Solutions of Known Concentrations Using the $C_{18}$ Adsorption-HPLC-GC Procedure[a,b]**

| Experiment no. | 1 | 2 | 3 | 4 | 5 | 6 | 7 | 8 | 9 | 10 |
|---|---|---|---|---|---|---|---|---|---|---|
| Concentration of each PAH (ng/liter) | 80 | 80 | 80 | 22 | 80 | 80 | 107 | 600 | 1000 | 965 |
| Number of liters forced through $C_{18}$ column | 14.5 | 14.3 | 14.2 | 18.2 | 14.0 | 12.2 | 13.8 | 6.4 | 6.2 | 10.2 |
| Number of μg of each PAH for quantitative recovery | 1.16 | 1.14 | 1.14 | 0.392 | 1.12 | 0.978 | 1.48 | 3.84 | 6.20 | 9.84 |
| Glass-fiber filter | None | S&S 29 | S&S 30 | S&S 30 | Whatman GF/B | Whatman GF/F | Whatman GF/F | Whatman GF/F | None | None |
| Type $C_{18}$ packing[c] | WC | WC | WC | WC | WC | WC | WC | WC | WP | WP |
| 1-Methyl-naphthalene | — | — | — | — | — | — | — | — | — | — |
| Fluorene | 76 ± 3 | 64 ± 6 | 77 ± 11 | 42 ± 8 | 72 ± 18 | 70 ± 20 | 69 ± 4 | 93 ± 16 | 89 ± 13 | 76 ± 10 |
| Dibenzofuran | — | — | — | — | — | — | — | 83 ± 10 | 80 ± 8 | — |
| 1-Methyl-4-chloronaphthalene | — | — | — | — | — | — | — | — | — | 95 ± 2 |
| Phenanthrene | 87 ± 3 | 101 ± 2 | 87 ± 2 | 105 ± 21 | 85 ± 2 | 90 ± 4 | — | — | — | — |
| 2-Methylanthracene | 81 ± 24 | 64 ± 5 | 72 ± 11 | 115 ± 18 | 61 ± 9 | 87 ± 13 | 74 ± 2 | — | — | — |
| 1-Methylphenanthrene | 87 ± 10 | 78 ± 13 | 82 ± 13 | 99 ± 68 | 87 ± 22 | 111 ± 12 | 86 ± 10 | — | — | — |
| 9-Methylanthracene | 103 ± 21 | 92 ± 13 | 87 ± 24 | — | 101 ± 17 | 110 ± 18 | 106 ± 12 | — | — | — |
| Fluoranthene | 96 ± 18 | 101 ± 16 | 87 ± 20 | 81 ± 5 | 67 ± 4 | 76 ± 9 | 88 ± 14 | — | — | — |
| Pyrene | 65 ± 16 | 84 ± 28 | 79 ± 35 | 92 ± 55 | 36 ± 10 | 100 ± 11 | 106 ± 14 | — | — | — |
| 9-Chlorophenanthrene | — | — | — | — | — | — | — | 88 ± 7 | 86 ± 5 | — |

[a] Reprinted with permission from Oyler et al. (1978). Anal. Chem. **50**, 837–842. Copyright 1978, American Chemical Society.
[b] The errors given in this table reflect the standard deviation in the determination of the weight of material injected into the GC plus the error involved in the measurement of the volumes of the HPLC fractions.
[c] WC, Waters Corasil II; WP, Waters Porasil B.

sampling volumes were very large (from 6.2 to 18.2 liters), stainless-steel tanks of 19.5-liter volume were pressurized to 125 psi with high-purity nitrogen in order to force the water sample through the filter and concentrator column. Aqueous solutions of known PAH concentrations were prepared by adding different amounts of PAH stock solutions in acetonitrile to the water contained in the pressure tanks. The filter assembly and concentrator column were connected to the outlet of the tank, and the pressure was adjusted to cause a flow of from 20 to 25 ml/min. The loaded concentrator column was then positioned on-line ahead of the analytical $C_{18}$ reverse-phase column (3.9-mm i.d. × 30 cm, or 3.2-mm i.d. × 25 cm; 10-μm particle) and desorbed using a linear gradient from 50 to 90% acetonitrile/water in 30 min at a flow rate of 1.2 ml/min. Detection was by UV absorption at 254 and 280 nm. Hand-collected LC peaks were analyzed by GC and gas chromatography/mass spectroscopy (GCMS).

Table IX lists the recoveries for various PAH concentrations, sampling volumes, and in-line fiber filters. The results indicate that submicron filters (0.7 μm), such as Whatman GF/F, have no deleterious effect on PAH recovery. However, the 1.0-μm Whatman GF/B filter did lower the recovery of some of the PAHs.

The various chlorination reaction products were generated by exposing the aqueous PAH solutions contained in the pressure tanks to sodium hypochlorite and sulfuric acid (to attain different pH's). These chlorination products are listed in Table X and were identified by GCMS, primarily from methylene chloride extracts of the individual LC fractions. Table X illustrates the tendency of the PAHs to undergo chlorination reactions under conditions typical of those encountered during disinfection processes. This tendency is particularly enhanced at low pH.

## C. Column Switching, On-Line

Ogan, *et al.* (1978) mounted a concentrator column (2.6-mm i.d. × 10 cm), packed with pellicular $C_{18}$ SIL-X-II (40 μm), in place of the loop in an injection valve in order to determine trace amounts of aqueous aromatic hydrocarbons. Figure 35 shows a schematic of the extraction column in the loop injector. This particular arrangement utilizes two pumps, one to apply sample to the concentrator column and one to desorb the concentrator onto the analytical column to effect separation. A mobile phase comprising 38:15:47 acetonitrile/methanol/ water was used at 1 ml/min together with fluorescence detection ($\lambda_{ex}$ = 365 nm, $\lambda_{em}$ = 455 nm, where $\lambda_{ex}$ and $\lambda_{em}$ are the excitation and emission wavelengths, respectively). The analytical column was packed with 10-μm $C_{18}$ reverse-phase.

Although only some 80 ml of sample are actually enriched, about 500 ml of sample are required in order to saturate all active sites in the chromatograph itself with PAHs prior to the enrichment step. Methanol was added to the aqueous

## TABLE X
### Summary of Aqueous Chlorination Studies[a]

| Starting PAH | $Cl_2$ (mg/liter) | PAH (ng/liter) | Reaction time (hr) | pH | Products identified (% yield) |
|---|---|---|---|---|---|
| 1-Methylnaphthalene | 24.0 | 531 | 3.0 | 3.8 | Monochloro-1-methylnaphthalene |
|  | 20.4 | 336 | 3.0 | 4.1 | Monochloro-1-methylnaphthalene (73 ± 5) |
| Fluorene | 1.205 | 334 | 0.5 | 7.0 | Fluorene (73 ± 4) |
|  | 18.8 | 819 | 3.0 | 4.1 | Fluorene |
|  | 23.5 | 773 | 3.0 | 3.4 | Fluorene, monochlorofluorene |
|  | 21.3 | 333 | 3.0 | 3.35 | Fluorene (~5), monochlorofluorene (52 ± 4) |
| Anthracene | 12.9 | 1000 | 3.7 | 4.0 | Anthracene (3) |
|  |  |  |  |  | Anthraquinone (90 ± 11) |
|  | 0.0 | 1042 | 4.0 | 4.4 | Anthracene |
|  | 12.4 | 965 | 3.75 | 6.5 | Anthraquinone (78 ± 9) |
| Phenanthrene | 2.0 | 552 | 0.08 | 7.1 | Anthraquinone (61 ± 16) |
|  | 3.2 | 820 | 0.5 | 7.1 | Phenanthrene |
|  | 3.7 | 236 | 0.5 | 6.8 | Phenanthrene (77 ± 14) |
|  | 26.3 | 233 | 3.0 | 6.0 | Phenanthrene (86 ± 4), monochlorophenanthrene (4 ± 1) |
|  | 19.3 | 820 | 3.0 | 4.1 | Monochlorophenanthrene |
|  | 19.5 | 239 | 3.0 | 4.2 | Phenanthrene (9 ± 4), monochlorophenanthrene (38 ± 5) |
|  | 20.0 | 118 | 3.0 | 4.05 | Monochlorophenanthrene (39 ± 5) |
| 1-Methylphenanthrene | 3.1 | 925 | 0.5 | 6.9 | 1-Methylphenanthrene |
|  | 21.0 | 994 | 3.0 | 4.0 | Monochloro-1-methylphenanthrene |
|  | 25.6 | 178 | 3.0 | 4.0 | Monochloro-1-methylphenanthrene (~8) |
| Fluoranthene | 3.4 | 824 | 0.5 | 6.8 | Fluoranthene |
|  | 22.0 | 239 | 3.0 | 5.9 | Fluoranthene (63 ± 3) |
|  | 17.7 | 824 | 3.0 | 4.1 | Fluoranthene chlorohydrin |
|  | 23.9 | 239 | 3.0 | 4.03 | Fluoranthene (42 ± 3), monochlorofluoranthene (32 ± 1) |

[a] Reprinted with permission from Oyler et al. (1978). Anal. Chem. **50**, 837–842. Copyright 1978, American Chemical Society.

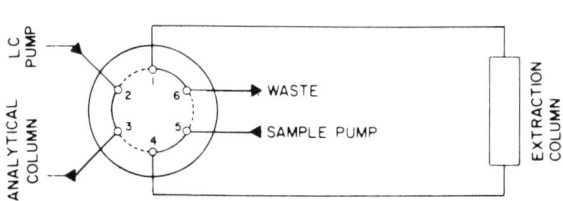

**Fig. 35.** Connection of the extraction column to the loop-injection valve. The solvent path is indicated for the valve in the Fill position (—) and the Inject position (- - -). [Reproduced from Ogan et al. (1978). *J. Chromatogr. Sci.*; by permission, Preston Publications, Inc.]

samples to a final concentration of 20%. This was done in order to improve the solubilities of the PAHs in the aqueous solution so that adsorption of PAHs onto glass and metal chromatographic surfaces would be decreased. Furthermore, the addition of methanol (also present at 15% in the mobile phase) may have enhanced the interaction of PAHs with the concentrator column packing by causing its hydrophobic surface to become wetted.

In order to ensure saturation by the sample solution of all active sites within the sample-pump surfaces, 400 ml of the sample were first pumped through it with the injection valve in the inject position. (In this configuration, sample goes directly to waste without passage through the concentrator column.) Afterward, the valve was turned to the fill position and 100 ml (80 ml of actual aqueous sample) were pumped through the concentrator column. The sample pump was

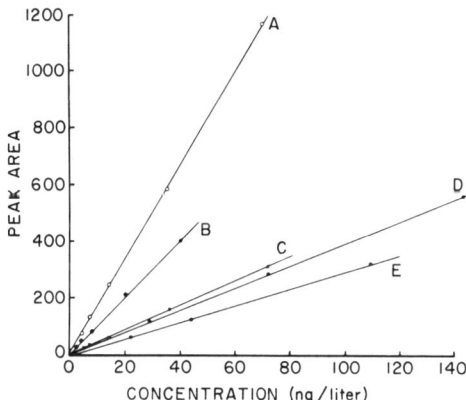

**Fig. 36.** Peak area versus PAH concentration for 100-ml samples pumped through the extraction column: A, benzo[*a*]pyrene; B, perylene; C, fluoranthene; D, benzo[*ghi*]perylene; E, 7H-benz-[*de*]anthracen-7-one. [Reproduced from Ogan *et al.* (1978). *J. Chromatogr. Sci.*, by permission, Preston Publications, Inc.]

then stopped and the valve rotated to inject, whereby the mobile phase from the second pump back-flushed the concentrator column onto the analytical column.

Plots of peak area versus concentration for each of five different PAHs in 100-ml sampling volumes are shown in Fig. 36. Excellent linearity for all five compounds is obvious.

Figure 37 plots peak area versus sampling volumes. [Concentrations ranged from 44 to 8 ppt (ng/liter) and were different for each PAH, but were held constant for each PAH.] The linearity in response for all PAHs, except benzanthrone, indicated that these PAHs can be recovered quantitatively from sampling volumes at least up to 400 ml. However, there was evidently breakthrough for benzanthrone for a 400-ml sampling volume.

Eisenbeiss et al. (1978) also studied the on-line enrichment of various PAHs (listed in Table XI) in a manner very similar to that used later by Ogan et al. (1978), described above. The analytical column was packed with LiChrosorb RP-18 (5 μm); the concentrator column (4.0-mm i.d. × 5 cm) was packed with a

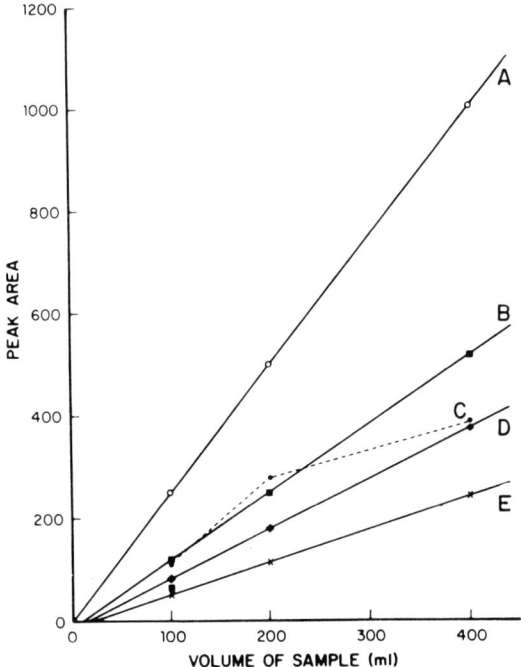

**Fig. 37.** Peak area versus sample volume for the aqueous concentrations of: A, benzo[a]pyrene (14 ppt); B, benzo[ghi]perylene (29 ppt); C, 7H-benz[de]anthracen-7-one (44 ppt); D, perylene (8 ppt); E, fluoranthene (14 ppt). [Reproduced from Ogan et al. (1978). J. Chromatogr. Sci., by permission, Preston Publications, Inc.]

"special" Merck packing and was mounted in place of the loop on a Valco injector. The enrichment was carried out at 4.0 ml/min after first diluting the samples with isopropanol to a concentration between 17.5 and 20% alcohol. This was added in order to keep the concentrator packing wetted during the enrichment step. After completion of sample loading, the injection valve was moved from position 1 to position 2, which inserted the precolumn in-line ahead of the analytical column. Desorption and back-flushing of the concentrator column were carried out by means of a second pump with an isocratic mobile phase of 85% methanol/water. Fluorescence was used for detection with $\lambda_{ex} = 360$ nm (or 367) and $\lambda_{em} = 470$ nm.

A 10-μl aliquot of a standard PAH mixture was used to spike a 500-ml aqueous sample (containing isopropanol at 17.5%). This produced a resultant solution of 200 ppb fluoranthene and 40 ppb each of the remaining PAHs listed in Table XI. The solution was enriched, then the injection valve was moved to position 2. The resultant chromatogram is shown in Fig. 38. A 10-μl injection of the same PAH standard used to spike the water sample was directly injected for comparison purposes (Fig. 39). The recovery efficiencies are the ratios of the peak heights in these two chromatograms; they are listed in Table XI.

Euston and Baker (1979) described an automated system for the concentration

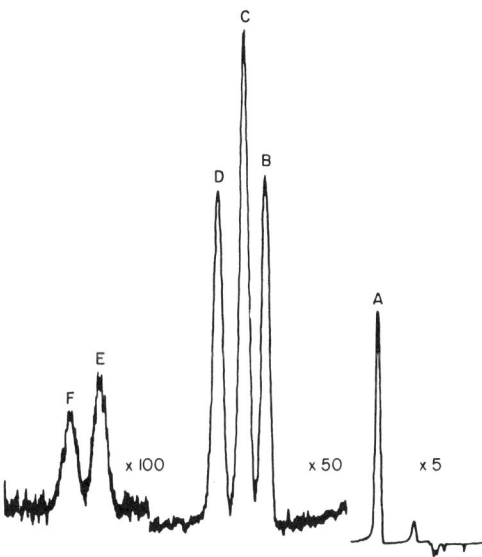

**Fig. 38.** Chromatogram resulting from enrichment of 500 ml of 17.5% isopropanol/water solution containing the six labeled PAHs (listed in Table XI) (for PAH concentrations see text). Detection by fluorescence: $\lambda_{em} = 470$ nm; $\lambda_{ex} = 367$ nm (Eisenbeiss et al., 1978).

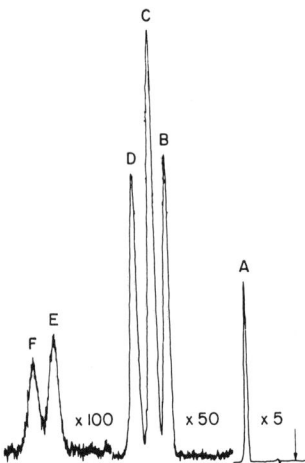

**Fig. 39.** Chromatogram of a direct injection of 10 μl of the PAH standard solution with the conditions as in Fig. 38 (see Table XI) (Eisenbeiss et al., 1978).

of various PAHs from dilute solutions. The concentrator column (2.1-mm i.d. × 5 cm) was packed with Perisorb RP; the analytical column (4.6-mm i.d. × 25 cm) was packed with LiChrosorb RP-18 (5 μm). The columns were mounted onto an automated six-port valve in the same manner as described earlier by Ogan et al. (1978) and shown here in Fig. 35. After loading, the precolumn was back-flushed onto the analytical column and separation was effected using an isocratic mobile phase consisting of 82.5% acetonitrile in water. Up to 100-ml sample volumes were enriched, containing 0.25 ppb of each of 9 different PAHs. The repeatability of retention time was excellent (~0.08% RSD); the precision

**TABLE XI**

**Recovery for the Enrichment Relative to Direct Injection[a]**

| Curve label[b] | PAH compound | Recovery rate (%) |
|---|---|---|
| A | Fluoranthene | 100 |
| B | Benzo[b]fluoranthene | 85 |
| C | Benzo[k]fluoranthene | 75 |
| D | Benzo[a]pyrene | 76 |
| E | Benzo[ghi]perylene | 86 |
| F | Indeno[1,2,3-cd]pyrene | 85 |

[a] Eisenbeiss et al. (1978).
[b] Figures 38 and 39.

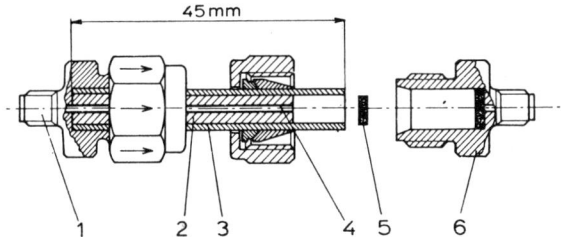

**Fig. 40.** Precolumn design: (1) Swagelok, ¼ × ¹⁄₁₆ in.; (2) variable-length PTFE rod, 4.6-mm o.d.; (3) stainless-steel tube, ¼ in. × 45 mm; (4) stainless-steel capillary, ¹⁄₁₆-in. o.d., 0.25-mm i.d.; (5) PTFE or stainless-steel 20-μm frit; (6) Swagelok, ¼ × ¹⁄₁₆ in. (Van Vliet et al., 1979).

of the raw area measurement ranged from 3.0% RSD (phenanthrene) to 9.8% RSD (chrysene).

Preliminary experiments to investigate such basic parameters as precolumn (concentrator) design (length), particle size of precolumn adsorbent, and pumping rate during the loading process were conducted by Van Vliet et al. (1979). PCBs (Arochlor 1254 and 1260) and phthalate esters [di-n-butyl (DBP) and di-(2-ethylhexyl) (DEHP)] were used as model nonpolar compounds and chloroanilines (4-mono- and 3,4-dichloro) were selected as a more polar model-compound type.

The precolumn, shown in Fig. 40, consisted of stainless-steel tubing (4.6-mm i.d. × 4.5 cm) into which a variable-length PTFE rod with an outer diameter of 4.6 mm could snugly fit. (Sometimes 2.9-mm-i.d. stainless-steel tubes were used with 2.9-mm-diameter PTFE rods.) The length of the plug of stationary phase was varied by varying the length of the PTFE rod, which was drilled down its long axis to accept a piece of stainless-steel capillary tubing (¹⁄₁₆-in. o.d. × 0.25-mm i.d.), which held the packing material. Either a 0.5- or 2.0-μm outlet frit and a 20-μm inlet frit with conventional Swagelok fittings completed the precolumn assembly. 5-, 10-, or 50-μm RP-18 packing materials were used to fill the precolumns.

The separation column (4.6-mm i.d. × 12.5 cm) was packed with LiChrosorb RP-18 (5 μm). A valve was mounted between the separator column and the precolumn so that enrichment occurred on the precolumn with the valve in the drain position; desorption and separation of the sample occurred by rotating the valve to the on-column position. A step gradient from 75 to 85 to 95% methanol/water at a flow rate of 2.0 ml/min effected separation.

The dependence of precolumn performance on the particle size of the stationary phase is demonstrated in Table XII using 5-, 10-, and 50-μm (and larger) packings in 4.6-mm i.d. × 2-mm precolumns. Replacing the 5-μm material with the 10-μm material resulted in slight deterioration; however, the 50-μm material (used in Sep-Pak cartridges) resulted in very poor performance.

## TABLE XII

**Dependence of Precolumn Performance on Particle Size of Stationary Phase**[a,b]

| Stationary phase | Particle size ($\mu$m) | Peak height (mm/$\mu$g) | | n |
|---|---|---|---|---|
| | | DBP | DEHP | |
| LiChrosorb RP-18 | 5 | 98.3 | 33.4 | 9 |
| | 10 | 92.3 | 31.3 | 6 |
| $C_{18}$ (Sep-Pak) | >50 | 74.5 | 15.0 | 7 |

[a] Van Vliet et al. (1979).
[b] Conditions: trace enrichment on a 2 × 4.6-mm i.d. precolumn from ~20-ml sample volumes containing 50 ng of each phthalate per milliliter.

Band broadening due to the insertion of a 2-mm-long precolumn packed with a plug of 5-$\mu$m LiChrosorb RP-18 resulted in only 3% increase compared to a 10-$\mu$l injection without it. Utilizing this size precolumn for all additional work, various sample-loading rates were investigated next. The adsorption efficiency of the concentrator column remained constant over the loading-rate range specified in Table XIII.

Finally, storage of loaded precolumns from 18 to 60 hr did not seriously detract from performance. Both phtalates used as test adsorbates showed between 0 and 7% losses ($n$ = 10).

The migration of DEHP and DBP from PVC medical foil into demineralized water with time was studied by exposing 10 g of PVC to 1 liter of water. Sampling volumes enriched ranged from 225 to 925 ml. The results are listed in Table XIV and show a general increase in phthalate concentration in the water with increasing exposure time.

## TABLE XIII

**Dependence of Preconcentration Efficiency on Pumping Speed**[a,b]

| Pumping speed (ml/min) | Peak height (mm/$\mu$g) | |
|---|---|---|
| | DBP | DEHP |
| 5.0 | 78.7 | 28.7 |
| 12.5 | 68.0 | 28.4 |
| 25.0 | 77.5 | 28.7 |

[a] Van Vliet et al. (1979).
[b] Conditions: trace enrichment from 25–100-ml sample volumes containing 17 ng of each phthalate per milliliter.

TABLE XIV

Migration of Phthalate Esters from Poly(vinyl chloride) into Demineralized Water[a,b]

| Contact time (hr) | Preconcentrated volume (ml) | Phthalate concn (ppb) | |
|---|---|---|---|
| | | DEHP | DBP |
| 0 | 925 | <0.09 | <0.04 |
| 1 | 425 | 0.2 | <0.07 |
| 5 | 425 | 1.8 | 0.2 |
| 24 | 375 | — | 0.5 |
| 26 | 350 | 2.0 | 0.3 |
| 96 | 350 | 3.2 | 0.5 |
| 600 | 225 | 2.4 | — |

[a] Van Vliet et al. (1979).
[b] Conditions: demineralized water contacted with 10 g poly(vinyl chloride) per liter water.

Using a loading rate of 1.5 ml/min, 4-mono- and 3,4-dichloroaniline were also enriched. Desorption and separation were done isocratically at 45% methanol/water, detection by UV absorption at 243 nm. These fairly polar compounds with low capacity factors demonstrated relatively rapid breakthrough on the precolumn (caused by passage of too large a sample volume). The effects of enriching various volumes (from 2 to 20 ml) are shown in Fig. 41 (chromatograms A and B). In addition, chromatogram C in the same figure shows the linearity of buildup on the precolumn from different sample concentrations when the sampling volume is kept small (2 ml) so that breakthrough does not occur.

Electrochemical detection was used for the determination of PCP enriched from river water (Mayer and Shoup, 1981). The concentrator column (4.0-mm i.d. × 3 cm) was packed with 5-μm BAS Biophase (RP-18) material and mounted in place of the loop on a pneumatically controlled injection valve. The analytical column was also packed with the same BAS material as the concentrator column.

River-water samples were loaded at 3 ml/min for 6 min with the valve in the load position. Upon switching to inject, the mobile phase flushed the adsorbates onto the analytical column for 2 min. This procedure desorbed the precolumn and allowed another sample to be enriched while the previous sample was being chromatographed, by returning the injector to the load position.

The mobile phase consisted of 40% acetonitrile, 60% 0.2 $M$ NaClO$_4$/0.005 $M$ citrate at pH 5.0. (At this particular pH, PCP is partially ionized and its ensuing capacity factor is purposely reduced, allowing for faster turnaround times with a PCP retention time of 16 min.) An example of both UV absorption at 254 nm and electrochemical detection is shown in Fig. 42 of the chromatogram obtained from enriching 20 ml of a 13 ppb PCP-spiked river-water sample.

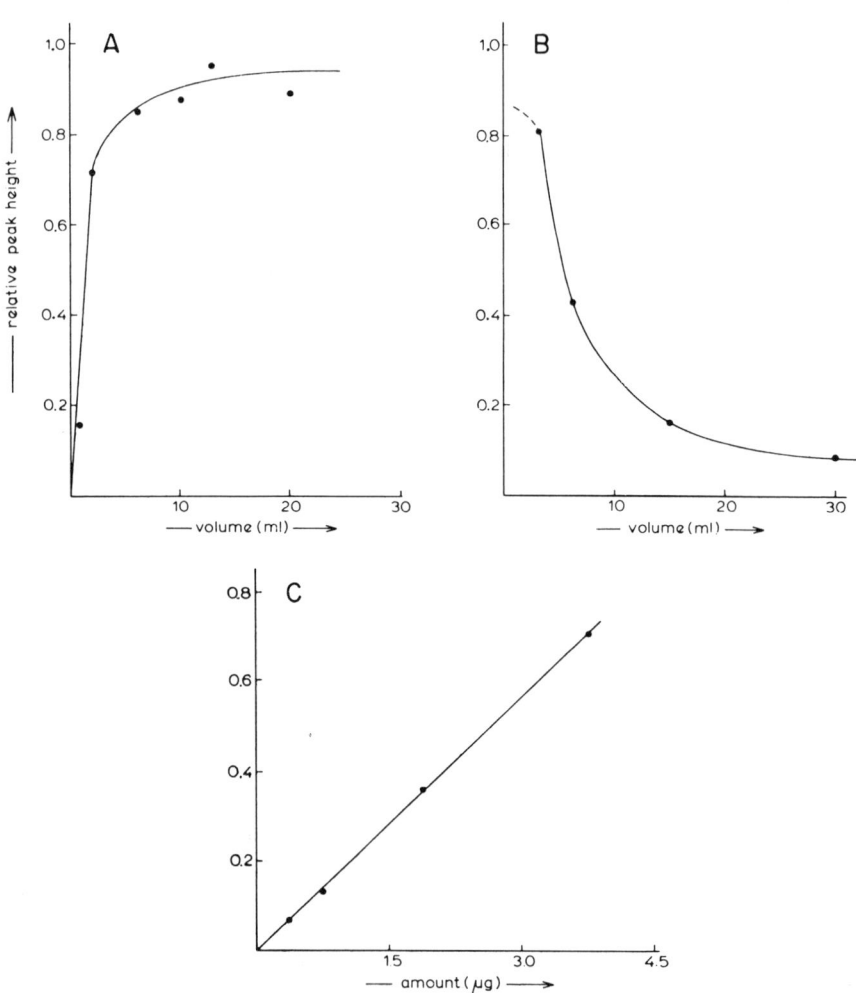

**Fig. 41.** Trace enrichment of 4-monochloroaniline: (A) from 2- to 20-ml sample volumes having identical concentrations; (B) from different sample volumes containing equal amounts of the analyte; (C) calibration curve recorded after trace enrichment from 2-ml volumes having different concentrations (Van Vliet et al., 1979).

Werkhoven-Goewie et al. (1981) also investigated the on-line enrichment of chlorinated phenols on short precolumns using various reverse-phase $C_{18}$ (and one $C_{22}$) packings and carbonaceous materials (including pyrocarbon-modified silica, carbon black, and activated carbon). These investigators sought to predict the breakthrough volume (maximum enrichment factor) utilizing, as model compounds, phenols with varying degrees of chlorination (mono- through pentachlorophenol).

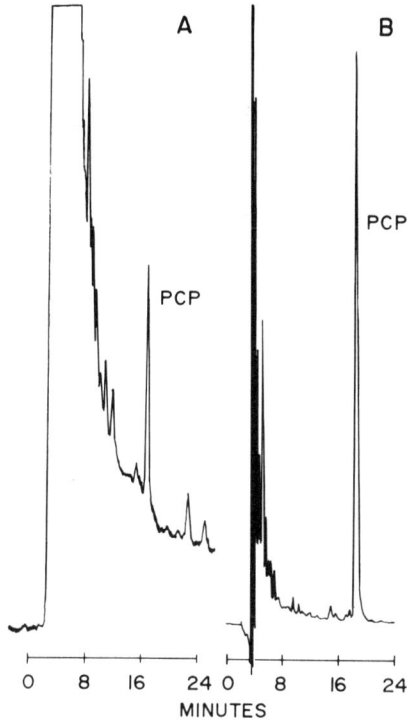

**Fig. 42.** Both UV (A) and amperometric (B) detectors in tandem following enrichment of 20 ml of 13 ppb PCP-spiked Wabash River water: (A) 254 nm, 0.1 AUFS; (B) 0.9 V, 50 nA (Mayer and Shoup, 1981).

For the breakthrough study, variable-length precolumns (4.6-mm i.d.) were mounted in the loop position of a six-port valve (no analytical column was used). Breakthrough was monitored by UV absorption at 220 nm, using loading rates of 5 or 20 ml/min. Aqueous sample solutions of chlorinated phenols (100 ppb) were first pumped through the chromatographic system, straight to the detector, with the valve in the bypass position, in order to permit complete filling (and active-site saturation) of the pump with the various sampling solutions prior to the enrichment step. After a steady baseline was observed on the detector, the valve was moved to the inject position to allow enrichment on the precolumn to occur. It was necessary to add phosphoric acid to the tetra- and pentachlorophenol sample solutions in order to suppress their ionization (increase $k'$). (No effect of acid addition on $k'$ values was demonstrated for the less acidic mono-, di-, and trisubstituted chlorophenols.)

Utilizing an analytical column (5.0-mm i.d. × 10 cm) packed with Hypersil ODS (5 μm) in conjunction with a 4.6-mm i.d. × 2.2-mm precolumn packed with LiChrosorb RP-18 (5 μm), no noticeable band broadening occurred for the on-line enrichment of 150 ml of a solution of acidified distilled water containing tri-, tetra-, and pentachlorophenols (10 ppb each) relative to an 18-µl injection (containing 500 ng of each chlorophenol). For pentachlorophenol, the enrichment of 250 ml of an aqueous sample solution resulted in a peak width the same as that obtained with a 100-µl loop injection.

Enrichment volumes from 50 to 250 ml and UV-absorption detection at 220 nm provided detection limits (for tri-, tetra-, and pentachlorophenols) between 0.05 and 3.0 ppb.

## III. CLINICAL APPLICATIONS

### A. Direct Injection

Frei (1978) investigated the problem of interferences which are concentrated along with the analyte(s) of interest during the trace-enrichment step. The contamination proved sufficiently severe that both the resolving power of gradient elution and postcolumn chemical derivatization (fluorescence tagging) were necessary. In addition, band broadening due to large injection volumes (to 300 ml) was investigated.

Injections of the same total amount of peptide (nonapeptide standards were used as analytes) but in varying injection volumes showed only a 10–20% band broadening for a 300-ml injection relative to a 100-µl injection volume. A Nucleosil-$C_8$ (5 µm) column (4.0-mm i.d. × 15 cm) and a 20% acetonitrile/pH-7.0-water mobile phase at 1.0 ml/min and UV detection at 210 nm comprised the chromatographic setup.

Frei cautioned that any large injection volumes should consist of aqueous buffer and not of the mobile phase used for chromatographic separation in order to minimize the band broadening. Figure 43 compares peak heights for four nonapeptides ranging in $k'$ from 1.4 to 10.8, dissolved in both mobile phase and buffered water. Significant band broadening was encountered, even for relatively small injection volumes (340 µl) when the samples were dissolved in the mobile phase, whereas no detectable broadening occurred for the same injection volume (i.e., 340 µl) in aqueous mobile phase (both relative to a 34-µl injection volume). This problem of increased bandwidth was particularly noticeable for compounds with low retention ($k'$) since they begin migrating down the column during the enrichment step. For a protein with a $k'$ value of 10.8, the retention on the column was sufficiently strong so that no noticeable bandwidth increase occurred, regardless of the sample solution employed.

Erni et al. (1976) described an LC system incorporating an eight-port valve, used to generate step gradients, which could also be used as an introduction system for large sample volumes. With this valve it was possible to introduce any

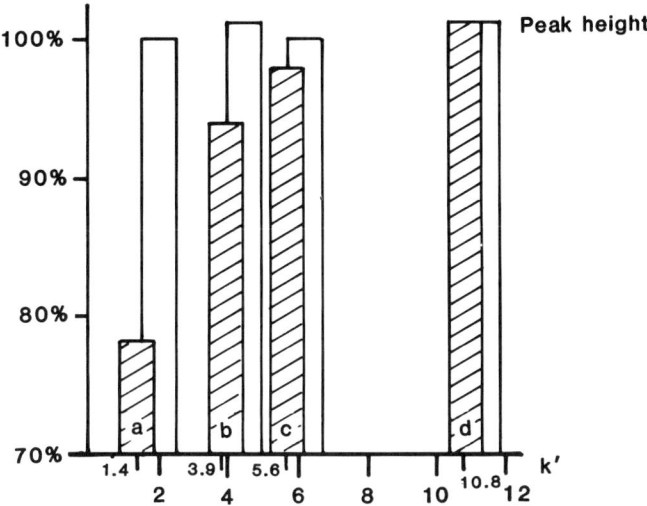

**Fig. 43.** Effect of injection volume on peak height (34-μl injection taken as 100%): (□) 340-μl injection from aqueous sample; (▨) 340-μl injection of sample dissolved in mobile phase; (a)–(d) represent different peptides ranging in $k'$ from 1.4 to 10.8 (Frei, 1978).

volume of sample solution into the chromatographic system by replacing one of the solvent reservoirs with the sample. The chromatographic system was flushed with water prior to sample enrichment; then after the injection volume was loaded a solvent step of medium polarity, intermediate between the sample injection solution and the elution of the analyte, allowed a preliminary front cut of the sample. It was possible to determine 2 ppb of the ergot alkaloid dihydroergocornine with an injection volume of 165 ml (5% reproducibility). The detection limit was 100 ppt (signal-to-noise ratio unspecified).

In a similar study to one described earlier (Frei, 1978), Schauwecker et al. (1977) investigated large injection volumes in conjunction with gradient elution and postcolumn derivatization, with the aim of improving detection limits, for the analysis of ergot alkaloids and proteins.

Some basic studies on the influence of injection volume on peak height and peak area were conducted using aqueous solutions of dihydroergocristine as a model. Injection volumes from 50 μl to 4.38 ml were made using loops of different sizes; larger injection volumes were pumped onto the column (Nucleosil $C_{18}$ or Merck RP-8, 3.0- and 4.0-mm i.d. × 3, 5, 10, and 15 cm). Detection was by UV absorption at 280 nm. A (40:60) 0.1 $M$ ammonium carbonate solution/acetonitrile mobile phase was used at a flow rate of 0.97 ml/min to effect chromatographic separation.

Table XV lists the effects of injection volume on peak heights and peak areas

### TABLE XV

**Influence of Injection Volume on Corrected Peak Area ($A_k$) and Corrected Peak Height ($H_k$) for Dihydroergocristine Mesilate[a,b]**

|  | $A_k$ | | $H_k$ | |
|---|---|---|---|---|
| Volume injected | Arbitrary units | Percentage[c] | Arbitrary units | Percentage[c] |
| 50 µl | 28.0 | 100 | 21.1 | 100 |
| 343 µl | 26.9 | 96.1 | 20.2 | 95.7 |
| 4384 µl | 23.9 | 85.4 | 17.7 | 83.9 |
| ~7 ml (pump) | 25.0 | 89.3 | 15.6 | 73.9 |
| 59 ml (pump) | 25.0 | 89.3 | 14.5 | 68.7 |
| 165 ml (pump) | 22.5 | 80.3 | 12.7 | 59.6 |

[a] Schauwecker et al. (1977).
[b] Chromatographic conditions: Nucleosil $C_{18}$, 5-µm column (3.0-mm i.d. × 10 cm); acetonitrile/0.1 M ammonium carbonate (40:60); 0.97 ml/min; 280 nm detection.
[c] Based on the smallest injection volume (50 µl) being taken as 100%.

of dihydroergocornine mesilate, taking the smallest injection volume (50 µl) as the comparison standard (100%). The overall concentration of dihydroergocornine mesilate per injection was kept constant, and the peak heights and areas were normalized to unit weight of model compound by dividing the observed heights and areas by the sample concentration times the injection volume for each. (These normalized values, $A_k$ and $H_k$, are referred to as "corrected" values in Table XV.) The values of $A_k$ should have remained constant; however, they decreased 10% for an injection volume of 59 ml and 20% for a 165-ml volume (relative to a 50-µl injection). It was presumed that the large injection volumes influenced the retention properties of the packing material to produce the observed results. The corrected peak heights $H_k$ showed significant reductions of up to 40% for the 165-ml injection. Nevertheless, these results did not represent drastic loss of column efficiency and were more than offset by lowering of the detection limit by 1000 or more.

The dihydroergocristine mesilate levels in urine samples were also determined. Due to the more complex sample matrix, step-gradient elution from 0.1 M ammonium carbonate (4 min) to 40% acetonitrile (10 min) and then 60% acetonitrile (5 min) was generated at 70°C. An upper sample volume of 1.78 ml (loop) was set due to the increased background. The separation of 2.74 ppm dihydroergocristine mesilate from urine is shown in Fig. 44. This separation is notable in that the alkaloid is separated from all other interferences in the urine without having to resort to use of a preliminary extraction or other urine pretreatment techniques. Turnaround time for this system was 13–14 min. The detection limit for this alkaloid in urine (1.78 ml) was 100 ppb (3:1 signal-to-noise ratio).

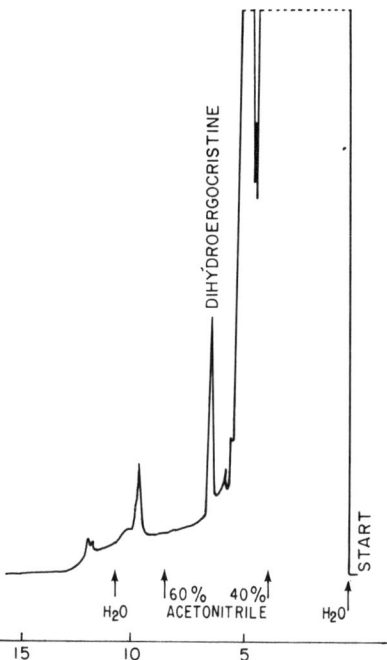

**Fig. 44.** Chromatogram of 2.74 ppm dihydroergocristine mesilate in urine. Conditions: Nucleosil $C_{18}$ column (4.0-mm i.d. × 15 cm); step gradient with acetonitrile/0.1 $M$ ammonium carbonate; 3 ml/min; 237-nm detection (Schauwecker et al., 1977).

The advantages of postcolumn derivatization as a measure of improving detectability and selectivity were investigated using Fluram as a fluorigenic agent on the nonapeptide oxytocin. Figure 45 compares a 20-μl injection to a 1.777-ml injection of oxytocin solutions onto a Merck RP-8, 5-μm column (4.0-mm i.d. × 25 cm) using acetonitrile/water (20:80) at 1.2 ml/min. Chromatograms of UV absorption at 210 nm and fluorescence after derivatization are shown. The absolute amount of oxytocin injected was the same for both injection volumes. Most of the interferences encountered by the UV detector are transparent to the fluorescence detector.

Cyclosporine A is a novel immunosuppressive peptide derived from fungi. It is characterized as an extremely hydrophobic cyclic peptide. It is administered to patients who are recipients of organ transplants. Measuring the drug in body fluids has proved particularly difficult.

Cyclosporine A was determined by direct injection of 1.77-ml samples onto a Merck RP-8, 5-μm column (4.0-mm i.d. × 3.0 cm) using 0.1 $M$ ammonium carbonate/acetonitrile step gradient at 4.0 ml/min, 70°C, with detection by UV absorption at 215 nm. Since more serious interferences are encountered at this

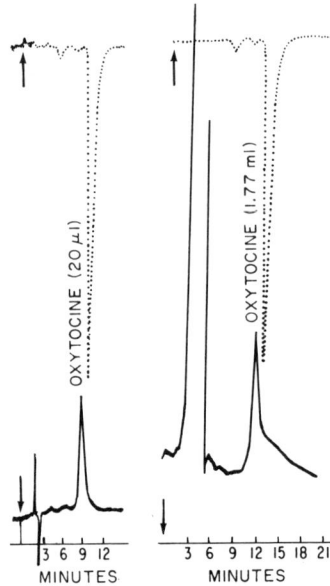

**Fig. 45.** Chromatograms of oxytocin in buffered solutions with different injection volumes (loop, 20 μl and 1.777 ml) on a column of 25-cm length: (—) UV detection at 210 nm prior to derivatization; (···), fluorescence detection after derivatization with Fluram. Conditions: solution of 30 mg Fluram in 100 ml acetonitrile; flow rate, 0.19 ml/min. Detection conditions: $\lambda_{ex} \approx 360$ nm; $\lambda_{em} \approx 470$ nm. Chromatographic conditions: Merck RP-8, 5 μm; column, 4-mm i.d.; mobile phase, acetonitrile/water (pH 7) (20:80); flow rate, 1.2 ml/min; column thermostated at 24°C. The absolute amount of oxytocine injected was the same for both injection volumes (Schauwecker et al., 1977).

**TABLE XVI**

**Quantitative Results for Trace Enrichment of Cyclosporine A**[a]

| Sample cyclosporine A | Retention time (min) | Peak area (arbitrary units) | Peak height (arbitrary units) |
|---|---|---|---|
| 18.1 ppm (H$_2$O) | 6.33 ± 0.02 | 333.4 ± 0.5 | 180 ± 4 |
| 18.1 ppm (urine)[b] | 11.04 ± 0.04 | 328.4 ± 1.0 | 403 ± 3 |
| 360 ppb (H$_2$O) | 6.34 ± 0.05 | 3.32 ± 0.14 | 2.6 |
| 360 ppb (urine)[c] | 13.43 ± 0.03 | 3.14 ± 0.05 | 2.2 |
| 314 ppb (urine)[d] | 13.22 ± 0.08 | 1.36 ± 0.32 | ~1.5 |

[a] Schauwecker et al. (1977)
[b] 55% Acetonitrile for separation step.
[c] Deproteinated before addition of cyclosporine A.
[d] Deproteinated after addition of cyclosporine A.

wavelength, the gradient was used to provide a cleaner separation of the cyclosporine A. For comparison purposes, aqueous samples of this protein were also run under the same chromatographic conditions. Peak heights and areas for cyclosporine A isolated from water and urine samples are compared in Table XVI. It can be seen that although the retention times are doubled for the cyclosporine A in urine, the peak heights and areas for the same concentrations (18.1 ppm and 360 ppb) are in agreement. Due to adsorption of the cyclosporine A onto the deproteinized fraction, a loss of cyclosporine A is evident for the 314-ppb determination.

## B. Column Switching, Off-Line

Lawrence and Allwood (1980) also developed an HPLC technique for measuring cyclosporine A in body fluids. Their procedure employs off-line isolation of the drug onto disposable cartridges prior to desorption and HPLC analysis. In this way, the compound of interest is extracted from biological fluids and preliminary separation from interferences is performed. This reduction of background has allowed for the trace analysis of cyclosporine A.

In order to release protein-bound drug, 3 ml of methanol was added to 1 ml of serum. After standing, 2 ml of the supernatant was then diluted with 1 ml of water and passed through a Sep-Pak $C_{18}$ cartridge. After loading, the cartridge was then rinsed with 5 ml of water. The more polar interferences are desorbed next using 5 ml of 75% methanol/water. Cyclosporine A was recovered from the cartridge by rinsing with methanol, discarding the first 0.5 ml and retaining the next 1 ml. This procedure recovered 90% of the cyclosporine A present in the sample.

Using 95:5 methanol/water at a flow rate of 1 ml/min on a μBondapak $C_{18}$ column (4.0-mm i.d. × 30 cm), cyclosporine A eluted at 4 min and was detected by UV absorption at 220 nm. Quantitative results using this procedure were in close agreement with the more traditional radioimmunoassay technique, even though some interference from serum constituents was encountered.

Dupont and DeJager (1981) also used off-line enrichment of the drug cefoperazone onto Sep-Pak $C_{18}$ cartridges from body fluids. This drug is hardly hydrophobic; in fact, it is sufficiently hydrophilic that it cannot be extracted from body fluids by organic solvents.

The Sep-Pak cartridges were prewashed with methanol and then water. A 2-ml sample volume was enriched. After a wash with water to elute unadsorbed substances, the cefoperazone was recovered with a rinse of 2 ml of 50:50 methanol/water. An injection of 25 μl was made onto a μBondapak $C_{18}$ column (4.6-mm i.d. × 25 cm). Using 50:50 methanol/water as mobile phase at a flow rate of 1.5 ml/min, the retention time was 2.3 min (UV detection at 228 nm). Serum protein was found to pass through the cartridge without adsorption during the enrichment step. Recovery of cefoperazone from aqueous solutions ranging from 1 to 4 ppm cefoperazone was 101.6–98.5%; recovery from serum ranging from 1 to 10 ppm cefoperazone averaged 92%. The limit of sensitivity of 250 ppb is adequate for clinical use.

## C. Column Switching, On-Line

Not only because of the relatively high solubility of the drug metabolite endralazine pyruvate (EP) in aqueous media, but also due to the sensitivity of this biochemical to hydrolysis in acidic media, liquid/liquid extraction methods could not be used for extraction of EP from body fluids. Consequently, Erni *et al.* (1981) developed an on-line enrichment technique for this drug metabolite.

The chromatographic system consisted of a guard (concentrator) column (4.6-mm i.d. × 6 cm) packed with LiChrosorb RP-2 (10 μm) and an analytical

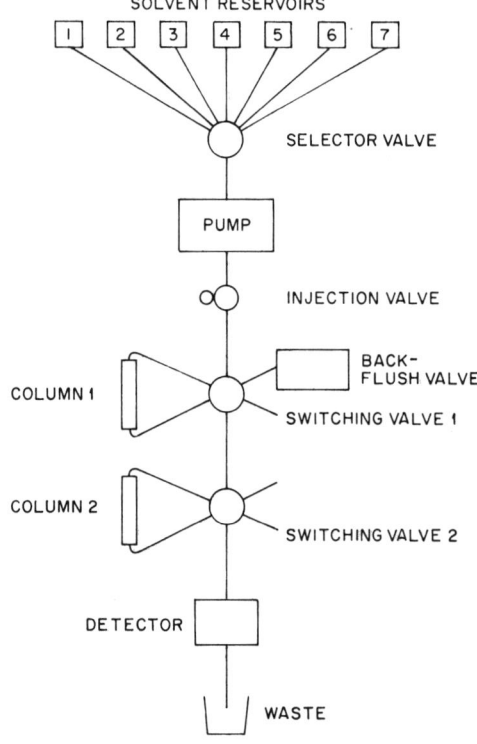

**Fig. 46.** Diagram of the column-switching system with a step-gradient-elution solvent delivery system (Erni *et al.*, 1981).

(separator) column (4.6-mm i.d. × 25 cm) packed with LiChrosorb RP-8 (5 μm). Both columns were mounted in the loops of two Valco air-operated switching valves. This arrangement permitted the introduction of various "cuts" (i.e., front cut, heart cut, or end cut) during desorption of the concentrator column, to be introduced onto the analytical column. For instance, very often the first part of a reverse-phase chromatogram contains polar compounds which can cause tailing and interferences for the analysis of the compounds of interest. An end-cut technique would eliminate those interferences by bypassing the separator column and diverting the first part of the chromatogram to waste. A diagram of the column switching system is shown in Fig. 46.

Urine samples were prepared for analysis for EP content by first adjusting the

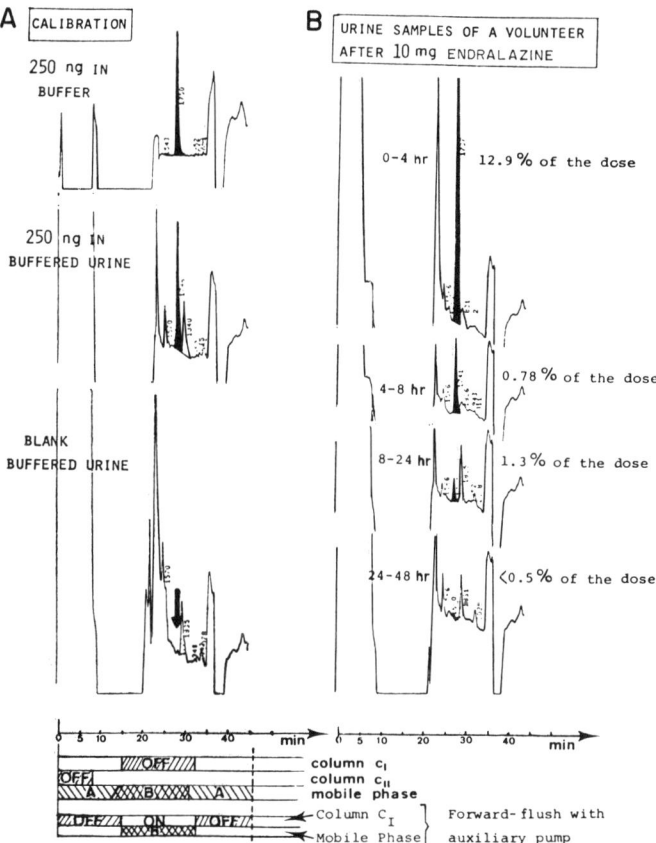

**Fig. 47.** Analysis of urine samples using column switching and a gradient system: (A) calibration chromatograms; (B) urine samples of a volunteer who received orally 10 mg of endralazine (Erni et al., 1981).

pH to 8.0 and stabilizing with Na-EDTA solution to prevent metal-catalyzed oxidation (3 ml urine were diluted to 10 ml). Injection volumes of 500 µl were made from the stabilized urine solutions (150 µl undiluted urine injected). A heart cut containing the drug metabolite was obtained by eluting the concentrator column first with 0–1.5% acetonitrile/buffered aqueous solution. Then the fraction containing the EP was transferred to the separator column and eluted by means of a step gradient to 20% acetonitrile. During the elution of EP from the separator column, the concentrator was flushed with the same 20% acetonitrile solution by means of an auxiliary pump. At the end of the run, both columns were reconditioned with the same initial mobile phase (0–1.5% acetonitrile).

Figure 47 shows the chromatograms of the urinary output of EP using the on-line enrichment technique, after oral administration of a 10-mg dose of endralazine to a human volunteer. The technique had a peak-area reproducibility better than 5% RSD for $n = 6$ at 250 ppb. The linear concentration range was 0–2500 ppb. A detection limit of 20 ppb for a 500-µl injection volume was given.

Lankelma and Poppe (1978) used a two-valve chromatographic system together with a concentrator and separator column for the on-line concentration of methotrexate (Mtx) in plasma. This drug has been used for over 20 years in cancer therapy, and its determination in body fluids is of great importance. The chromatographic system is shown in Fig. 48. The amount of sample present in

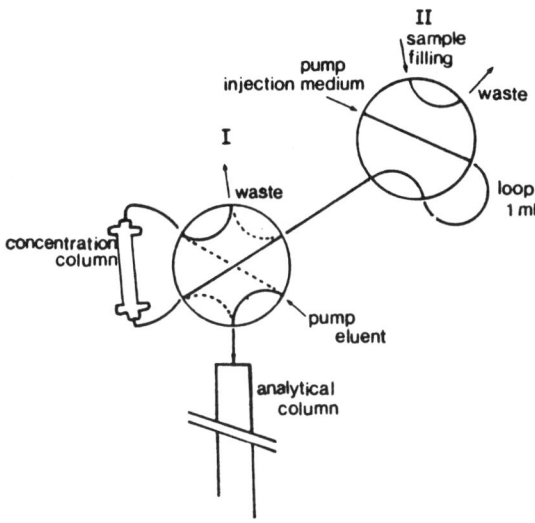

**Fig. 48.** Concentration system with two valves (I and II): solid line, concentration; broken line, injection (Lankelma and Poppe, 1978).

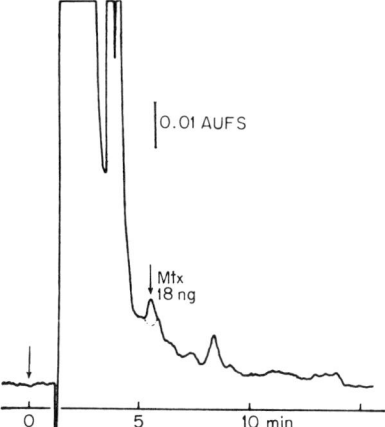

**Fig. 49.** Injection of deproteinized plasma, spiked with 18 ng of Mtx directly on the analytical column: injection volume, 1 ml; analytical column, Partisil SAX; particle size, 10 μm (Lankelma and Poppe, 1978).

the loop of valve II is enriched onto the concentrator column (in the loop of valve I). After switching valve I to inject, the concentrator column is back-flushed onto the separator column.

The concentrator column (3.0-mm i.d. × 4.6 cm) was packed with Merck RP-8 (10 μm); the separator column (4.6-mm i.d. × 25 cm) was packed with an anion-exchange material, Partisil SAX (Whatman). A flow rate of 2.5 ml/min of a 20% methanol/phosphate buffer solution was used to desorb the concentrator column and as mobile phase for the separator column; UV absorption at 306 nm was used for detection.

Figure 49 shows the chromatogram resulting from the direct injection of 1 ml of deproteinized plasma spiked to 18 ppb with Mtx. Because the baseline did not stabilize before the elution of the analyte of interest, Mtx quantitation posed a problem.

Figure 50 demonstrates the power of the front-cut technique for removing the interferences which coelute. After injection of the 1 ml of deproteinized plasma containing 36 ppb Mtx onto the concentrator column, various volumes of water were pumped through the concentrator column in order to elute as much of the polar interferences as possible prior to back-flushing the Mtx fraction onto the separator column. Prolonged elution with water does reduce the interferences; however, a rinse volume of 30 ml or more also causes elution of the drug. A rinse of 20 ml appears optimal; comparison of the chromatogram to that shown in Fig. 49 shows the usefulness of the cleanup step in reducing background interferences.

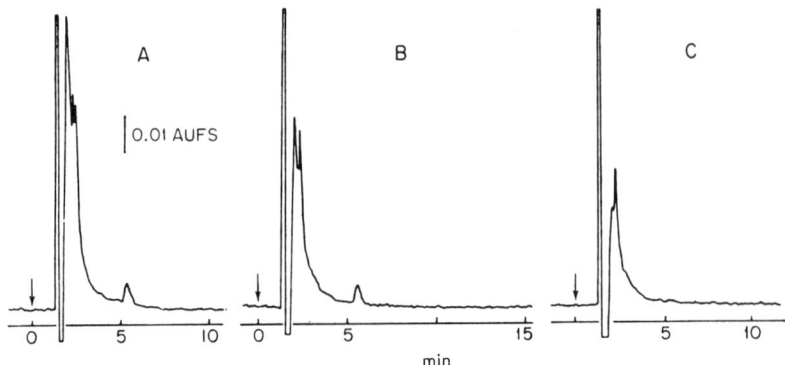

**Fig. 50.** Injections of 1 ml of deproteinized plasma spiked to 36 ppb with Mtx using the on-line concentration technique: (A) 11-ml, (B) 20-ml, and (C) 30-ml water rinses of the loaded concentrator column before switching to the position for injection on the separator column (Lankelma and Poppe, 1978).

Koch and Kissinger (1980a) developed a sensitive and selective method for the determination of serotonin (5-HT) and 5-hydroxyindoleacetic acid (5-HIAA) in cerebrospinal fluid and small regions of the brain. The method utilized a brain-tissue-homogenate cleanup on gravity-fed isolation columns (Amberlite CG-50 for 5-HT and Sephadex G-10 for 5-HIAA).

The sample-enrichment setup consisted of a 2-ml injection loop on the first of two six-port injection valves, with a second six-port valve containing the concentrator and separator columns. The arrangement and operation of this system are the same as for that of Lankelma and Poppe (1978) shown in Fig. 48.

Two-milliliter injection volumes (loop) from the isolation columns were used for enrichment of 5-HT and 5-HIAA onto the concentrator column (4.6-mm i.d. × 3.0 cm) packed with Brownlee Labs RP-8 material. After washing the column with ammonium acetate buffer at 1.8 ml/min, the concentrator column was back-flushed onto the separator column (4.0-mm i.d. × 30 cm) packed with µBondapak $C_{18}$. Methanol/ammonium acetate buffer (15:85) at 1 ml/min was used for desorption from the concentrator and as the mobile phase on the separator columns. An electrochemical detector monitored the chromatographic effluent at a potential of +0.5 V (Ag/AgCl reference).

Typical concentrations of 5-HT and 5-HIAA found in various regions of rat and sheep brain are listed in Table XVII. Figure 51 shows the chromatogram obtained by carrying an amount of brain homogenate equivalent to 1 mg of brain tissue through the procedure.

The incorporation of an enrichment step in the sample-extraction scheme overcame the problem of inefficient sample utilization inherent in the combination of gravity-fed isolation columns and analytical HPLC columns. Since the

## TABLE XVII

**Regional Brain Content of 5-HT and 5-HIAA Found Using the On-Line Sample-Enrichment Procedure**[a]

| Region | 5-HT[b] | 5-HIAA[b] | Region | 5-HT[b] | 5-HIAA[b] |
|---|---|---|---|---|---|
| Rat | | | Sheep | | |
| Whole brain | 392 ± 26 | 157 ± 10 | Pineal gland | 2070 ± 139 | 17,300 ± 1120 |
| Hypothalamus | 661 ± 44 | 400 ± 28 | Amygdala | 266 ± 18 | 442 ± 29 |
| Hippocampus | 435 ± 30 | 200 ± 12 | Hippocampus | 104 ± 7 | 267 ± 17 |
| Medial forebrain bundle | 156 ± 10 | 210 ± 14 | Raphe | 682 ± 46 | 2420 ± 157 |
| Amygdala | 163 ± 11 | 165 ± 11 | Thalamus | 690 ± 47 | 1200 ± 78 |
| Pineal gland | 2300 ± 149 | 863 ± 56 | | | |

[a]Reprinted with permission from Koch and Kissinger (1980a). Copyright 1980, Pergamon Press, Ltd.
[b]Nanograms per gram of wet tissue.

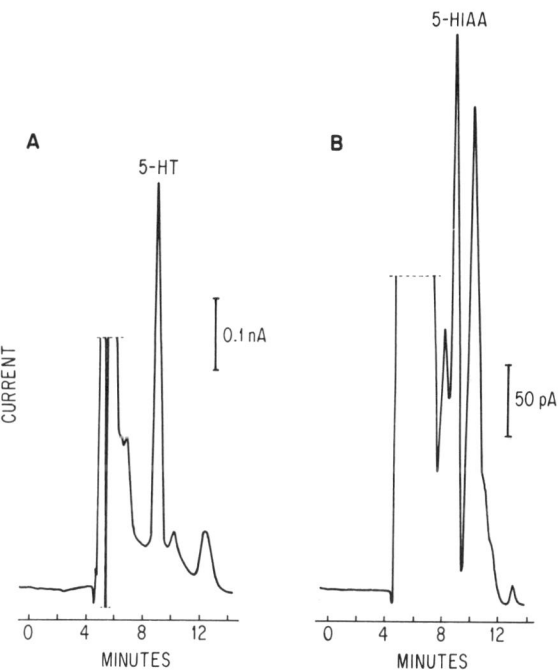

**Fig. 51.** Determination of 5-HT and 5-HIAA in the equivalent of 1 mg of brain: (A) serotonin, (B) 5-hydroxyindoleacetic acid. [Reprinted with permission from Koch and Kissinger (1980a). Copyright 1980, Pergamon Press, Ltd.]

former provide bandwidth volumes much greater than conventional HPLC injection volumes, most of the sample would go to waste if large injection volumes were not realizable.

Recoveries were 89.5% for 5-HT and 80.9% for 5-HIAA. The enrichment was linear for concentrations ranging from 5 pg/ml to 5 ng/ml (5 ppt to 5 ppb). The detection limits at 5 : 1 signal-to-noise ratio were 1.1 ppt for 5-HT and 7.5 ppt for 5-HIAA. The total analysis time was from 1–2 hr per sample. This on-line enrichment method has been used to assay the equivalent of 1 mg of raw brain-tissue homogenate and results indicate excellent promise of applicability to brain slices and punchouts.

In addition, Koch and Kissinger (1980b) also determined 5-HT and 5-HIAA in human serum and plasma in a manner nearly identical to that described above (Koch and Kissinger, 1980a).

## IV. PHARMACOLOGICAL APPLICATIONS

### A. Direct Injection

As an alternative to bioassay, Krummen and Frei (1977) developed an HPLC technique using trace enrichment for the determination of various nonapeptide drugs (oxytocin, demoxytocin, lypressin, ornipressin, and felypressin). These compounds are of varied therapeutic effect ranging from use as nasal sprays to uterine-contraction stimulants. Reverse-phase $C_{18}$ (Nucleosil 5 and 10 μm) and $C_8$ (Merck 5 and 10 μm) packings were used in either 3.0- or 4.0-mm i.d. × 25 cm columns (10 μm) or 4.0-mm i.d. × 15 cm columns (5 μm); 20% acetonitrile/pH-7.0 buffered water was used as the isocratic mobile phase at flow rates between 1.8 and 4.0 ml/min. Detection was by UV absorption at 210, 215, or 220 nm.

Liquid pharmaceutical formulations were injected (35–750 μl) directly into the LC without any sample preparation. Due to the low dosage of some forms (e.g., oxytocin ampules at 1.0 IU/ml) and also due to a need to investigate by-products of the drugs, the largest injection volumes possible were desirable. In addition, the effects of aqueous and mobile-phase injection solutions on band broadening (in conjunction with large injection volumes) were investigated. This was done indirectly by measuring peak height; values are listed in Table XVIII for 340-μl injection volumes of aqueous and mobile-phase solutions relative to a 34-μl injection volume (taken as 100%). For the aqueous injections no dependence of peak heights on $k'$ was observed, even for low-$k'$ compounds. On the other hand, pronounced lowering of peak heights occurred for mobile-phase injections of compounds with $k'$ values less than 5.0.

Figure 52 compares 50–750-μl injection volumes of a commercial oxytocin

## TABLE XVIII

### Effect of a Large Injection Volume on Peak Height[a,b]

| Substance | $k'$ | Peak height[c] ($n = 3$) | |
|---|---|---|---|
| | | 340-µl injection of aqueous solution | 340-µl injection of mobile-phase solution |
| Lypressin and ornipressin | 1.4 | 100 ± 3 | 78 ± 3 |
| Oxytocin | 3.9 | 101 ± 2 | 94 ± 1 |
| Felypressin | 5.6 | 100 ± 2 | 98 ± 1 |
| Demoxytocin | 10.8 | 101 ± 1 | 101 ± 1 |

[a] Krummen and Frei (1977).

[b] The amount injected is kept constant for each substance. Column: RP 8 (10 µm), 25 cm × 3 mm.

[c] Expressed as a percentage of that given by a 34-µl injection of a solution of 10-fold higher concentration (taken as 100%).

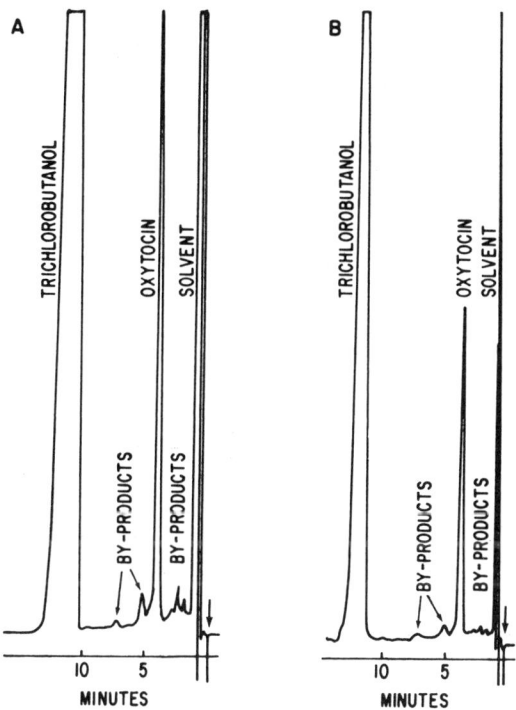

**Fig. 52.** Chromatograms of oxytocin solution (10 IU/ml) on a column (25 cm × 4 mm) of Nucleosil $C_{18}$ (10 µm). Mobile phase, 20% acetonitrile in phosphate buffer of pH 7; flow rate, 4.0 ml/min. (A) Injection volume 750 µl, UV monitor at 210 nm (sensitivity 0.1 AUFS); (B) injection volume 50 µl, UV monitor sensitivity 0.02 AUFS (Krummen and Frei, 1977).

liquid dosage form without any detectable band broadening. The larger injection volume provided a particularly more effective means of determining any trace-level components and by-products in the formulation which would have been marginally detectable using a smaller (50 µl or less), more conventional injection volume.

Dissolution rate is often used as a quality-control test for many drug-containing tablets, and the concentration of the active ingredient in the dissolution medium is often determined by UV spectroscopy. However, the sensitivity for low-dosage tablets is lacking with this analytical approach. In fact, the dosage level of some tablets is even too low to be adequately analyzed using conventional HPLC techniques and injection sizes. As a means of solving these problems Finlay et al. (1979) developed an enrichment technique for determining less than 1.0 mg/tablet of betamethasone using water as the dissolution medium.

The HPLC column (4.5-mm i.d. × 20 cm) was packed with 10-µm Spherisorb-ODS and was thermostated at 60°C; 40–55% methanol/water was used as mobile phase at 2 ml/min with detection by UV absorption at 254 nm.

The effect of injection volume on peak height was assessed by injecting 10–1500 µl of aqueous betamethasone solution. (The total amount of drug per

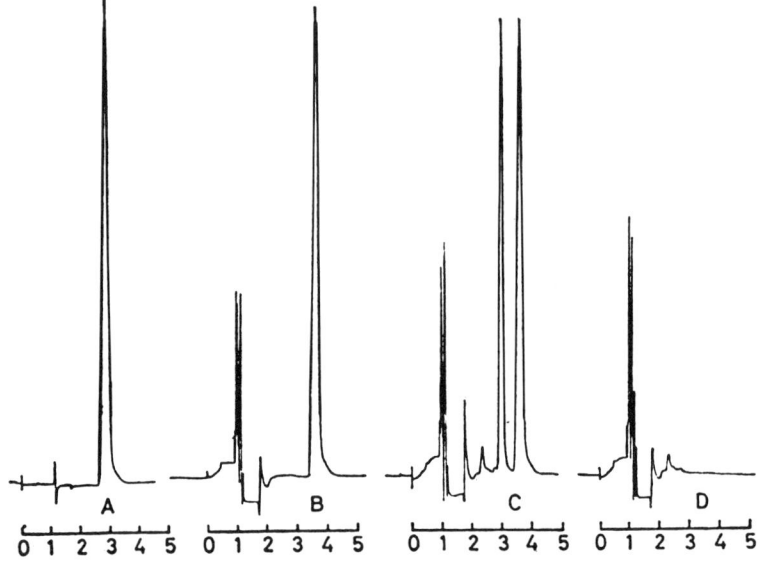

**Fig. 53.** Chromatograms of: (A) betamethasone (38 µg/ml) in water; loop injection, nominal 10 µl; (B) betamethasone (0.25 µg/ml) in water; loop injection, nominal 1500 µl; (C) betamethasone (0.25 µg/ml) and prednisone (0.15 µg/ml) in water, filtered through MF-Millipore 1.2 µm; loop injection, nominal 1500 µl; (D) water, filtered through MF-Millipore 1.2 µm; loop injection, nominal 1500 µl (Finlay et al., 1979).

injection was kept constant.) Peak heights per microgram of betamethasone injected were calculated for each injection volume and showed a less than 10% reduction per microgram of drug for the 1500-$\mu$l injection relative to the 10-$\mu$l injection.

It was discovered that after filtering the solutions of the tablets, several small peaks, not observable without filtering, became apparent. These were attributed to contaminants leaching from the filter and were eliminated by rejecting the first 10 ml of filtrate. Figure 53 compares chromatograms of the same total amount of injected betamethasone from 10-$\mu$l (A) and 1500-$\mu$l (B) injection volumes of aqueous betamethasone solutions. The quickly eluting contaminant peaks shown in chromatogram C were traceable to a blank extract of the filter (D).

## B. Column Switching, Off-Line

Trace enrichment on a silica column was used by Popl *et al.* (1975) in order to concentrate aromatics and polar contaminants present at trace levels from white oils and *n*-alkanes. Because of their carcinogenic activity, the content of polycyclic aromatics becomes an important factor when considering the use made of white oils, i.e., not only for medical purposes affecting the skin (cosmetic and topical applications), but more importantly for use in the food industry.

The concentrator column (1.0-cm i.d. × 90 cm) was packed with 34 g of Woelm silica gel and a flow rate of 1.7 ml/min was used to enrich the samples. Medicinal oil for pharmaceutical use (1 liter, 0.854 g/ml density) and dearomatized *n*-alkanes (765 ml, $C_{10}$–$C_{17}$, 0.755 g/ml density) were enriched separately. The breakthrough of anthracene in the eluate was monitored by means of fluorescence measurements taken on the collected fractions. (The column demonstrated retention of anthracene at 1 ppm for up to 1300 ml loading volume of the medicinal oil.) The loaded column was washed with 150 ml of *n*-pentane and the aromatic fraction was recovered by desorption with 100 ml ether. Evaporation of the eluate yielded 1.052 g of the aromatic concentrate, appearing as a yellow, viscous oil. The *n*-alkane sample yielded 0.0257 g of aromatic concentrate, appearing as a white powder.

The aromatic concentrates were further purified on a basic alumina column to remove polar components. Final chromatographic separation of the purified aromatic fractions was carried out on an alumina column (3.0-mm i.d. × 100 cm) or by gel-permeation chromatography (GPC) (1.3-cm i.d. × 300 cm).

Results showed that about 85% of mono- and diaromatic hydrocarbons in the medicinal oil were not retained by the silica-gel concentrator column during loading and rinsing processes. The remaining 15% were eluted from the concentrator column by means of the ether along with the tri- and polyaromatics and polar compounds. The dearomatized *n*-alkane sample contained only large aromatics and polar compounds. This procedure allowed the determination of cer-

**TABLE XIX**

Quantity of Determined Aromatics[a]

| Aromatic type | Sample A[b] | Sample B[c] |
|---|---|---|
| Anthracene | — | 1 |
| Phenanthrene | 140 | 30 |
| Alkylfluoranthenes | 10 | 1 |
| Benzo[k]fluoranthene | 1 | 1 |
| Triphenylene | 1.5 | — |
| Alkyldibenzothiophene | 180 | — |
| Alkylaryldisulfone | — | 39,000 |

[a] Reprinted with permission from Popl et al. (1975). Anal. Chem. **47**, 1947–1950. Copyright 1975, American Chemical Society.
[b] Medicinal oil (ppb).
[c] n-Alkanes, $C_{10}$–$C_{17}$ (ppb).

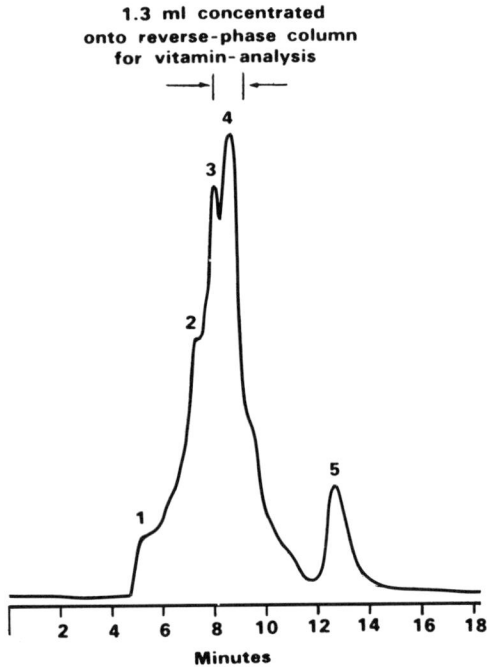

**Fig. 54.** On-column concentration technique for separation of B vitamins in food protein supplement, GF chromatographic separation of protein supplement. Sample size, 100 μl. Conditions: column, Micro-Pak TSK 2000SW (30 cm × 7.5 mm); flow rate, 1.2 ml/min; mobile phase, 10% methanol/90% water containing 0.1 $M$ $KH_2PO_4$ and 0.01 $M$ 1-heptanesulfonic acid; 254 nm, 2.0 AUFS. Peaks: 1–3, proteins; 4, vitamins; 5, unknown. [Reproduced from Majors (1980). J. Chromatogr. Sci., by permission, Preston Publications, Inc.]

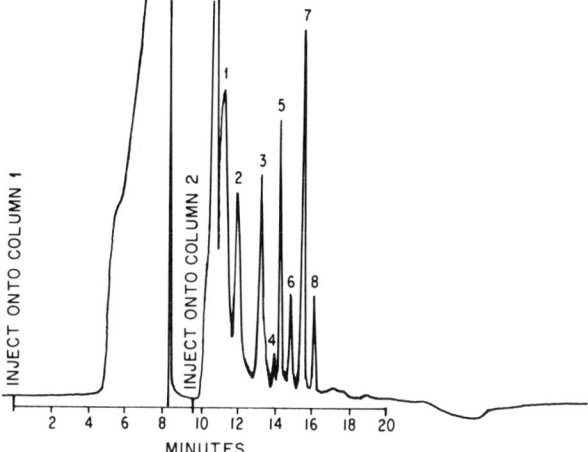

**Fig. 55.** On-column concentration technique for separation of B vitamins in food protein supplement. Composite chromatogram with reverse-phase analysis of vitamin fraction by on-column concentration method. Sample size, 1.3 ml. Conditions: column 1, Micro-Pak TSK 2000SW; column 2, Micro-Pak MCH-10 (30 cm × 4 mm); flow rate, 2 ml/min; mobile phase, 10–80% methanol in water containing 0.1 $M$ $KH_2PO_4$ and 0.01 $M$ 1-heptanesulfonic acid in 15 min; UV detection, 254 nm, 0.5 AUFS. Peaks: 1, niacin; 2, unknown; 3, pyrodoxine ($B_6$); 4, unknown; 5, unknown; 6, thiamine ($B_1$); 7, riboflavin ($B_2$); 8, unknown. [Reproduced from Majors (1980). *J. Chromatogr. Sci.*, by permission, Preston Publications, Inc.]

tain aromatics down to the 1.0 ppb level. Table XIX lists some of the aromatic hydrocarbons which were identified by fluorescence and phosphorescence spectrometry and chromatographic retention times and gives their concentrations in the white oils.

## C. Column Switching, On-Line

Erni and Frie (1978) and Majors (1980) used an on-line "multidimensional" LC/LC system for the analyses of glycosides in plant extracts and vitamins in a protein food supplement, respectively. In these approaches, an injection valve, located after the primary column, diverted sample either to waste or to the secondary column. When the fractions of interest began to elute, the entire chromatographic flow coming from column 1 was diverted to column 2. The multidimensional aspect can pair similar or very dissimilar column packings in the primary and secondary columns. The above examples used a gel-permeation packing in the primary column and reverse-phase $C_{18}$ in the secondary column. However, the chromatographic mobile phases must be compatible. Furthermore, the mobile phase must be sufficiently weak in elution strength so that an enrichment can occur on the head of the secondary column during the sample introduction step.

Majors' primary column (7.5-mm i.d. × 30 cm) contained an aqueous gel-filtration (GF) packing (Micro-Pak TSK 2000 SW); the secondary column (4.0-mm i.d. × 30 cm) contained a reverse-phase packing (Micro-Pak MCH). The aqueous GF column mobile phase consisted of 10% methanol, 90% aqueous $KH_2PO_4$/heptanesulfonic acid solution. Flow rates of 1.2 and 2.0 ml/min were used for the GF and the reverse-phase separations, respectively; UV absorption at 254 nm served as detection for both chromatographic systems.

Figure 54 shows the GF separation of a 100-μl aqueous injection of the water-soluble portion of a protein food supplement, fortified with vitamins to the 10 ppm level. From injection of standards, it was determined that the vitamins of interest were contained in fraction 4 (1.3 ml). This peak was diverted (heart cut) onto the secondary reverse-phase column. Fractions 1, 2, and 3 contained the lower molecular weight proteins. These would have been irreversibly adsorbed onto the reverse-phase column if a direct injection of the protein supplement extract had been attempted. Figure 55 shows the resulting chromatogram obtained from the on-column concentration of peak 4 from the GF primary column. Elution of the vitamins was completed in under 17 min utilizing a 15-min gradient from 10 to 80% methanol.

## REFERENCES

Bristol, D. W. (1980). *J. Chromatogr.* **188,** 193–204.
Chesler, S. N., Gump, B. H., Hertz, H. S., May, W. E., Dyszel, S. M., and Enagonio, D. P. (1976). Trace hydrocarbon analysis: The National Bureau of Standards Prince William Sound/Northeastern Gulf of Alaska baseline study, January. *NBS Tech. Note (U.S.)* **889.**
Creed, C. G. (1976). *Res./Dev.* **27,** 40–44.
Dupont, D. G., and DeJager, R. L. (1981). *J. Liq. Chromatogr.* **4,** 123–128.
Eisenbeiss, F., Hein, H., Joester, R., and Naundorf, G. (1978). *Chromatogr. Newsl.* **6,** 8–12.
Erni, F., and Frei, R. W. (1978). *J. Chromatogr.* **149,** 561–569.
Erni, F., Frei, R. W., and Lindner, W. (1976). *J. Chromatogr.* **125,** 265–274.
Erni, F., Keller, H. P., Morin, C., and Schmitt, M. (1981). *J. Chromatogr.* **204,** 65–76.
Euston, C. B., and Baker, D. R. (1979). *Am. Lab. (Fairfield, Conn.)* **11,** 91–100.
Finlay, G., Jeffries, J. P., and Sharma, S. P. (1979). *J. Pharm. Pharmacol.* **31,** 634–635.
Frei, R. W. (1978). *Int. J. Environ. Anal. Chem.* **5,** 143–155.
Hertz, H. S., May, W. E., Chesler, S. N., and Gump, B. H. (1976). *Environ. Sci. Technol.* **10,** 900–903.
Huber, J. F. K., and Becker, R. R. (1977). *J. Chromatogr.* **142,** 765–776.
Ishii, D., Hibi, K., Asai, K., and Nagaya, M. (1978). *J. Chromatogr.* **152,** 341–348.
Karger, B. L., Snyder, L. R., and Horvath, C. (1973). "An Introduction to Separation Science," pp. 127–129. Wiley, New York.
Karger, B. L., Martin, M., and Guiochon, G. (1974). *Anal. Chem.* **46,** 1640–1647.
Kirkland, J. J. (1975). *Chromatographia* **8,** 665–666.
Koch, D. D., and Kissinger, P. T. (1980a). *Life Sci.* **26,** 1099–1107.
Koch, D. D., and Kissinger, P. T. (1980b). *Anal. Chem.* **52,** 27–29.
Krummen, K., and Frei, R. W. (1977). *J. Chromatogr.* **132,** 429–436.

Kummert, R., Molnar-Kubica, E., and Giger, W. (1978). *Anal. Chem.* **50,** 1637–1639.
Lankelma, J., and Poppe, H. (1978). *J. Chromatogr.* **149,** 587–598.
Lawrence, R., and Allwood, M. C. (1980). *J. Pharm. Pharmacol.* **32** (Suppl.), 100-p.
Little, J. N., and Fallick, G. J. (1975). *J. Chromatogr.* **112,** 389–397.
Majors, R. E. (1980). *J. Chromatogr. Sci.* **18,** 571–579.
May, W. E., Chesler, S. N., Cram, S. P., Gump, B. H., Hertz, H. S., Enagonio, D. P., and Dyszel, S. M. (1975). *J. Chromatogr. Sci.* **13,** 535–540.
May, W. E., Wasik, S. P., and Freeman, D. H. (1978). *Anal. Chem.* **50,** 997–1000.
Mayer, G., and Shoup, R. E. (1981). *Current Separations* **3** (May), 4–6. Bioanalytical Systems, Inc., W. Lafayette, Indiana.
Ogan, K., Katz, E., and Slavin, W. (1978). *J. Chromatogr. Sci.* **16,** 517–522.
Otsuki, A. (1977). *J. Chromatogr.* **133,** 402–407.
Otsuki, A., and Takaku, T. (1979). *Anal. Chem.* **51,** 833–835.
Oyler, A. R., Bodenner, D. L., Welch, K. J., Liukkonen, R. J., Carlson, R. M., Kopperman, H. L., and Caple, R. (1978). *Anal. Chem.* **50,** 837–842.
Popl, M., Stejskal, M., and Mostecky, J. (1975). *Anal. Chem.* **47,** 1947–1950.
Riggin, R. M., and Howard, C. C. (1979). *Anal. Chem.* **51,** 210–214.
Sampson, R. L. (1977). *Am. Lab. (Fairfield, Conn.)* **9**(5), 44–49.
Saner, W. A., and Gilbert, J. (1980). *J. Liq. Chromatogr.* **3,** 1753–1765.
Saner, W. A., Jadamec, J. R., Sager, R. W., and Killeen, T. J. (1979). *Anal. Chem.* **51,** 2180–2188.
Schauwecker, P., Frei, R. W., and Erni, F. (1977). *J. Chromatogr.* **136,** 63–72.
Van Vliet, H. P. M., Bootsman, T. C., Frei, R. W., and Brinkman, V. A. T. (1979). *J. Chromatogr.* **185,** 483–495.
Waters Associates (1976). "Pesticide Residues in Drinking Water," November. Milford, Massachusetts.
Waters Associates (1978a). "Half-Hour Determination Method for Chlorophenoxy Acids and Esters Using Liquid Chromatography," May. Milford, Massachusetts.
Waters Associates (1978b). "Rapid Sample Preparation for Analysis of PCB's in Water," September. Milford, Massachusetts.
Werkhoven-Goewie, C. E., Brinkman, V. A. Th., and Frei, R. W. (1981). *Anal. Chem.* **53,** 2072–2080.
West, S. D., and Parka, S. J. (1981). *J. Agric. Food Chem.* **29,** 223–226.

# HPLC ANALYSIS OF POLAR SUBSTANCES ON UNMODIFIED SILICA

### J. B. Green and P. L. Grizzle*

U.S. Department of Energy
Bartlesville Energy Technology Center
Bartlesville, Oklahoma

|  |  |  |
|---|---|---|
| I. | Introduction | 223 |
| II. | Literature Survey | 224 |
| III. | Nonaqueous Systems Containing Aliphatic Carboxylic Acids | 227 |
|  | A. Experimental | 227 |
|  | B. Results and Discussion | 231 |
| IV. | Nonaqueous Systems Containing Amines | 238 |
|  | A. Experimental | 240 |
|  | B. Results and Discussion | 241 |
| V. | Coordinating Amine- and Carboxylic Acid-Based Separations for Optimum Resolution of Complex Samples | 257 |
|  | A. Experimental | 257 |
|  | B. Results and Discussion | 258 |
| VI. | Conclusions | 264 |
|  | References | 265 |

## I. INTRODUCTION

Although other factors have contributed, much of the current popularity and rapid growth of high-performance liquid chromatography (HPLC) has resulted directly from the development of microparticulate bonded-phase packing materials, especially those used in the reverse-phase mode. Even as early as 1978, many laboratories reported that over 75% of their separations were performed on

---

*Present address: Sun Exploration and Production Co., P.O. Box 936, Richardson, Texas 75080.

covalently derivatized silica columns (Snyder and Kirkland, 1979). Today, during informal conversations with equipment vendors and other people in the field, one frequently hears estimates as high as 95% for use of bonded-phase columns. Thus, an article on separations on unmodified silica seems a bit out of the mainstream of activity. Recently, others writing about what is traditionally called liquid–solid chromatography (LSC) have appeared to be increasingly defensive or apologetic toward what has become or is rapidly becoming known as an inferior separation method.

However, current research employing mobile phases containing additives designed to promote special separation selectivity and enhance reproducibility of LSC may well reverse this trend toward obsolescence. For example, mobile phases containing amines have yielded separations of carbohydrates on plain silica equal to those of the best covalently bonded $NH_2$–silica columns. Also, aqueous mobile phases containing cetyltrimethylammonium bromide have been shown to dynamically coat plain silica, thus imparting it with a hydrophobic surface similar to many commercially available reverse-phase packings. The key to these types of separations is obviously the special additive, usually present in low concentration, which imparts the change in the silica surface. Thus, this article will focus mainly on the types of additives tried to date, their effects on selectivity, and areas of application.

Although some general qualitative comments can be made about the mechanisms of these separations, insufficient data exist for a detailed description. Thus a "show-and-tell" rather than a theoretical approach will be employed. Many of the examples are separations of fossil-fuel liquids; however, fossil fuels contain such a wide variety of compounds that, hopefully, readers working in other areas will be able to apply the separations shown here to their own work.

The discussion that follows is divided into four sections: a literature review section that covers all types of mobile-phase additives (Section II), followed by three sections containing original work on carboxylic acid additives (Section III), amine additives (Section IV), and separation schemes involving multiple silica columns and additives (Section V).

## II. LITERATURE SURVEY

Traditionally, LSC has been limited to low- or medium-polarity solutes and nonaqueous solvent systems. However, as shown in Table I, more recent work has often involved ionic solutes and aqueous mobile phases. One approach involves the addition of cationic or electrically neutral surfactants in the mobile phase which effects a hydrophobic character to the surface; for example,

## TABLE I
## Polar Mobile-Phase Additives

| Typical bulk solvent | Additive | Type of solute tried | Probable interaction mechanism | Reference |
|---|---|---|---|---|
| Aqueous–alcoholic | Cetyltrimethylammonium bromide | Neutral, anionic | Hydrophobic; also ion pairing with anions | Knox and Laird (1976); Ghaemi and Wall (1979); Hansen (1981); Hansen et al. (1981) |
| | Nonionic surfactants, e.g., Tweens | Widely varied; anionic, cationic, and neutral | Hydrophobic; ion-pairing agents may be added as in conventional reverse-phase chromatography | Wall (1980); Ghaemi and Wall (1980); Ghaemi and Wall (1981) |
| Aqueous–organic | Strong inorganic and organic acids | Ionic | A complex combination of partition, ion exchange, and adsorption depending upon pH, solute, etc. | Svendsen and Greibrokk (1981) |
| | Polyamines | Carbohydrates | Adsorption onto a silica surface coated with amine | Aitzetmuller (1978); Wheals and White (1979); Hendrix et al. (1981) |
| Alkane–chloroalkane–alcoholic | Aliphatic carboxylic acids | Nonionic acids and bases | Adsorption onto a silica surface coated with bound carboxylic acid | Lawrence and Leduc (1978); Green (1981); Green and Hoff (1981) |
| Chloroalkane–alcoholic | Potassium salt–crown ether complexes | Sulfonic acids | Formation and adsorption of potassium–crown ether sulfonates | Brugman and Kraak (1981) |
| Alkane–ether–alcoholic | Amines and ammonium hydroxides | Nonionic acids and bases | Adsorption onto a silica surface coated with amine | Section IV of this article |
| Variable | Water | Nonionic-polar | Partition into an adsorbed water phase or other interactions with adsorbed water | Paanakker et al. (1978); Guillemin (1980); Scott and Traiman (1980) |

$$\text{silica} + X^{\oplus}\,[\text{H}_2\text{O}] \rightleftharpoons \text{silica—X} + \text{H}^{\oplus}$$
$$+$$
$$Y$$
$$\updownarrow \qquad\qquad (1)$$
$$\text{silica —X(Y)}$$

X = cetyltrimethylammonium ion, Y = solute

It should be noted that the system shown in Eq. (1) also has a built-in ion-pairing ability for anionic solutes. Of course, these systems are designed to be competitive with bonded-phase ODS, $C_8$, etc., columns. They may have certain advantages, however, including the ability to vary the organic loading on the column (via control of the surfactant concentration) and better column-to-column reproducibility. With regard to the latter point, Hansen et al. (1981) compared the behavior of 14 different silicas in the system shown in Eq. (1) and found only 1 to be significantly different. Thus, dynamic coating may provide a reverse-phase system largely independent of the origin of column material.

Svendsen and Greibrokk (1981) used dilute solutions of organic and inorganic acids as mobile phases in chromatography of biogenic amines on silica. Solute retention varied mainly with pH and the structure of the acid used in the mobile phase. Similarly, several workers have reported success in using amine-loaded aqueous phases for separations of carbohydrates (Table I). In both systems, the most probable retention mechanism involves solute interaction with silica-bound acid or amine.

Nonaqueous systems employing mobile phases containing acids and bases most likely work via solute-adsorbed acid or base interactions also. Those systems are described in Sections III–V in this article. A novel example of this type of interaction in a nonaqueous system was shown by Brugman and Kraak (1981). They used potassium salt–crown ether complexes in mobile phases to separate sulfonic acids. Competition between silica-bound and free complex for the sulfonic acids governed retention in that system.

Finally, water can be viewed as a mobile-phase additive in LSC. In the past, water has been viewed only as a deactivating agent which occupied sites on the silica surface and made them unavailable to solutes (Snyder, 1968). However, the interaction of solutes with adsorbed water has become recognized (Paanakker et al. 1978; Snyder and Kirkland, 1979; Scott and Traiman, 1980). Guillemin (1980) dramatically demonstrated the ability of adsorbed water to retain solutes in his separation of phenols on silica using distilled water as the mobile phase.

The work of Scott and Kucera (1978, 1979) and Scott (1980) supports the concept of polar mobile-phase additives being adsorbed on silica and interacting with solutes. However, this viewpoint is not universally accepted. For example, Snyder and Poppe (1980) have written a lengthy review claiming that in nearly all cases, competition of solutes and mobile phase for silica adsorption sites is the dominant feature governing retention in LSC. Fortunately, a detailed understanding of the mechanism is not an absolute prerequisite for effectively using polar additives in LSC.

## III. NONAQUEOUS SYSTEMS CONTAINING ALIPHATIC CARBOXYLIC ACIDS

In 1978, Lawrence and Leduc showed that the addition of small amounts of acetic, propanoic, or butyric acid to mobile phases of medium polarity greatly improves the peak shape and chromatographic efficiency of phenols, phenoxyalkanoic acids, and triazine compounds on silica. Although, as pointed out in their paper, addition of carboxylic acids to mobile phases is not new, their study represents one of the few in which the effects of acid concentration and structure on retention are investigated.

Using Lawrence and Leduc (1978) as a starting point, the potential of carboxylic acid–silica systems for separating polar materials in liquid fuels was investigated by Green (1981) and Green and Hoff (1981). Among other things, it was found that adding acids to the mobile phase *increased* retention of many basic solutes, thus indicating that the acid did more than just block sites on silica which caused peak tailing, as had been often reported previously. Subsequent work has focused on evaluating chlorinated and long- or branched-chain acids as mobile-phase additives and on further application of acid-containing systems to the separation of fuels.

### A. Experimental

Although much of the apparatus and procedure was the same as described in detail earlier (Green, 1981; Green and Hoff, 1981; Grizzle *et al.*, 1981), the essential features are repeated below for the sake of continuity in this article.

*1. Apparatus*

Two basic HPLC setups were used. The first was a standard two-pump system (Waters Associates, Milford, MA) with gradient controller and either two UV detectors in series, set at different wavelengths, or a single detector capable of simultaneously monitoring two different wavelengths, (e.g., SP-M230, Spectra Physics, Santa Clara, CA). The second system was a one-pump system with a low-pressure-gradient former ahead of the pump (Spectra Physics M8000). This system had the ability to mix three solvents to form gradients—a nearly essential feature for analyzing complex mixtures such as fuels. Dual-wavelength UV detection was employed in the second system also, and both systems had automatic sample injectors (WISP, Waters Associates).

The Spectra Physics liquid chromatograph was also used for preparative-scale runs. For preparative runs, a motorized six-way valve for fraction collection

(Altex, Berkeley, CA) and a special preparative UV detector with 0.1-cm-pathlength cell (Model UA-5, ISCO, Lincoln, NE) were added. Also, a pneumatic six-way valve (Valco, Houston, TX) with a 1-ml loop was used for sample injection instead of the WISP autosampler. Both six-way valves were controlled by the HPLC microprocessor for completely automatic operation.

Both analytical and preparative HPLC columns were packed in this laboratory after the method of Coq *et al.* (1975). The construction of the preparative column is described by Vogh and Thomson (1981).

## 2. Materials

Hexane, methanol, and halogenated solvents were Burdick & Jackson Labs, Muskegon, MI, "distilled-in-glass" grade and were used as received. Pure-grade heptane was obtained from Phillips Petroleum, Bartlesville, OK, and ethanol from U.S. Industrial Chemical Co., New York, NY. Acetic, propanoic, and decanoic acids were obtained from Fisher Scientific, Pittsburgh, PA; Mal-

**TABLE II**

**HPLC Chromatographic Conditions: Two-Pump System**

| Condition | Gradient number 1 | Gradient number 2 |
|---|---|---|
| Column length × diameter (cm) | 25.0 × 0.46 | 16.3 × 0.46 |
| Precolumn length × diameter (cm) | 7.0 × 0.46 | 7.0 × 0.46 |
| Column packing ($SiO_2$, Merck) | Si-60, 5-μm | Si-60, 10-μm |
| $N$ (average plates/m) | 10,000–15,000 | 8000–10,000 |
| Precolumn packing ($SiO_2$, Merck) | Si-60, 10-μm | Si-60, 10-μm |
| Flow rate (ml/min) | 2.0 | 2.0 |
| Average back pressure (psi) | 1200 | 1200 |
| Chart speed (cm/min) | 0.5 | 0.5 |
| Gradient conditions: initial %B → final %B | 1 → 100 | 6 → 100 |
| Gradient time (min), time at 100% B (min), reequilibration time (min) | 30, 2, 20 | 40, 0, 20 |
| Gradient curve | Linear | Linear |
| Solvent A | Heptane | 1,2-Dichloroethane containing 0.0335 $M$ acid[a] |
| Solvent B | Dichloromethane containing 0.0268 $M$ acid[a] and 0.3% (v/v) ethanol | 1,2-Dichloroethane containing 0.0335 $M$ acid[a] and 8.0% (v/v) ethanol |

[a] Either acetic, propanoic, 3-chloropropanoic, monochloracetic, decanoic, or no acid.

## TABLE III

### HPLC Chromatographic Conditions: Ternary-Gradient System

| | |
|---|---|
| Column | 15 cm × 0.46-cm i.d.; 10-μm Si-60 (E. Merck) |
| Precolumn | 7 cm × 0.46-cm i.d.; 10-μm Si-60 (E. Merck) |
| $N$ (average plates/m) | 10,000 |
| Flow rate | 2.0 ml/min |
| Chart speed | 0.5 cm/min |
| Temperature | 30.0°C |

Gradient (linear)[a]

| Time (min) | %A | %B | %C | Time (min) | %A | %B | %C |
|---|---|---|---|---|---|---|---|
| | | | | (Begin reequilibration cycle) | | | |
| 0 | 99 | 1 | 0 | 71 | 0 | 100 | 0 |
| 28 | 6.6 | 93.4 | 0 | 75 | 0 | 100 | 0 |
| 30 | 0 | 93.0 | 7.0 | 77 | 99 | 1 | 0 |
| 70 | 0 | 0 | 100 | 95 | 99 | 1 | 0 |

[a] A, Heptane; B, dichloromethane containing 0.2% (v/v) ethanol and 0.0268 $M$ propanoic acid; C, 92% dichloromethane, 8% (v/v) ethanol containing 0.0335 $M$ propanoic acid.

linckrodt, St. Louis, MO; and Aldrich, Milwaukee, WI, respectively. Purification of the heptane, ethanol, and acids has been described earlier (Green, 1981).

In addition, 2- and 3-chloropropanoic and 4-chlorobutyric acids were obtained from Aldrich, and monochloroacetic acid was purchased from J. T. Baker, Phillipsburg, NJ. Initially, purification via formation of ammonium salts was attempted since that method had worked well with the nonhalogenated acids. However, using UV absorbance as the quality criterion, very little improvement was obtained with that method.

An acceptable grade of 3-chloropropanoic acid was finally obtained from vacuum distillation over a 30-cm glass-helices-packed column followed by two recrystallizations from heptane. The same procedure was applied to monochloroacetic acid, but a grade suitable only for preliminary experiments was obtained. No satisfactory method for purifying 2-chloropropanoic and 4-chlorobutyric acids (both liquids) has been discovered as yet.

### 3. Procedure

Tables II–IV list the experimental parameters for the dual-pump, ternary-gradient, and ternary-gradient–preparative-scale separations, respectively. Comparison of Tables II and III reveals that use of the ternary gradient essentially enables the sequential exposure of a single injection to both gradients discussed in Table II, thus avoiding multiple runs on each sample.

Acid and base concentrates of the various fuels were isolated by either nonaqueous ion-exchange chromatography or liquid–liquid extraction into aque-

## TABLE IV

### HPLC Chromatographic Conditions: Preparative Ternary-Gradient System

| | |
|---|---|
| Column | 30 cm × 2.5-cm i.d. 316 ss |
| Packing | Woelm TLC grade (~20-μm) silica |
| $N$ (average plates/m) | 4000 |
| Flow rate | 28 ml/min |
| Chart speed | 0.5 cm/min |
| Temperature | 35.0 ± 0.1°C |

#### Gradient (linear)[a]

| Time (min) | %A | %B | %C | Time (min) | %A | %B | %C |
|---|---|---|---|---|---|---|---|
| | | | | (Begin reequilibration cycle) | | | |
| 0 | 97 | 3 | 0 | 51 | 0 | 100 | 0 |
| 4 | 93 | 7 | 0 | 54 | 0 | 100 | 0 |
| 28 | 20 | 80 | 0 | 57 | 97 | 3 | 0 |
| 32 | 0 | 85 | 15 | 66 | 97 | 3 | 0 |
| 43 | 0 | 0 | 100 | | | | |
| 48 | 0 | 0 | 100 | | | | |

| Fraction collection times | | Backflush conditions | |
|---|---|---|---|
| Fraction number | Time (min) | | |
| 1 | 4–10.0 | Solvent | Pyridine |
| 2 | 10.1–22.5 | Flow rate (ml/min) | 28 |
| 3 | 22.6–34.0 | Total volume (liters) | 1–2 |
| 4 | 34.1–61.0 | Temperature (°C) | 80 |

[a] A = Hexane; B = dichloromethane containing 1.0% (v/v) methanol and 0.15% (v/v) propanoic acid; C = dichloromethane containing 12.5% (v/v) methanol and 0.2% (v/v) propanoic acid.

ous–alcoholic media. Both procedures have been described elsewhere (Green and Hoff, 1981; Guerin et al., 1981). The first method involves passing the sample dissolved in a nonaqueous solvent through two columns packed with macroporous ion-exchange resins. The first column contains an anion resin in the hydroxide form, the second contains a cation-exchange resin in the hydrogen form. Hydrocarbons and other neutrals elute, and the acids and bases are retained on the first and second columns, respectively. They are recovered by unpacking the columns and extracting the resins. The second method is similar in principle, except that the hydrogen and hydroxide ions are dissolved in 2 : 1 methanol–water instead of being bound to ion-exchange resins. Thus, the sample is dissolved in diethyl ether and extracted with 1–2 $M$ NaOH and 1–2 $M$ HCl. The NaOH extract is neutralized and back-extracted to recover the acids, and the HCl extract is similarly treated to recover the bases.

Propanoic acid was removed from fractions obtained from preparative runs by bubbling in ammonia gas and filtering the resulting ammonium propionate solid. Dissolution of the fraction in the least polar solvent possible prior to exposure to gaseous ammonia facilitated the most rapid and complete formation of the salt. Often the sample required two or three exposures and filtrations to effect complete removal of the propanoic acid.

## B. Results and Discussion

Figure 1 summarizes the main effects of adding carboxylic acids to mobile phases in separations on silica. Obviously, many of the components (numbers 1–9; see Table V) are affected very little by the addition of acetic acid to the mobile phase. However, components with a basic character (numbers 10–17) are generally more retained, and in some cases (numbers 16 and 17) much more retained. In fact, addition of carboxylic acids makes the order of elution of basic solutes correspond more closely to their inherent basicity (Green, 1981). This accounts for the large shift in the retention of $N$-methylaniline and $N,N$-dimethylaniline, since N-alkylation increases the basicity of anilines. The peak shape of many solutes improves with the addition of carboxylic acids to the mobile phase. However, this feature is less obvious in Fig. 1 because most of the solutes shown chromatograph well in almost any type of system.

Figure 2 shows how a ternary gradient can be used to resolve mixtures containing solutes of widely varying polarity (see also Table VI). The synthetic mixture shown contains all of the compounds in Fig. 1 plus a set of pyridine analogs and two dinitrogen compounds. The solutes fall into four general categories: (1) hydrocarbons + N-alkylated pyrroles, (2) pyrroles and $N$-arylanilines, (3) anilines, and (4) pyridine homologs and dinitrogen compounds. Addition of the propanoic acid to the mobile phase improves the chemical class nature of the separation by increasing the retention difference between the pyrrole and aniline classes. Other types of compounds can overlap and complicate the picture (Green, 1981), but the separation pattern in Fig. 2 still represents a very effective one for base extracts from fuels and potentially for other types of samples. Analytical-scale chromatograms can yield useful qualitative information and, for simple mixtures such as the one shown, quantitative data. Preparative-scale separations (see below) yield chemically unique fractions for spectroscopic analysis and/or further separations.

Figure 3 shows some examples of the type of separation shown in Fig. 2 applied to real samples. Figure 3A shows an abundance of aniline- and pyridine-type components typical of a relatively low-boiling coal-liquid distillate. The residue bases from the same coal liquid (Fig. 3B) show a greater predominance of pyridine benzologs than aminoarenes. Also, the >370°C sample contains predominantly polycyclic azaarenes as evidenced by the generally decreased

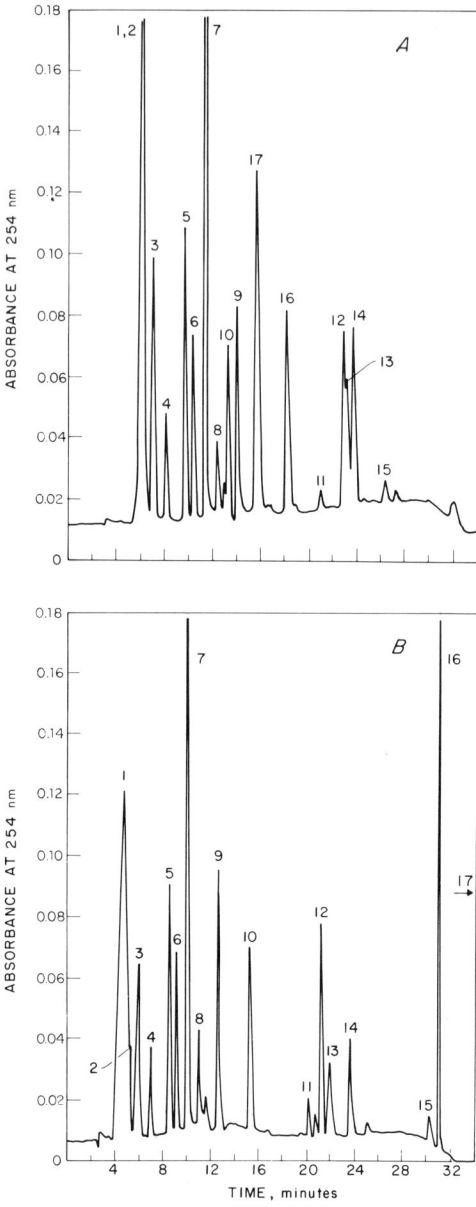

**Fig. 1.** Separations of a synthetic mixture of nitrogen-containing compounds, showing the effect of adding a carboxylic acid to the mobile phase. (A) No acid added; (B) 0.0268 $M$ acetic acid. See Table II, gradient 1 for experimental conditions and Table V for peak identities. Absolute amounts of each solute injected range from 0.3 to 1.0 µg.

## TABLE V

### Peak Identities for Figs. 1, 11, 12, and 15

| Peak no. | Compound | Peak no. | Compound |
|---|---|---|---|
| 1 | Chrysene | 10 | N-Phenylbenzylamine |
| 2 | Benzo[ghi]perylene | 11 | 2,6-Dimethylaniline |
| 3 | 9-Ethylcarbazole | 12 | 6-Aminochrysene |
| 4 | 1-Methylindole | 13 | p-Chloroaniline |
| 5 | N-Phenyl-1-naphthylamine | 14 | 2-Aminoanthracene |
| 6 | Diphenylamine | 15 | Aniline |
| 7 | Dibenzo[a,c]phenazine | 16 | N-Methylaniline |
| 8 | 3-Methylindole | 17 | N,N-Dimethylaniline |
| 9 | 13H-Dibenzo[a,i]carbazole | | |

retention and resolution of the pyridine class. This behavior results from the decreasing basicity of pyridine benzologs as the number of condensed aromatic rings increases (see Fig. 2), caused by increased resonance delocalization of the lone pair of electrons on the nitrogen atom. Current work with more acidic mobile-phase additives (see below) is aimed at improving resolution of these higher boiling bases. Finally, Fig. 3C shows the separation of a shale-oil base extract from which only the very volatile (<200°C) material has been removed. Since the UV response of the higher ring azaarenes is much greater per unit weight than that of most other compounds, they are by far the most prominent feature of the chromatogram. Also, a great similarity is evident between Figs. 3B

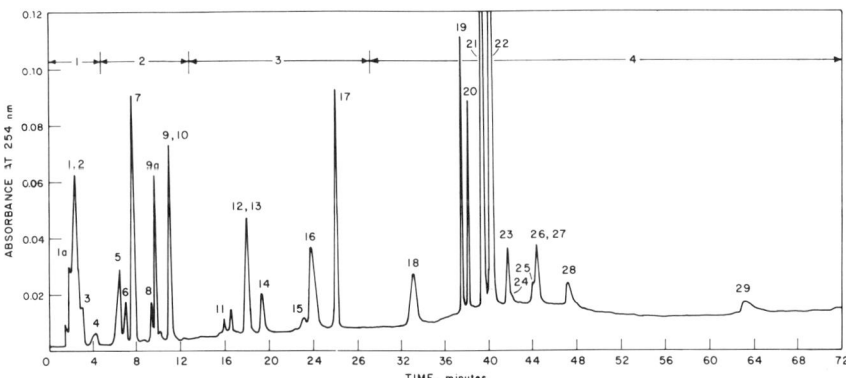

**Fig. 2.** Compound-class separation of basic solutes using a ternary gradient. See Table III for experimental conditions and Tables V and VI for peak identities. Absolute amounts of each solute injected range from 0.2 to 0.5 μg.

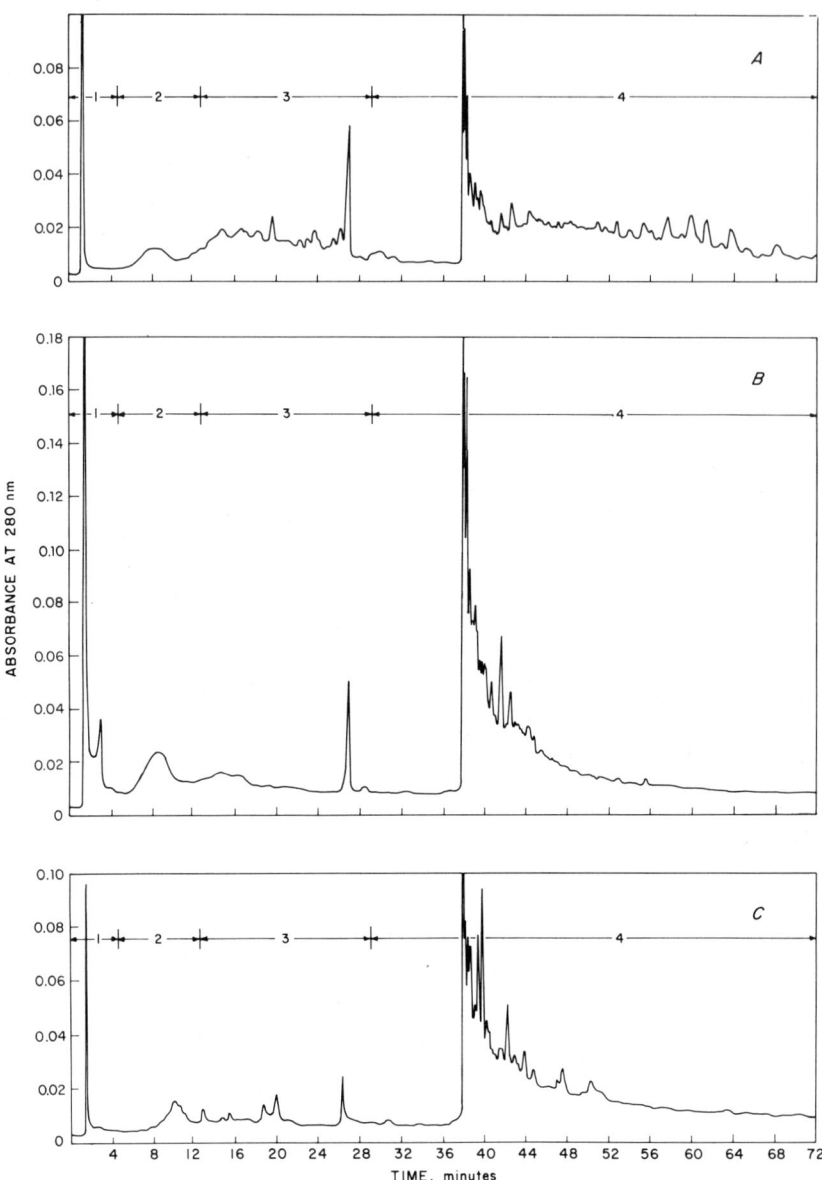

**Fig. 3.** Ternary-gradient separation of coal-liquid and shale-oil base concentrates. See Table III for conditions. (A) 124 μg western Kentucky coal-liquid bases from 200–370°C distillate; (B) 66 μg western Kentucky >370°C residue bases. (C) 96 μg >200°C strong-bases concentrate from shale oil. See text for explanation of retention regions 1–4.

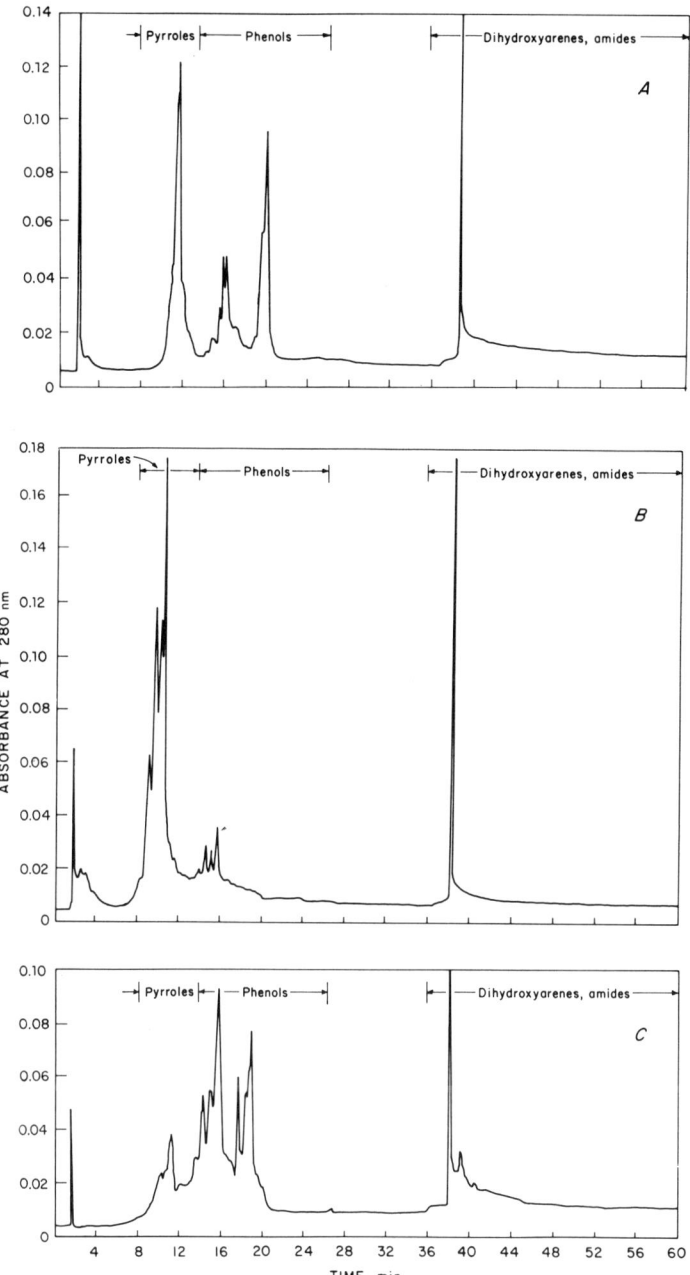

**Fig. 4.** Ternary-gradient separation of acids isolated from shale oil by two different methods. See Table III for chromatographic conditions. (A) Strong and (B) weak acids isolated by nonaqueous ion-exchange chromatography; (C) total acids isolated from extraction with 1 $N$ NaOH. Amounts injected in (A)–(C) were 80, 40, and 89 µg, respectively.

and C. This superficial similarity is largely due to inadequate resolution of the individual compounds (several hundred in each liquid), and the previously mentioned bias of the UV detector toward large chromophores present in both extracts. Thus, LC separations, with the present state of the art, have only limited usefulness for fingerprinting complex mixtures.

Figure 4 shows that carboxylic acid-containing mobile phases also perform reasonably well with acidic solutes. The shale-oil sample was the same one from which the bases shown in Fig. 3C were isolated. As indicated in the Fig. 4 legend, acids were isolated by both nonaqueous ion-exchange chromatography and (on a different aliquot) liquid–liquid extraction. The ion exchange procedure yielded a strong-acid fraction (Fig. 4A), composed largely of phenols with minor amounts of pyrroles and what could either be polyhydroxylated compounds or amides, and a weak-acid concentrate (Fig. 4B) made up mostly of pyrroles with a minor amount of phenols that escaped retention on the first pass through the ion-exchange columns. The extraction acids (Fig. 4C) are nearly devoid of pyrroles due to their very weak acidity. In fact, even the first group of compounds in Fig. 4C, whose retention ($\simeq 10$ min) corresponds to either pyrroles or hindered phenols, is believed to be mostly phenols, because the ratio of peak area at 280 nm to that at 254 nm is more characteristic of phenols than members of the pyrrole family (pyrrole, indole, carbazole, etc.).

Thus, whenever different fuel extracts have significant differences in chemical classes present, this LC method is quite useful as a qualitative tool. However, as pointed out in the discussion of Fig. 3, structural variability within a given class of compounds is often difficult to spot by this method. In spite of several limitations, separations based on carboxylic acid-containing mobile phases and unmodified silica greatly outperform all bonded-phase systems tried so far. Reverse-phase chromatography merely yields a "great smear" elution pattern based mainly on alkyl-chain length and degree of aromaticity. Also, normal-phase separations on cyano- and amine-bonded phases give poor correlation to functional groups present in solutes compared to those in Figs. 3 and 4. Examples of CN and $NH_2$ bonded-phase separations are presented in Section IV.

Figure 5 shows a preparative-scale separation of bases from a high-boiling petroleum distillate. Petroleum samples contain somewhat different types of compounds than synfuels (coal liquids and shale oils) and are generally more complex in terms of number of isomers present and variation in length of alkyl chains attached to aromatic rings. Nevertheless, a reasonable chromatogram is obtained with what is essentially a scaled-up version of the separations shown in Figs. 2–4. Characterization of the five fractions (including an 80°C pyridine backflush to elute strongly retained material) via IR spectrometry and additional separation techniques is in progress. The mass balance from the separation of a total of 1.6 g of base concentrate in 16 injections is shown in Table VII. It should be noted that the 100-mg injections were extremely conservative for the column

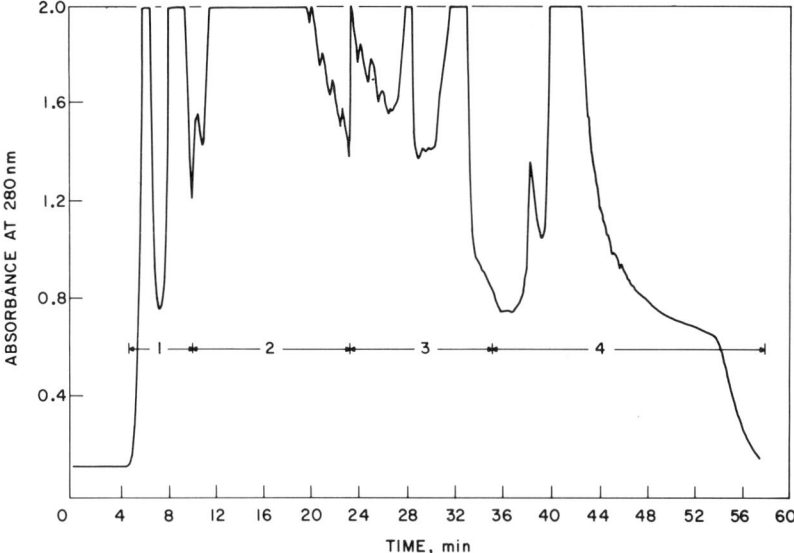

**Fig. 5.** Preparative separation of bases from a 535–675°C distillate of a Wilmington, CA, crude oil. Approximately 100 mg was injected in 1 ml. See Table IV for conditions and Table VII for mass balance of fractions. Regions 1–4 indicate cut points for fractions.

size used (Table IV). However, with the completely automated apparatus, the experiment still took less than 24 hr and yielded fractions of the highest quality possible with the column and separation techniques available.

As mentioned earlier, work is currently under way to determine if using more acidic mobile-phase additives can further improve the resolution of basic materials. Figure 6 shows results obtained with 3-chloropropanoic acid on middle- and high-boiling SRC-II coal-liquid bases. Obviously, the resolution is greatly superior with the added acid than without. The same samples have also been chromatographed using acetic, propanoic, and decanoic acids in the mobile phase (Green, 1981). Especially for the 425°+ bases, resolution using 3-chloropropanoic acid was noticeably superior to that obtained using any of the other acids. Preliminary results with monochloroacetic acid showed considerably increased retention of bases over even 3-chloropropanoic acid-containing mobile phases. However, some of the more basic solutes tailed badly. Thus, it appears that a point of diminishing returns may be reached in the search for more acidic additives, possibly resulting from a kinetically slow interaction between relatively strong acids and bases.

In conclusion, carboxylic acid mobile-phase additives can be used to improve resolution between entire chemical classes of compounds or of solutes within a chemical class. In general, this technique gives the greatest improvement in

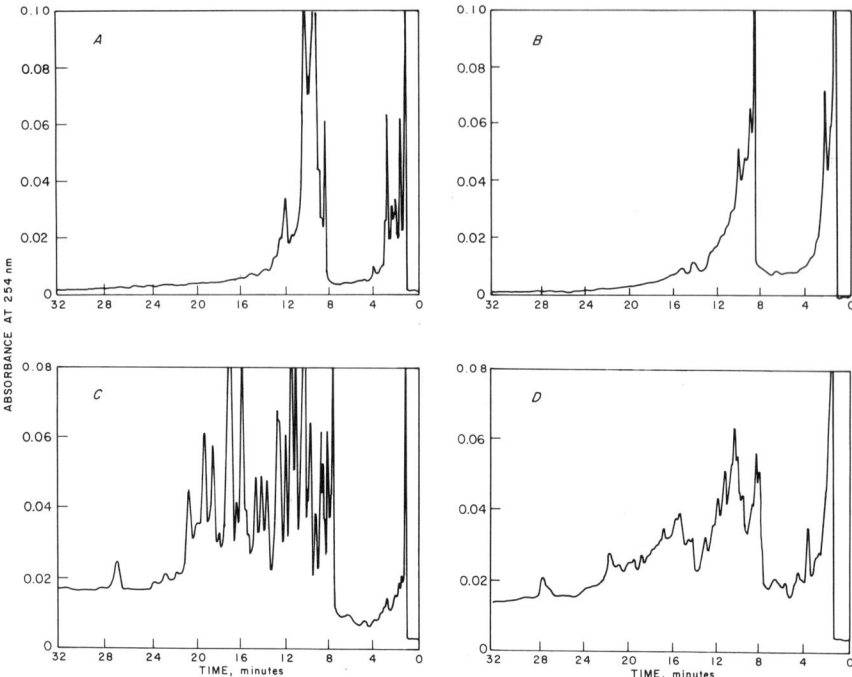

**Fig. 6.** A comparison of resolution of high-boiling SRC-II coal-liquid bases with and without 3-chloropropanoic acid in the mobile phase. (A) 7.6 μg 325–425°C and (B) 9.4 μg 425°C+ bases without acid; (C) 32 μg 325–425°C and (D) 31 μg 425°C+ bases with 0.0335 $M$ acid in the mobile phase. See Table II, gradient 2 for experimental conditions.

separations of basic solutes but can be used with a wide variety of materials. This separation approach shows promise for application on mixtures found in energy, environmental, and many other areas.

## IV. NONAQUEOUS SYSTEMS CONTAINING AMINES

Dissatisfaction with the resolution between groups of acidic solutes (see Section III) and the difficulty in removing carboxylic acids from fractions obtained from preparative runs of acid concentrates of fuels prompted the research reported in this section. Also, it was hoped that the amine-based separations might have a different enough selectivity from the carboxylic acid-spiked systems to allow the two modes to be used together to achieve optimum resolution of complex mixtures. To some degree, this hope has been realized. However,

## TABLE VI

### Peak Identities for Figs. 2 and 8

| Peak no. | Compound | Peak no. | Compound |
|---|---|---|---|
| 1a | Triphenylamine | 23 | Isoquinoline |
| 9a | Carbazole | 24 | Tetramethylpyrazine |
| 1–17 | Same as Table V | 25 | 3-Methylpyridine |
| 18 | 2-Azafluoranthene | 26 | 2,6-Dimethylquinoline |
| 19 | 1-Azapyrene | 27 | 5-Aminoindole |
| 20 | 4-Azafluorene | 28 | 2-Methylpyridine |
| 21 | Acridine | 29 | Benzimidazole |
| 22 | 2-Methylacridine | | |

selectivity using amine-loaded mobile phases turned out to be considerably different from what was originally envisioned.

As with carboxylic acids, adding amines to nonaqueous mobile phases in LSC is not new. For example, Snyder (1968) mentions amine-containing solvents in his discussion of "the basic solvent anomaly." Also, Guillemin (1980) and Colin et al. (1981) have reported using alkylamines in separations of basic materials on silica to improve peak shape. However, the effects of amines on the retention behavior of widely varying solutes has not been systematically investigated.

Acidic solutes are usually separated on silicas bonded with amino-, cyano-, or long-chain hydrocarbon functionalities (Schabron et al., 1978). Thus, for comparison, experiments were also conducted with some bonded-phase columns.

## TABLE VII

### Preliminary Results from a Preparative Separation of Bases from a 535–675°C Distillate of a Wilmington, CA Crude Oil

| Fraction | Weight percent | Major compound type |
|---|---|---|
| 1 | 6.9 | Hydrocarbons, $N$-alkylpyrroles |
| 2 | 24.0 | Pyrrolic types, unknown |
| 3 | 22.2 | Thiazoles, unknown |
| 4 | 45.3 | Pyridine benzologs |
| Backflush | 1.8 | Unknown |
| Total | 100.2 | |

## A. Experimental

### 1. Apparatus

A standard two-pump gradient HPLC system with controller (Waters Associates), dual-wavelength detector (Spectra Physics M230 or Waters M440), and autoinjector (Waters WISP) was used throughout the preliminary work reported here, except that a fluorescence detector (Schoffel FS 970, Kratos, Westwood, NJ) was used whenever the mobile phase contained pyridine.

TABLE VIII

**Amines Involved in the Preliminary Study**

| Name | Supplier[a] | Purification method | Approximate UV cutoff (nm)[b] |
|---|---|---|---|
| tert-Butylamine | A | Distillation[c] | 253 |
| Isopropylamine | E | None | 267 |
| n-Butylamine | F | None | 273 |
| Triethylamine | A | Distillation, salt formation[d] | 250 |
| Monoethanolamine | F | None | 268 |
| Triethanolamine | F | None | 274 |
| Morpholine | A | Distillation, salt formation | 250 |
| 2-Methoxyethylamine | A | Distillation | 264 |
| N,N,N,N-Tetra(hydroxyethyl)ethylenediamine | E | None | 275 |
| Hydroxylamine | B | Neutralization[e] | 313 |
| N,N-Diethylhydroxylamine | A | Distillation, salt formation | 260 |
| Ethylenediamine | F | Triple distillation | 266 |
| N,N,N,N-Tetramethylethylenediamine | F | Double distillation | 292 |
| Piperazine | A | Recrystallization[f] | 263 |
| Hexamethylenetetramine | A | None | 295 |
| Pyridine | BJ | None | 330 |
| Tetramethylammonium hydroxide | A | None | 270 |
| Tetrabutylammonium hydroxide | MCB | None | 247 |
| Tetrabutylammonium bromide | A | None | 233 |

[a] A, Aldrich Chemical, Milwaukee, WI; E, Eastman Kodak, Rochester, NY; F, Fisher Scientific, Pittsburgh, PA; B, J. T. Baker, Phillipsburg, NJ; BJ, Burdick and Jackson, Muskegon, MI; MCB, MCB, Cincinnati, OH.

[b] Neat liquid or 20% (w/w) solution.

[c] Vacuum or atmospheric distillation over a 30-cm glass-helices column.

[d] Dissolution of the amine in ethyl ether; addition of a slight excess of 98% sulfuric acid; recovery of the sulfate salt; recrystallization of the salt (usually in tetrahydrofuran); regeneration of the amine by addition of a gross excess of KOH; followed by distillation to recover the amine.

[e] Hydroxylamine hydrochloride was treated with MgO in methanol to form the free amine and insoluble $MgCl_2$.

[f] Piperazine was recrystallized from ethyl acetate.

## 2. Materials

Table VIII lists the amines, their suppliers, and purification methods used in this study. During the preliminary screening of amines, rigorous purification was generally not done as long as a reasonable baseline could be obtained at 280 nm. In general, primary amines were cleanest; tertiary amines were highly contaminated and difficult to purify. As indicated by the spacing in Table VIII, there were five groups of amines investigated: alkylated, oxygen-containing, those containing multiple nitrogen atoms, aromatic, and quaternary ammonium salts.

The bulk mobile-phase components—hexane, methyl *tert*-butyl ether, methanol, and ethyl acetate—were purchased from either Burdick and Jackson or Fisher and were used as received. Silica for columns was obtained from E. Merck, Darmstadt, West Germany (Lichrosorb Si-60, 10 μm); Whatman, Clifton, NJ (Partisil-10, 10 μm); and DuPont, Wilmington, DE (Zorbox, 7–8 μm). the columns were packed downward using carbon tetrachloride as the slurry medium and pentane as the driving solvent after the method of Coq *et al.* (1975).

For the purpose of comparison, experiments were also run on covalently bonded cyano and amino columns. The cyano column was purchased prepacked from DuPont; $NH_2$ columns were packed in this laboratory using Merck and DuPont materials and the same packing procedure used for silica columns.

## 3. Procedure

Most of the experimentation was carried out under conditions set forth in Table IX. However, use of tetraalkylammonium salts necessitated the more polar mobile-phase systems described in Table X. As shown in Table XI, pyridine was unique in that it was not added to the mobile phase in small amounts, but instead was one of the bulk components of the mobile phase. Finally, conditions for separations with bonded-phase columns and the control runs (no amines) on silica are described in Table XII. It should be pointed out that the silica column and precolumn used in the control experiments had never been exposed to any acidic or basic mobile-phase additives. This is important due to substantial "memory effect" resulting from exposure to mobile phases containing polar, strongly retained additives.

All chromatographic systems were characterized with a minimum set of acidic, basic, and neutral solutes. In addition, real samples were chromatographed in some of the more promising systems.

## B. Results and Discussion

### 1. Work with Bonded Phases

As mentioned in the introduction to this section, one of the main thrusts of the work with basic mobile-phase additives was to improve separations of acidic

TABLE IX
Typical Conditions for Preliminary Experiments with Amine-Spiked Mobile Phases

| Parameter | Primary | Typical entries | | |
|---|---|---|---|---|
| | | Alternate 1[a] | Alternate 2[b] |
|---|---|---|---|
| Column packing (length × i.d., cm) | DuPont (25 × 0.46) | Partisil-10 (15 × 0.46) | Li-60 (15 × 0.46) |
| Precolumn packing (length × i.d., cm) | None | Li-60 (7 × 0.46) | — |
| Flow rate (ml/min) | 2.0 | — | — |
| Chart speed (cm/min) | 0.5 | — | — |
| Back pressure (psig) | 500 | — | — |
| Gradient parameters: | | | |
| A solvent | Hexane | — | — |
| B solvent | Ethyl acetate containing 0.1–1.0% (w/v) amine | Methyl tert-butyl ether containing 0.1–2.0% (w/v) amine + 0–2% MeOH | Glyme or dimethoxymethane containing 0.1–1.0% (w/v) amine |
| Gradient shape | Linear | — | — |
| [B], initial → final % (v/v) | 2 → 75 | 4 → 100 | 2 → 100 |
| Time (min): programming, hold at end of program, reequilibration | 33, 2, 18 | — | — |
| Detection: AUFS × nm | 0.32 × 280 | 0.32 × 280+<br>0.32 × 254 | — |

[a] Dash indicates parameter is unchanged from primary condition.
[b] Dash indicates parameter is unchanged from alternate 1 condition.

## TABLE X
### Appropriate Conditions for Tetraalkylammonium Salt-Containing Systems

| Parameter | Typical entries | | |
|---|---|---|---|
| | Primary | Alternate 1[a] | Alternate 2[b] |
| Column packing (length × i.d., cm) | DuPont (25 × 0.46) | Partisil-10 (15 × 0.46) | Li-60 (15 × 0.46) |
| Precolumn packing (length × i.d., cm) | None | Li-60 (7 × 0.46) | — |
| Flow rate (ml/min) | 2.0 | — | — |
| Chart speed (cm/min) | 0.5 | — | — |
| Back pressure (psig) | 500–1000 | — | — |
| Gradient parameters: | | | |
| A solvent | Methyl *tert*-butyl ether (MTBE) | MTBE containing 0.01–0.2% amine | — |
| B solvent | 60% MTBE, 40% MeOH containing 0.01–0.02% amine | 60% MTBE, 40% MeOH | MeOH |
| Gradient shape | Linear | Convex | — |
| [B], initial → final % (v/v) | 4 → 100 | 1 → 100 | — |
| Time (min): programming, hold at end of program, reequilibration | 48, 2, 20 | 48, 2, 18 | — |
| Detection: AUFS × nm | 0.2 × 254+ | 0.2 × 280+ | — |
| | 0.2 × 280 | 0.2 × 313 | |

[a] Dash indicates parameter is unchanged from primary condition.
[b] Dash indicates parameter is unchanged from alternate 1 condition.

**TABLE XI**

**Conditions for Separations with Pyridine**

| Parameter | Typical entries | |
|---|---|---|
| | Primary | Alternate |
| Column packing (length × i.d., cm) | Li-60 (15 × 0.46) | — |
| Precolumn packing (length × i.d., cm) | Li-60 ( 7 × 0.46) | — |
| Flow rate (ml/min) | 2.0 | — |
| Chart speed (cm/min) | 0.5 | — |
| Back pressure (psig) | 800 | — |
| Gradient parameters: | | |
|   A solvent | Heptane | Hexane |
|   B solvent | Pyridine | 95:5 Pyridine/isopropanol |
|   Gradient shape | Linear | — |
|   [B], initial → final % (v/v) | 1 → 20 | 2 → 100 |
|   Time (min): programming, hold at end of program, reequilibration | 35, 2, 18 | 30, 3, 20 |
| Detection: | | |
|   Fluorescence | $\lambda_{ex} = 330$ nm, $\lambda_{em} \geq 370$ nm | — |
|   UV | 0.1 AUFS × 365 nm | — |

compounds in fuels. To obtain a basis for evaluating the various systems, initial experiments involved separations with state-of-the-art bonded-phase columns, and plain silica with conventional mobile phases. Figure 7 shows results obtained on a mixture of pyrrolic and phenolic compounds—two major types found in fuels. Obviously, the amino-silica column yielded the superior separation. However, separation of the pyrrole and phenol families is incomplete even on that column. On the other hand, the separation on plain silica gave almost no resolution between any of the acidic solutes.

As discussed in Section III, mobile phases containing carboxylic acids give superior group-type separations of basic solutes. A convincing demonstration of this superiority is evident in comparing the separations shown in Fig. 8 to the one shown earlier in Fig. 2. Obviously, the well-defined fractions of Fig. 2 are completely absent in any of the Fig. 8 separations. Of the three systems, the plain silica gives the best group-type separation for bases. However, peak shape for the more basic pyridine benzologs is poor without any type of polar mobile-phase additive. Also, reproducibility of retention times is poor whenever no polar additives are present.

In addition to incomplete separation of pyrrolic and phenolic compounds, the amino-silica column also suffered from a rapid chemical degradation under conditions described in Table XII. Figure 9 shows separations of a group of anilines on a new column and the same column after it had been used continuously for 1

## TABLE XII

### Conditions for Cyano and Amino Bonded-Phase and Conventional (No Amines) Silica Separations

| Parameter | Column packing (length × i.d., cm) | | |
|---|---|---|---|
| | DuPont CN (25 × 0.46) | DuPont NH$_2$ (25 × 0.46) | Partisil-10 SiO$_2$ (15 × 0.46) |
| Precolumn packing, (length × i.d., cm) | None | None | Li-60 (7 × 0.46) |
| Flow rate (ml/min) | 2.0 | 2.0 | 2.0 |
| Chart speed (cm/min) | 0.5 | 0.5 | 0.5 |
| Back pressure (psig) | 600 | 400 | 400 |
| Gradient parameters: | | | |
| A solvent | Hexane | Hexane | Hexane |
| B solvent | Ethyl acetate | Ethyl acetate | Ethyl acetate |
| Gradient shape | Linear | Linear | Linear |
| [B], initial → final % (v/v) | 1 → 25 | 2 → 75 | 2 → 75 |
| Time (min): programming, hold at end of program, reequilibration | 30, 0, 15 | 33, 2, 18 | 33, 2, 18 |
| Detection: AUFS × nm | 0.32 × 280 | 0.32 × 280 | 0.32 × 280 |

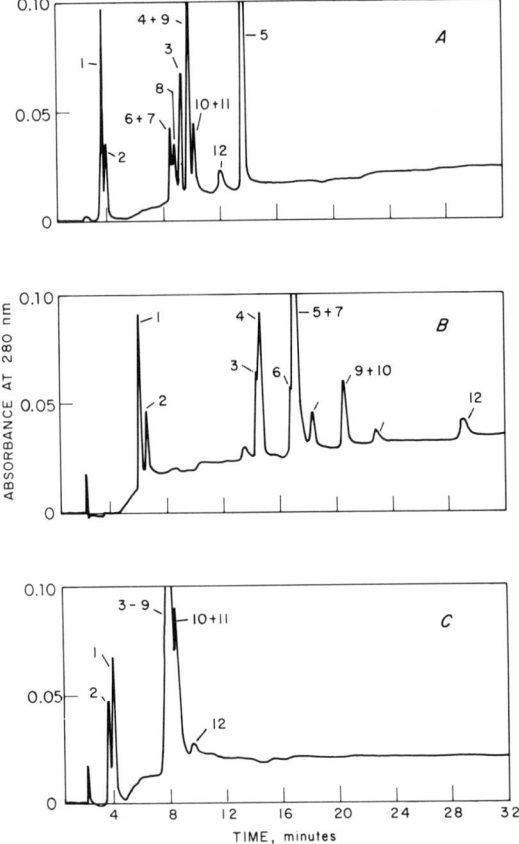

**Fig. 7.** Separations of a synthetic blend of pyrrolic and phenolic compounds on (A) cyano and (B) amino bonded-phase columns and (C) plain silica. See Table XII for chromatographic conditions and Table XIII for peak identities. Absolute amounts of each solute injected ranged from 0.5 to 1.2 μg.

week. The instability of these columns has been noted previously (Wheals and White, 1979; Karlesky et al., 1981) and was observed with both Merck and DuPont packing materials.

### 2. Aliphatic Amines

Given the rather unimpressive performance of the conventional normal-phase systems, experimentation with the various amines was initiated. Preliminary work with these systems (Tables VIII–XI) led to the following conclusions:

1. The degree of alkylation, the type of oxygen functionalities present, and the number of nitrogen atoms per amine molecule are all important in determin-

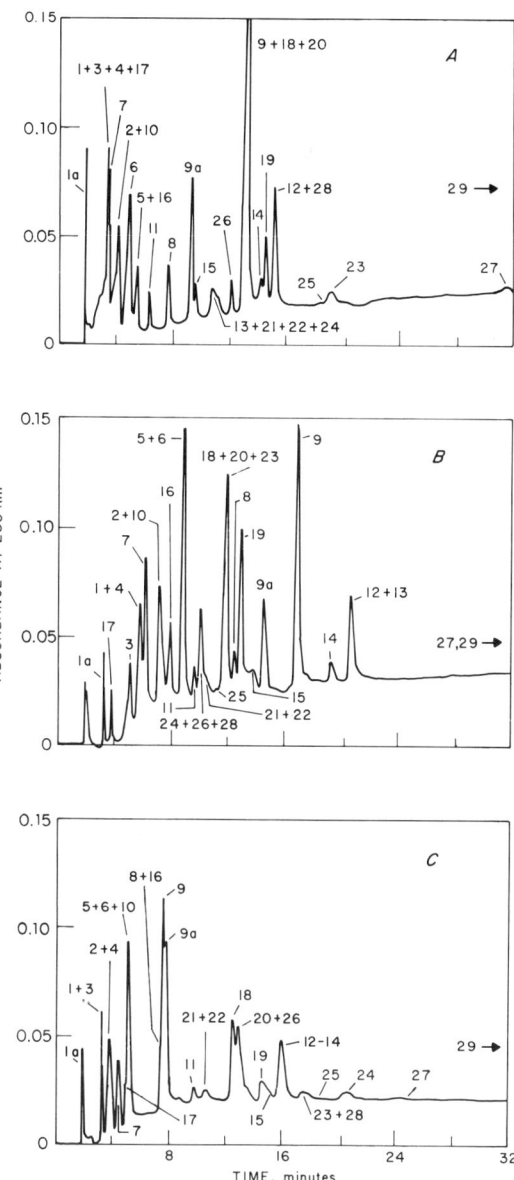

**Fig. 8.** Separations of a synthetic blend of nitrogen-containing compounds on (A) cyano and (B) amino bonded-phase columns and (C) plain silica. See Table XII for chromatographic conditions and Tables V and VI for peak identities (cf. Fig. 2). From 0.2 to 0.5 μg of each solute was injected.

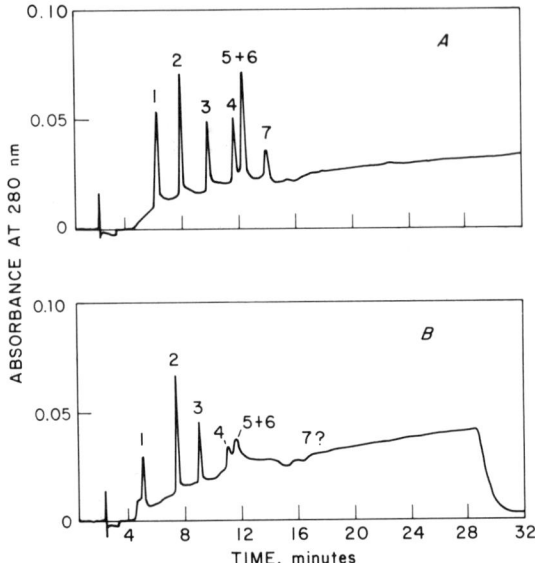

**Fig. 9.** Effect of column age on separation quality for $NH_2$-silica bonded-phase columns: (A) new column; (B) 1-week-old column. Peak identities: (1) 1,2,3,4-tetraphenylnaphthalene; (3) 2,4,6-trimethylaniline; (7) *p*-ethylaniline; (2), (4), (5), and (6) 2,6-, 2,5-, 2,3-, and 2,4-dimethylaniline, respectively.

ing the effect of a given amine as a mobile-phase additive. As an example of the effect of alkylation, tertiary amines, although the most basic, usually are the least effective in retaining acidic compounds. To show the effect of oxygen, phenols yield sharp, symmetrical peaks when either mono- or triethanolamine is added to the mobile phase, but are grossly tailing, barely distinguishable peaks whenever the mobile phase contains 2-methoxyethylamine. Finally, amines such as ethylenediamine that contain more than one nitrogen and at least two N–H functionalities retain acidic solutes as well as the covalently derivatized amino-silicas.

2. As chromatographic systems, amine-containing mobile phases and plain silica are generally highly efficient and quite reproducible unless the amine is grossly impure.

3. Tetraalkylammonium hydroxides are most effective for class separation of acidic materials.

4. Aromatic amines provide interesting separations, but their toxicity and high UV cutoff limit their utility.

5. Materials prefractionated with mobile phases containing carboxylic acids can be further separated in systems with amine-containing phases.

The following discussion will illustrate a portion of the work which led to these conclusions.

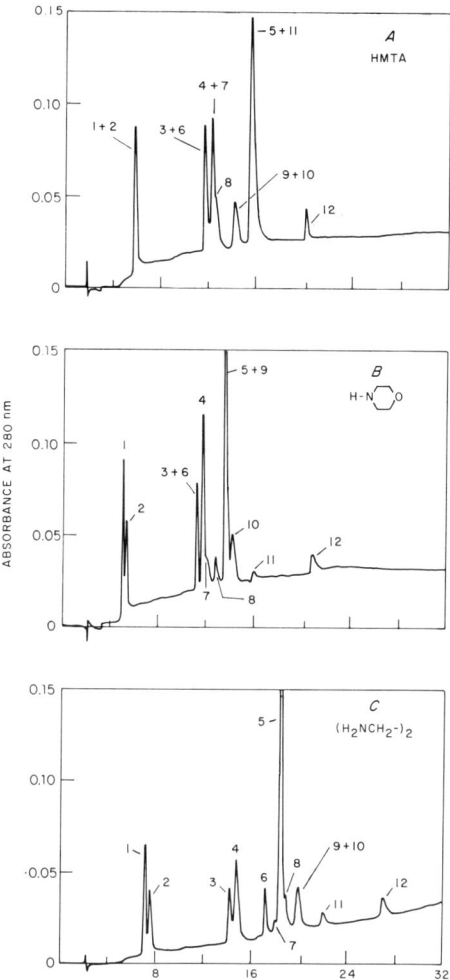

**Fig. 10.** Effects of different aliphatic amines on separations of pyrrolic and phenolic compound types on silica (cf. Fig. 7). See Table XIII for peak identities; for experimental conditions, see primary entries in Table IX. The B solvents used were: (A) ethyl acetate containing 1.0 g/liter hexamethylenetetramine; (B) ethyl acetate containing 0.2% (v/v) morpholine; and (C) glyme (1,2-dimethoxyethane) containing 0.1% (v/v) ethylenediamine.

Figure 10 shows the effect of adding three different amines to the same bulk mobile-phase system for separation of the mixture of pyrroles and phenols shown in Fig. 7. Obviously, complete separation of the two classes is not achieved in any of the three systems. However, it is interesting to note the close resemblance of the chromatogram obtained by adding ethylenediamine to the mobile phase (Fig. 10C) to the one obtained using the covalently derivatized amine column (Fig. 7B). The system containing hexamethylenetetramine (Fig. 10A) yielded good peak symmetry yet poor retention of acidic compounds. In light of its highly symmetrical structure

hexamethylenetetramine was expected to cover the silica surface with one nitrogen atom sticking up above the surface. This may explain the good peak symmetry obtained; the poor retention of acids may be due to inductive effects of the other nitrogen atoms on the one exposed, thus lowering its effective basicity. Systems containing morpholine (Fig. 10B) show considerably different behavior than those spiked with 2-methyoxyethylamine. Thus, the geometric shape of the amine as well as the functionalities present influence its behavior as a mobile-phase additive.

Figure 11 shows a chromatogram of the same mixture of nitrogen-containing

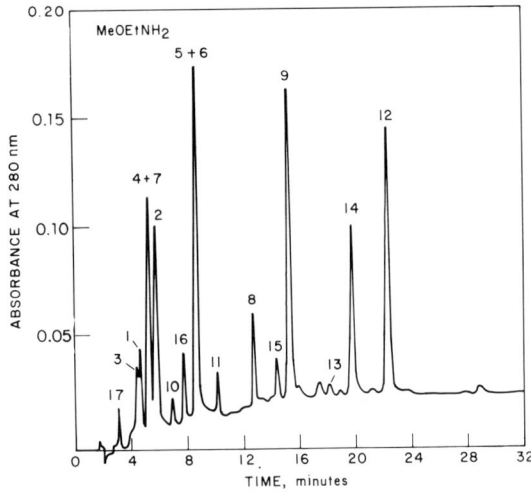

**Fig. 11.** Separation of nitrogen compounds using a mobile phase spiked with 2-methoxyethylamine (cf. Fig. 1). For peak identities see Table V; for experimental conditions see alternate 1 entries, Table IX. The B solvent in the gradient was methyl *tert*-butyl ether containing 0.5% (v/v) amine.

**TABLE XIII**

Peak Identities for Figs. 7, 10, and 13A

| Peak no. | Compound | Peak no. | Compound |
|---|---|---|---|
| 1 | 9-Methylcarbazole | 7 | 2,3-Dimethylphenol |
| 2 | Benzo[a]pyrene | 8 | o-Cresol |
| 3 | Indole | 9 | m-Ethylphenol |
| 4 | Carbazole | 10 | 3,4-Dimethylphenol |
| 5 | 13H-Dibenzo[a,i]carbazole | 11 | Phenol |
| 6 | 2,4,5-Trimethylphenol | 12 | 2-Naphthol |

compounds discussed earlier in Section III (Fig. 1). Obviously, separation selectivity in systems containing amines is much different than in those containing carboxylic acids. Several features of Fig. 11 merit attention. First, diarylamine compounds (peaks 5, 6, and 10) are clearly separated from pyrroles (peaks 8 and 9) when amine-containing mobile phases are used, whereas they are mixed in carboxylic acid-spiked systems (Fig. 1). Second, hydrocarbons (peaks 1 and 2) are surprisingly well retained in Fig. 11. Third, aniline homologs elute in the order of increasing *acidity* in the amine-containing system. For example, *N,N*-dimethylaniline (peak 17), the most basic aniline present, is the earliest eluting compound in Fig. 11. On the other hand, 6-aminochrysene, in which the lone pair of electrons on the nitrogen are extensively delocalized by the large chrysene

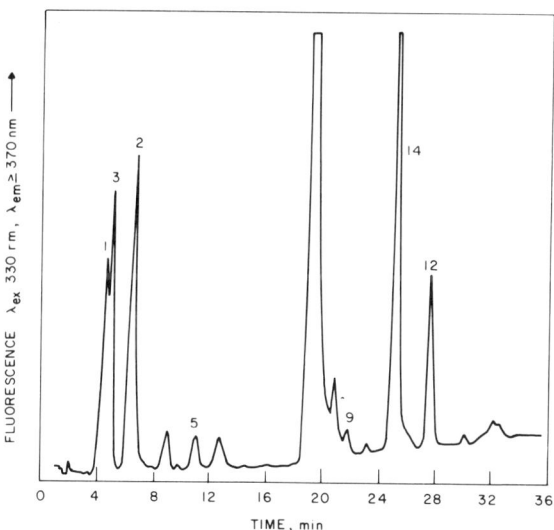

**Fig. 12.** Chromatogram of nitrogen compounds obtained with a pyridine-containing mobile phase. See Table V for peak identities, Table XI (primary entries) for experimental conditions.

moiety, is the last component to elute in the entire mixture. This is an important feature for work with fuels, since amino-polycyclic aromatic hydrocarbons (amino-PAHs) are important carcinogens in coal liquids (Guerin *et al.*, 1980). Thus, amine-containing mobile phases would be useful in the further separation of a prefractionated arylamine concentrate or a mixture of pyrroles and diarylamines, but not a gross mixture of many types of nitrogen-containing components.

### 3. Mobile Phases Containing Pyridine

Figure 12 shows the same mixture of nitrogen compounds as Figs. 1 and 11, separated with a heptane–pyridine gradient (Table XI). Since only 7 out of 17 compounds in the mixture were detected by fluorescence, it is difficult to compare Fig. 12 to Figs. 1 and 11. However, the peaks which were detected seem to follow the pattern of Fig. 11 fairly well. Isocratic work with heptane–pyridine concentration ratios between 100 and 10 and UV detection at 305 nm enabled detection of 7 additional compounds. Piecing those chromatograms together with Fig. 12 resulted in an elution pattern similar to that obtained with the aliphatic amines, e.g., Fig. 11. Additional work with pyridine benzologs as solutes also indicated similar behavior between pyridine and aliphatic amine-containing mobile phases.

### 4. Tetraalkylammonium Hydroxides

The potential of tetraalkylammonium hydroxides as mobile-phase additives for the separation of acidic materials is demonstrated in Figs. 13 and 14 (see also Tables XIV and XV). Stalling *et al.* (1981) have shown that these bases and alkali-metal hydroxides react with silanol groups to form tetraalkylammonium and alkali-metal silicates, respectively. They evaluated these substrates for preconcentration of phenols from fish tissue, and concluded that cesium silicate was best for that application. For solubility reasons, however, tetramethylammonium and tetrabutylammonium hydroxides are more suitable as mobile-phase additives in this work. Although tetramethylammonium hydroxide is slightly superior,

**TABLE XIV**

Peak Identities for Fig. 13B

| Peak no. | Compound | Peak no. | Compound |
|---|---|---|---|
| 1 | 2,4,6-Tri-*tert*-butylphenol | 6 | 2,3,5-Trimethylphenol |
| 2 | 2,3,4,5-Tetraphenylpyrrole | 7 | 3,4,5-Trimethylphenol |
| 3 | Benzanthrone | 8 | 3,5-Dimethylphenol |
| 4 | 4-Phenoxybutyronitrile | 9 | *p*-Ethylphenol |
| 5 | 2,6-Dimethylphenol | | |

## TABLE XV

### Peak Identities for Fig. 14A

| Peak no. | Compound | Peak no. | Compound |
|---|---|---|---|
| 1 | Dibenzofuran | 6 | Phenol |
| 2 | Benzophenone | 7 | 2-Naphthol |
| 3 | 2,4,5-Trimethylphenol | 8 | 1-Fluorenecarboxylic acid |
| 4 | o-Cresol | 9 | 1-Naphthoic acid |
| 5 | 3,4-Dimethylphenol | 10 | 2,3-Dihydroxynaphthalene (not eluted) |

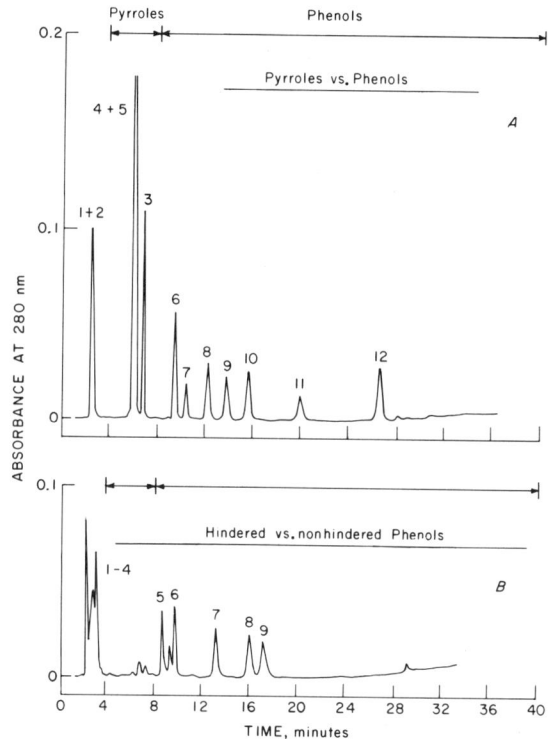

Fig. 13. Separations of synthetic mixtures of pyrroles and phenols with a tetramethylammonium hydroxide-spiked mobile-phase system. For peak identities see Tables XIII and XIV for (A) and (B), respectively. See primary entries, Table X for experimental conditions. Compare (A) with Figs. 7 and 10.

both salts give good separations of the pyrrole, phenol, and carboxylic acid classes. The very polar mobile phase (Table X) necessary for the elution of acidic solutes attests to their high affinity for silica coated with quaternary ammonium species. On the other hand, pyridine and other basic types without an active hydrogen elute easily with hexane. Figure 13A shows a successful separation of

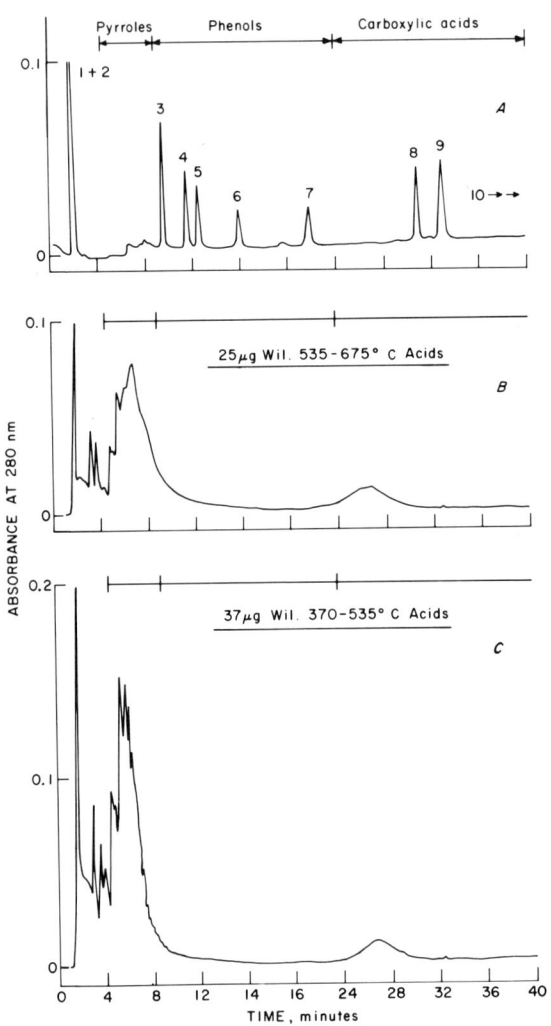

**Fig. 14.** Comparison of the retention of carboxylic acids in a synthetic mixture (A) to those in high-boiling distillates of Wilmington, CA crude oil (B and C). Absolute amounts in peaks in (A) range from 0.6 to 1.4 μg. See Table XV for peak identities in (A); experimental conditions were the same as for Fig. 13. The extensive alkylation in the petroleum acids obviously decreases their retention somewhat.

the same synthetic mixture of compounds attempted with bonded-phase and other amine-spiked systems (Figs. 7 and 10). In fact, the gradient profile can be changed to give an even wider split between the pyrrolic compounds (peaks 3-5, Fig. 13A) and 2,4,5-trimethylphenol (peak 6, Fig. 13A). Figure 13B shows that even 2,6-dimethylphenol (peak 5) is resolved from pyrroles. In general, Fig. 13 and Tables XIII and XIV show that retention of phenols increases with decreasing steric hindrance and alkyl substitution and with increasing aromaticity. At present, only amides—ordinarily present in small amounts in fuels—do not group cleanly into one fraction. They overlap between pyrroles and phenols, depending on the number of free hydrogens, alkyl substitution, etc.

Figure 14 shows an application of the tetramethylammonium hydroxide system to acid concentrates from two high-boiling distillates of a Wilmington, CA, crude oil. Although alkylation lowers the retention of the carboxylic acids somewhat in both boiling ranges, they are still clearly separated. The slight tailing of the pyrrolic compounds into the region characteristic of phenols is caused by the presence of either alcohols, extremely hindered phenols, or (most probably) amides.

Figure 14 shows the surprisingly symmetrical peaks obtained for carboxylic acids in this system. Increasing the concentration of tetraalkylammonium hydroxide in the mobile phase causes carboxylic acids to elute more rapidly, but has the opposite effect on phenols. Thus, a rather low concentration of salt (0.01–0.03% w/v) is usually optimum for good retention of phenols and good resolution between phenols and carboxylic acids.

The polar bulk mobile phase is a very important advantage of the tetraalkylammonium hydroxide-based separation. Acid concentrates frequently contain the most polar, and hence least soluble, constituents in the fuel. Thus, a separation utilizing hexane, for example, as the mobile phase is difficult or impossible to use on either an analytical or preparative scale.

One disappointing aspect of the tetraalkylammonium hydroxides as mobile-phase additives is the rather mediocre chromatographic efficiency obtained. Usually, the theoretical plate height is two or three times that normally observed for columns of a given particle size. Raising the temperature, adding other amines to the mobile phase, changing the concentration of the tetraalkylammonium hydroxide, and trying many different types of silica have all failed to improve significantly the chromatographic efficiency.

## 5. Summary: Future Research

Finally, the excellent stability of amine-spiked systems in general is demonstrated by the example shown in Fig. 15. The average decrease in retention of the components in the test mixture shown was 3.6% over a period of nearly 4 weeks of continual operation with the ethylenediamine-spiked mobile phase. Figure 15 indeed shows a sharp contrast to Fig. 9, thereby indicating the substantial superiority of dynamically coated as compared to covalently bonded amino-silica.

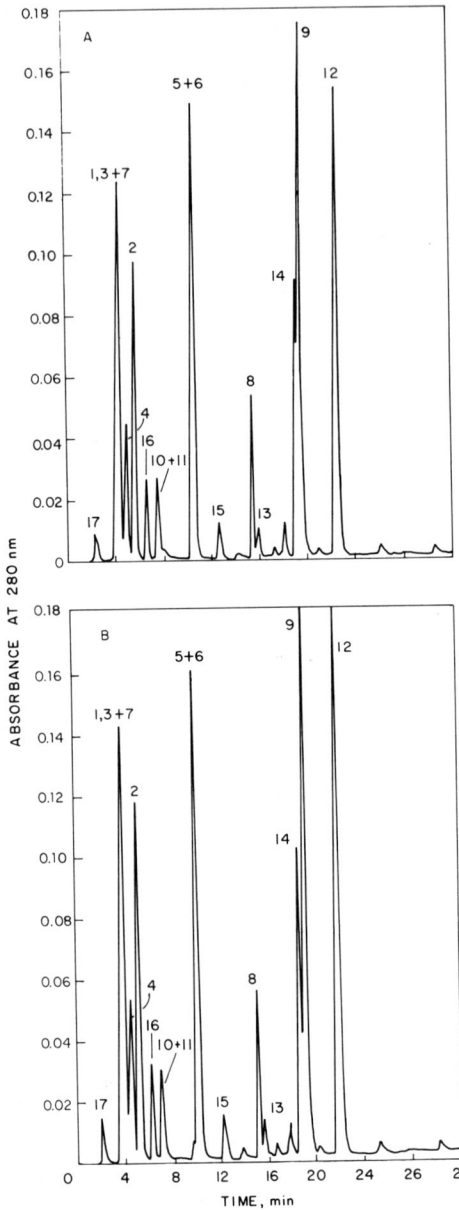

**Fig. 15.** Effect of column age on separation quality in a dynamically coated amine system: (A) 1-week-old column; (B) 4-week-old column. The average decrease in retention of peaks over 4 weeks was 3.6%. See Table V for peak identities and Table IX (alternate 1 entries) for experimental conditions. The B solvent was methyl *tert*-butyl ether containing 2.0% (v/v) methanol and 0.1% (v/v) ethylenediamine (cf. Fig. 9).

In summary, preliminary work with amine-spiked mobile phases has suggested a number of intriguing applications and unexpected conclusions. For example, separation of pyrrolic and phenolic chemical classes was possible only with polar mobile phases spiked with tetraalkylammonium hydroxides. In light of the large difference in inherent acidity between the two groups of compounds, one might have expected that separation to be trivial in any amine-containing system. Equally surprising was the strong retention of amino-PAHs on amine-loaded silica; an application of this finding is shown in the next section. Also, the similarity in chromatographic behavior between mobile phases containing pyridine and those with other amines was certainly unexpected in light of their large differences in structure and basicity.

Current research with amine-containing mobile phases in this laboratory is most intensive in the following areas:

(1)  scaling up separations to preparative scale to allow for a spectroscopic determination of the true applicability of these separations to real samples;

(2)  combining two different amine-based separations or an amine-based and carboxylic acid-based separation to obtain chemically unique fractions of fuels; and

(3)  rerunning four or five of the most promising amine systems under standardized conditions to assess at least qualitatively the applicability of different types of amines to different separation problems.

## V. COORDINATING AMINE- AND CARBOXYLIC ACID-BASED SEPARATIONS FOR OPTIMUM RESOLUTION OF COMPLEX SAMPLES

As alluded to in Sections III and IV, no one mobile-phase additive has been found which will resolve all compound classes in any complex sample. Thus, this section is a logical extension of work previously discussed in the article. For brevity, only one example—separation of a base concentrate from a coal liquid—will be discussed.

### A. Experimental

The SRC-II coal liquid was obtained from Pittsburg and Midway Coal Mining (Run No. 77 SR-12; Merriam, KS) and distilled on a spinning-band still (Perkin-Elmer, Norwalk, CT) into three distillates and a residue. The 325–425°C distillate was separated into acids, bases, and neutrals (Green and Hoff, 1981) using nonaqueous ion-exchange chromatography. The base concentrate was first separated on a preparative scale under conditions similar to those listed in Table IV

(Section III), using propanoic acid as the mobile-phase additive. Each fraction was checked for chromatographic overlap using conditions in Table III (Section III). Then, each fraction was separated twice on an analytical scale using mobile phases spiked with hexamethylenetetramine and ethylenediamine under conditions outlined in Table IX.

## B. Results and Discussion

Yield data and a sample chromatogram of the preparative separation using propanoic acid are shown in Table XVI and Fig. 16, respectively. The checks, unsurprisingly (see Fig. 16), for chromatographic purity indicated significant contamination of fraction 2A with components belonging to fractions 2 and 3. Since fraction 2A was an extremely small fraction (see Table XVI) and of unknown composition, this problem was deemed to be relatively unimportant and further analysis was continued. Overlap between other fractions was minor.

Figure 17A–F shows chromatograms of each subfraction rerun using a mobile-phase system containing ethylenediamine. First, it should be noted that unless each fraction had been previously separated, this separation would have been of very little value for this sample. However, since the sample had already been broken down into compound types, many of the separations from the ethylenediamine system can be readily interpreted.

Also, it should be noted that the small peak eluting at 16 min, most noticeable in Fig. 17A, is due to solvent impurities. Hence, Fig. 17A merely indicates the presence of a single group of compounds—aromatic hydrocarbons. These are impurities carried over from the initial isolation of the base extract. Fraction 2 (Fig. 17B) is much more interesting. The first group of peaks has not been elucidated; the second group is believed to be made up of diarylamine types (e.g., diphenylamine); and the third is composed of indoles and carbazoles.

**TABLE XVI**

**Summary of Results from Preparative Separation of SRC-II 325–425°C Bases**

| Fraction | Weight percent | Major compound type |
| --- | --- | --- |
| 1 | 3.8 | Hydrocarbons, N-alkylated pyrrolic compounds |
| 2 | 8.0 | Diarylamines, indoles, carbazoles |
| 2A | 2.8 | Mono- and diarylamines |
| 3 | 12.5 | Aniline types, hindered pyridines |
| 4 | 72.2 | Pyridine benzologs |
| Backflush | 4.0 | Amines, amides, bifunctionals, unknown |
| Total | 103.3 | |

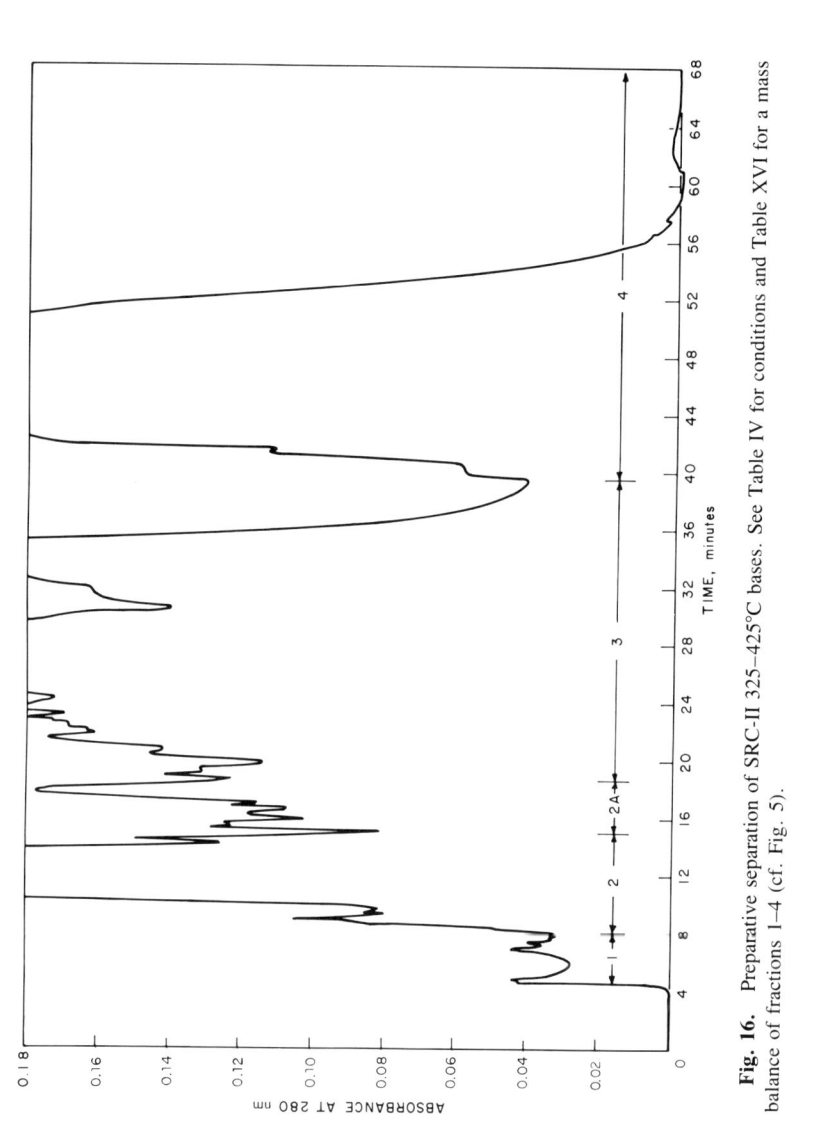

**Fig. 16.** Preparative separation of SRC-II 325–425°C bases. See Table IV for conditions and Table XVI for a mass balance of fractions 1–4 (cf. Fig. 5).

## TABLE XVII
### Guide for Matching Mobile Phases to Different Types of Solutes for Separations on Silica

| Solute type | Suggested additive(s) | Suggested bulk mobile-phase components |
|---|---|---|
| Acidic (general guides) | Amines or ammonium hydroxides | Alkanes–ethers–alcohols containing 0.01–0.3% amine |
| | Long-chain (≥4 carbons) aliphatic carboxylic acids, especially for strong or multisubstituted acids | Alkanes–dichloromethane containing 0.1–5% alcohol and 0.1–0.5% acid |
| Hydroxyarenes | Tetraalkylammonium hydroxides | Methyl *tert*-butyl ether containing 1–50% alcohol and 0.01–0.1% $R_4NOH$ |
| | Same as for hydroxyarenes | Same as above |
| Carboxylic acids and di- and trihydroxyarenes | Carboxylic acids—chain length, ≥4 carbons | Dichloromethane containing 0–5% alcohol and 0.2–0.5% acid |
| Pyrroles, amides, and other N–H compounds | Aliphatic amines, ethylenediamine, methoxyamines | Alkanes–methyl *tert*-butyl ether containing 0–5% alcohol and 0.1–0.3% amine |
| Neutral compounds (aldehydes, ketones, esters, sulfones, etc.) (general guides) | Carboxylic acids, 2–4 carbons. Usually, peak shape is improved but retention changes little | Alkanes–dichloromethane containing 0–1% alcohol and 0.1–0.2% acid |

| | | | |
|---|---|---|---|
| Basic (general guides) | | Carboxylic acids—optimum chain length is directly related to base strength of solutes | Alkanes–dichloromethane containing 0.1–15% alcohol and 0.2–0.3% acid |
| | | Aliphatic amines—especially for stronger bases and those bases containing N–H functions | Alkanes–methyl *tert*-butyl ether containing 0–10% alcohol and 0.1–0.2% amine |
| Azaarenes, diazaarenes, and azoles | Carboxylic acid chain length | Aromatic ring number | Dichloromethane containing 0.5–10% alcohol and 0.2–0.4% acid |
| | >4 | 1 pyridines, pyrazines, etc. | |
| | | 2 quinolines, benzothiazoles, etc. | |
| | Acetic or chlorinated acids | 3 and up | |
| Aminoarenes | | Carboxylic acids induce separations in the order of increasing basicity, whereas amines have the opposite effect. In general, acid additives are best for resolving isomers of anilines and aminonaphthalenes. Aliphatic oxygen-containing amines are best for amino-PAHs | Alkanes–dichloromethane containing 0.2–0.3% acid and 0–2% alcohol |
| | | | Alkanes–methyl *tert*-butyl ether containing 0.1–0.2% amine and 0–2% alcohol |
| N,N-Disubstituted amides | | Acetic or propanoic acid | Dichloromethane containing 1–10% alcohol and 0.2–0.3% acid |
| Strong bases, difunctional bases | | In general, ethylenediamine works well with these compounds | Alkanes–methyl *tert*-butyl ether containing 0.1–0.5% ethylenediamine and 0–10% alcohol |

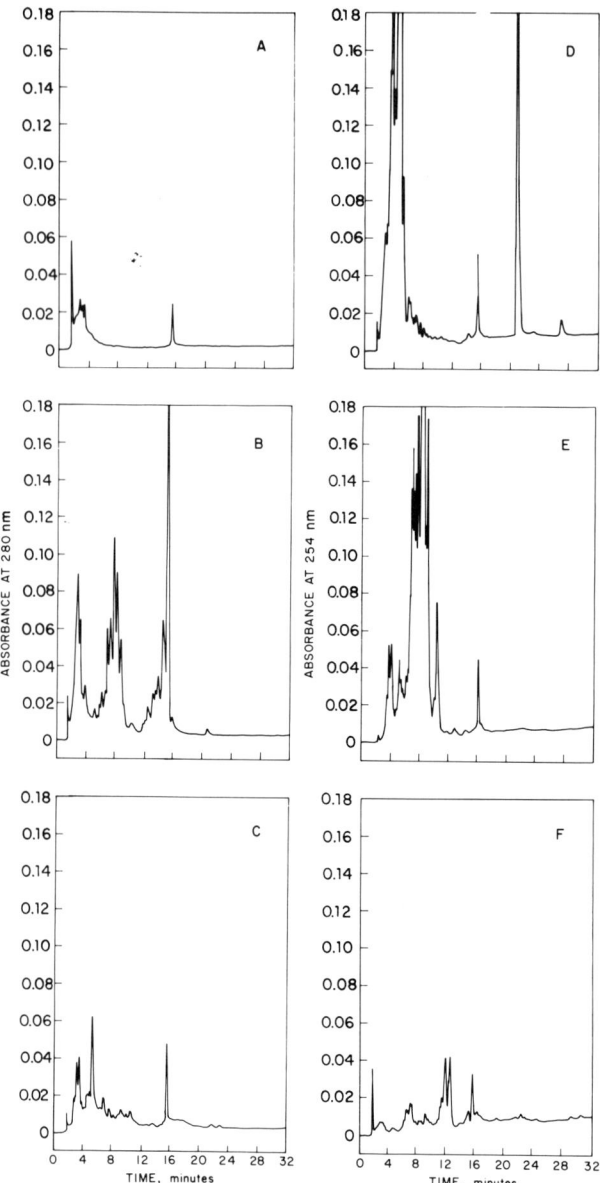

**Fig. 17.** Further separation of fractions shown in Fig. 16 using an ethylenediamine-spiked mobile phase. Conditions were the same as specified for Fig. 15. (A) 13 µg fraction 1; (B) 16 µg fraction 2; (C) 14 µg fraction 2A; (D) 21 µg fraction 3; (E) 27 µg fraction 4; (F) 30 µg backflushed material.

Interestingly, fraction 2A (Fig. 17C) contains very little pyrrolic-type compounds, expected to be present from overlap between fractions 2 and 2A, and instead appears to contain mono- and diarylamines. For example, a compound such as $N$-phenylbenzylamine, intermediate in retention between aniline and diphenylamine, would be expected to fall into fraction 2A. The two peaks at 21 and 27 min in Fig. 17D are believed to be 3- and 4-condensed-ring amino-PAHs, respectively. As mentioned in Section IV, this separation is quite significant because of the current interest in these types of compounds. The bulk of the compounds in fraction 3 are believed to be either analogous to 1,2,3,4-tetrahydroquinoline, which of course is an $N$,2-dialkylated aniline, or hindered azaarenes such as 7,8-benzoquinoline. The chromatogram of this fraction run with the hexamethylenetetramine-containing mobile phase (not shown) indicated little or no $N,N$-disubstituted anilines present in fraction 3. Fraction 4 (Fig. 17E) comprises most of the bases in the distillate (Table XVI) and is almost entirely alkylated pyridine benzologs. Although the hexamethylenetetramine-containing mobile phase (not shown) did a better job of resolving this fraction than shown in Fig. 17E, amine-containing mobile phases in general give mediocre resolution of pyridine-type compounds. Figure 17F shows the fraction obtained from back-flushing the silica column at the end of the preparative run with the propanoic acid-spiked mobile phase (see Table IV). Very little can be said at present concerning this chromatogram except that the ethylenediamine-spiked mobile phase elutes at least a portion of this fraction.

One additional point concerning Fig. 17 bears mentioning. Methyl *tert*-butyl ether containing 2.0% (v/v) methanol was substituted for the ethyl acetate used in most of the work with amines reported in Section IV. This change in solvents yielded much flatter baselines during the solvent gradient and the ability to monitor significantly lower UV wavelengths. Very little of the current work in this laboratory employs ethyl acetate as a mobile-phase component.

Although in a very early stage of development, the coordination of separations employing different types of mobile-phase additives shows promise for yielding high-quality fractions from complex samples and possibly resolving individual components in somewhat simpler samples. This technique is potentially quite useful as a trace qualitative tool for assessing compound types and could be used for trace quantitative work in cases where peaks could be positively identified and UV response factors obtained. Although the coupling of acid- and base-containing systems is discussed here, this is by no means the only combination possible. Virtually any compound which is strongly retained on silica is a possible mobile-phase additive. Detector compatibility, solubility in the bulk mobile phase, and the imagination of the experimenter are the only obvious limitations.

## VI. CONCLUSIONS

Aside from low absolute-detection limits, low potential for contamination, etc., a good trace-organic-analysis method must have a built-in qualitative identification feature. Of course, this is what has made GC–MS so popular today. One of the main advantages of the systems discussed here is that they yield retention patterns which are useful qualitative tools, unlike size-exclusion or reverse-phase chromatography which give a rough indication of molecular size at best. Multiple runs of the same sample on silica with mobile phases containing different additives often yield definitive assessments of the compound types present. Also, these more volatile nonaqueous mobile-phase systems are inherently more compatible with LC–MS interfaces than the aqueous–organic solvents currently in use. A compound-class separation coupled with a mass spectrometer would indeed be a powerful tool for analyzing complex mixtures of polar species.

From a quantitative standpoint, the separation techniques presented should be equal to other liquid chromatographic methods for low-level organic analysis. However, selection of the optimum detector is complicated by the frequent use of solvent gradients and by the wide variety of possible mobile-phase components. For example, application of electrochemical and fluorescence detectors, ideally suited to aqueous mobile phases, will be more limited. On the other hand, use of detectors which depend on solvent volatility, such as the various moving hot-wire detectors, will be favored.

Often a very important aspect of trace-organic analysis is removal of sample matrix and/or concentration of the sample components of interest. Especially if the desired compounds are of one chemical class or group, a preparative-scale separation similar to the examples shown here could be very effective in obtaining a purified concentrate prior to final analysis by spectroscopic or chromatographic means. Also, since the mobile-phase additive largely controls separation selectivity, separations often can be scaled up directly using silica from different manufacturers without fear of large shifts in retention, peak-order reversals, etc.

A possible stumbling block to application of these systems is their apparent complexity and the difficulty in predicting the ideal mobile-phase system for a given sample. Of course, this is still a relatively undeveloped research area and is actively being pursued in the authors' and other laboratories. Nevertheless, Table XVII may serve as an initial guide for matching mobile-phase additives to different types of compounds.

In conclusion, LSC chromatography encompasses much more than the separation of toluene, naphthalene, phenanthrene, and chrysene on silica using hexane as the mobile phase. The above separation is mentioned because it, or a very similar one, is frequently the *only* one mentioned for plain silica in column-vendors' literature. Hopefully, this article might entice a few of the many dedicated bonded-phase chromatographers to "take a step backward" and try plain silica whenever they encounter a separation which defies resolution on their favorite bonded phase.

## Acknowledgments

The authors thank D. E. Seizinger, J. W. Goetzinger, and J. S. Ball for reviewing the manuscript.

## REFERENCES

Aitzetmuller, K. (1978). *J. Chromatogr.* **176**, 354–358.
Brugman, W. J. T., and Kraak, J. C. (1981). *J. Chromatogr.* **205**, 70–78.
Colin, H., Schmitter, J.-M., and Guiochon, G. (1981). *Anal. Chem.* **53**, 625–631.
Coq, B., Gonnet, C., and Rocca, J.-L. (1975). *J. Chromatogr.* **106**, 249–262.
Ghaemi, Y., and Wall, R. A. (1979). *J. Chromatogr.* **174**, 51–59.
Ghaemi, Y., and Wall, R. A. (1980). *J. Chromatogr.* **198**, 397–405.
Ghaemi, Y., and Wall, R. A. (1981). *J. Chromatogr.* **212**, 271–281.
Green, J. B. (1981). *J. Chromatogr.* **209**, 211–229.
Green, J. B., and Hoff, R. J. (1981). *J. Chromatogr.* **209**, 231–250.
Grizzle, P. L., Green, J. B., Sanchez, V., Murgia, E., and Lubkowitz, J. (1981). *Prepr., Div. Pet. Chem., Am. Chem. Soc.* **26**(4), 839–850.
Guerin, M. R., Ho, C.-H., Rao, T. K., Clark, B. R., and Epler, J. L. (1980). *Environ. Res.* **23**, 42–53.
Guerin, M. R., Rubin, I. B., Rao, T. K., Clark, B. R., and Epler, J. L. (1981). *Fuel* **60**, 282–288.
Guillemin, C. L. (1980). *In* "Chromatography Review" (F. W. Karasek, ed.), Vol. VI, No. 1, pp. 6–8. Spectra Physics, Santa Clara, California.
Hansen, S. H. (1981). *J. Chromatogr.* **209**, 203–210.
Hansen, S. H., Helboe, P., Thomsen, M., and Lund, U. (1981). *J. Chromatogr.* **210**, 453–460.
Hendrix, D. L., Lee, R. E., Jr., Baust, J. G., and James, H. (1981). *J. Chromatogr.* **210**, 45–53.
Karlesky, D., Shelly, D. C., and Warner, I. (1981). *Anal. Chem.* **53**, 2146–2147.
Knox, J. H., and Laird, G. R. (1976). *J. Chromatogr.* **122**, 17–34.
Lawrence, J. F., and Leduc, R. (1978). *Anal. Chem.* **50**, 1161–1164.
Paanakker, J. E., Kraak, J. C., and Poppe, H. (1978). *J. Chromatogr.* **149**, 111–126.
Schabron, J. F., Hurtubise, R. J., and Silver, H. F. (1978). *Anal. Chem.* **50**, 1911–1917.
Scott, R. P. W. (1980). *J. Chromatogr. Sci.* **18**, 297–306.
Scott, R. P. W., and Kucera, P. (1978). *J. Chromatogr.* **149**, 93–110.
Scott, R. P. W., and Kucera, P. (1979). *J. Chromatogr.* **171**, 37–48.
Scott, R. P. W., and Traiman, S. (1980). *J. Chromatogr.* **196**, 193–205.
Snyder, L. R. (1968). "Principles of Adsorption Chromatography." Dekker, New York.
Snyder, L. R., and Kirkland, J. J. (1979). "Introduction to Modern Liquid Chromatography," 2nd ed. Wiley, New York.
Snyder, L. R., and Poppe, H. (1980). *J. Chromatogr.* **184**, 363–413.
Stalling, D. L., Petty, J. D., and Smith, L. M. (1981). *J. Chromatogr. Sci.* **19**, 18–26.
Svendsen, H., and Greibrokk, T. (1981). *J. Chromatogr.* **212**, 153–166.
Vogh, J. W., and Thomson, J. S. (1981). *Anal. Chem.* **53**, 1345–1350.
Wall, R. A. (1980). *J. Chromatogr.* **194**, 353–363.
Wheals, B. B., and White, P. C. (1979). *J. Chromatogr.* **176**, 421–425.

# INDEX

## A

Abate (herbicide), 172–173
Acetic acid, 227–229, 231, 237
N-Acetylretinylamine, 14
Adsorption chromatography, carotenes, 6
Aliphatic amines, 249
Aliphatic carboxylic acids, high-performance liquid chromatography, 227–238
Alkyamines, high-performance liquid chromatography, 239
N-Alkylation, 231
Alumina
  in high-performance liquid chromatography
    carotenoids, 5
    retinoids, 10–11, 14
    vitamin D, 22, 26
    vitamin E, 32
  in trace-enrichment techniques, 217
Amines
  high-performance liquid chromatography, 238–257
    coordination with carboxylic acid-based separations, 257–263
6-Aminochrysene, 251
Amino-polycyclic aromatic hydrocarbons (amino-PAHs), 252, 257, 263
Amino-silica column, high-performance liquid chromatography, 244–247
Anilines, 231, 244
Animal feeds
  high-performance liquid chromatography
    tocopherol acetate, 36
    vitamin D, 20, 24–26
Anion-exchange chromatography
  folacin, 57, 59–60
  vitamin $B_6$, 54

Antioxidants, steric exclusion chromatography, 124
API, *see* Atmospheric pressure ionization
API–MS, *see* Atmospheric pressure ionization mass spectrometry
Aqueous steric exclusion chromatography (gel filtration), 126, 132, 137, 140, 144, 220
Araboascorbic acid, *see* Isoascorbic acid
Aromatic amines, 248
Aromatic hydrocarbons, 188–189, 191–196, 217–219, 258, 263
N-Arylaniline, 231
Ascorbic acid (vitamin C), liquid chromatography, 43–47
Atmospheric pressure ionization (API), in liquid chromatography–mass spectrometry, 83–84
Atmospheric pressure ionization mass spectrometry (API–MS), 99, 101
Automated switching in SEC–LC techniques, 134–136
Azaarenes, 231

## B

Benzanthrone, 194
Benzene, 188–189
Benzidine, 174–176
Betamethasone, 216–217
BHA, steric exclusion chromatography, 124
BHT, steric exclusion chromatography, 124
Blood
  high-performance liquid chromatography
    ascorbic acid, 45
    folacin, 56, 59–60
    niacin, 51

267

Blood (cont.)
  riboflavin, 49
  vitamin A and retinoids, 13–15
  vitamin $B_6$, 53, 55
  vitamin D and metabolites, 27–29
  vitamin E, 36–37
  vitamin K, 42
  steric exclusion chromatography, 132, 137
  trace-enrichment analysis, 207–208, 210–212, 214
Brain tissue
  high-performance liquid chromatography
    ascorbic acid, 45
    folacin, 59
    thiamine, 49
  trace-enrichment analysis, 212–214
Bunker-C oil, 185–186
Butyric acid, 227, 229

## C

Carbazole, 258
Carboxylic acid
  high-performance liquid chromatography, 227–238, 248, 251, 254–255
  coordination with amine-based separations, 257–263
β-Carotene, 3, 5, 7, 132
Carotenoids
  liquid chromatography, 3–17
  steric exclusion chromatography, 130
Cation-exchange chromatography, folacin, 58
Cefoperazone, 207–208
Cereals
  high-performance liquid chromatography
    niacin, 51
    vitamin $B_6$, 55
Chemical ionization (CI)
  in gas chromatography–mass spectrometry, 70, 73
  in liquid chromatography–mass spectrometry, 77–78, 86–89, 94, 98–100
Chemical ionization mass spectrometry (CIMS), 78, 82, 93
Chlorinated phenols, 200–202
Chlorinated reaction products, 191–192
4-Chlorobutyric acid, 229
Chlorophenoxy herbicides, 177
2-Chloropropanoic acid, 229
3-Chloropropanoic acid, 229, 237–238
Cholecalciferol (vitamin $D_3$), 17–18, 20–22, 26–29, 38–39
  hydroxylated metabolites, 23
4,6-Cholestadienol, 22
Cholesterol phenylacetate, 38–39
CI, see Chemical ionization
CID, see Collision-induced dissociation
CIMS, see Chemical ionization mass spectrometry
Closed-column chromatography
  tocopherols and tocotrienols, 32
Coal liquids, 231, 234, 237–238, 252, 257–263
Cobalt corrinoids, 60
Collision-induced dissociation (CID), 91
Column chromatography
  carotenoid analysis, 5
  vitamin E, 32–33
Contamination, in trace-enrichment techniques, 154–157, 202
Corn oil, steric exclusion chromatography, 123–124, 131
p-Cresol, 157–159
Crude oil, see Petroleum
Cryptoxanthins, 6
Cyanocobalamin, see Vitamin $B_{12}$
Cyano-silica column, high-performance liquid chromatography, 245–247
Cyclosporine A, 205–207

## D

DBA, see Dibenzanthracene
DBP, see Dibutyl phthalate
DCB, see Dichlorobenzidine
DDT, steric exclusion chromatography, 120–121, 146–147
Decanoic acid, 228, 237
DEHP, see Di-(2-ethylhexyl) phthalate
Dehydroascorbic acid, 43–44, 47
7-Dehydrocholesterol, 17–18, 21–23
3-Dehydroretinol, 4, 10, 14
Demoxytocin, 214–215
DEP, see Diethyl phthalate
DHF, see 5,6-Dihydrofolic acid
Diarylamine, 251–252, 258, 263
Dibenzanthracene (DBA), 157
Dibutyl phthalate (DBP), 179–181, 197–199
Dichlorobenzidine (DCB), 174–176

Di-(2-ethylhexyl) phthalate (DEHP), 197–199
Diethyl phthalate (DEP), 179–181
Dihydroergocornine, 203–204
Dihydroergocristine, 203–205
5,6-Dihydrofolic acid (DHF), 56–59
Dimeric tocopherols, 32
$N,N$-Dimethylaniline, 231, 251
2,6-Dimethylphenol, 255
Dinitrogen compounds, 231
Dioxane, 33
Diphenylamine, 258
Direct injection technique
  in steric exclusion chromatography, 117–126
    problems, 143–145
  in trace enrichment, 163–175, 202–207, 214–217
Direct liquid introduction (DLI), in liquid chromatography–mass spectrometry, 78–83
Displacement chromatography, 153–154, 157–160
DLI, see Direct liquid introduction
Dursban (pesticide), 175–177

## E

Egg yolk, vitamin D and metabolites, 28
EH–MS, see Electrohydrodynamic ionization mass spectrometry
EI ionization, see Electron-impact ionization
Electrochemical detector, HPLC analysis of ascorbic acid, 44–45
Electrohydrodynamic ionization mass spectrometry (EH–MS), 99–100
Electron-impact (EI) ionization
  in gas chromatography–mass spectrometry, 70, 73
  in liquid chromatography–mass spectrometry, 77–78, 86–89, 94, 98–100
Elution chromatography, 153–154, 157–158
Emmerie–Engel reaction, 31–32
Endralazine pyruvate (EP), 208–210
5,6-Epoxyretinoic acid, 16–17
5,6-Epoxyretinoid, 4
Ergocalciferol (vitamin $D_2$), 17, 20, 26–29, 39
  hydroxylated metabolites, 23
Ergosterol, 17

Erythorbic acid, see Isoascorbic acid
D-Erythrohexono-1,4-lactono-2-ene, see Isoascorbic acid
Ethanol, 228
Ethoxyquin, 32
Ethyl acetate, 241
Ethylenediamine, 248–250, 258, 262–263

## F

Fat-soluble vitamins, liquid chromatography, 3–43
FD–MS, see Field-desorption mass spectrometry
Felypressin, 214–215
FI, see Field ionization
FID, see Flame-ionization detector
Field-desorption mass spectrometry (FD–MS), 99
Field ionization (FI), 98
Fish
  vitamin D, 19, 26
  vitamin E, 33
Flame-ionization detector (FID), 84, 87
Fluoranthene, 195–196
Fluorometric detector, column chromatography of vitamin E, 33–35
Fluridone (herbicide), 173–174
Folacin, 55–60
Folate, 55–56
Folic acid, 55–60
Foods
  high-performance liquid chromatography
    ascorbic acid, 43–47
    carotenoids, 5–7
    folacin, 56, 58
    niacin, 50–51
    riboflavin, 48–49
    thiamine, 48–49
    tocopherols and tocotrienols, 32–35
    vitamin A esters, 10–12
    vitamin $B_6$, 52–53, 55
    vitamin $B_{12}$, 60
    vitamin D, 19, 24–27
    vitamin K, 38, 41–42
  steric exclusion chromatography, 121–127, 130–131
Fortified foods
  β-carotene, 5
  vitamin D 19–20, 26

## G

Gas chromatography (GC), 69
  off-line multidimensional SEC techniques, 128–131, 133
Gas chromatography–mass spectrometry (GC–MS), 70–71, 73–74, 93–94, 264
  chlorination reactions of polycyclic aromatic hydrocarbons, 191
Gas–liquid chromatography (GLC)
  carotenoids, 5
  vitamin E, 32
GC, see Gas chromatography
GC–MS, see Gas chromatography–mass spectrometry
Gel filtration, see Aqueous steric exclusion chromatography
Gel-permeation chromatography (GPC), 119, 137–138, 144, 219
GFC (gel-filtration chromatography), see Aqueous steric exclusion chromatography
GLC, see Gas–liquid chromatography
Glycosides, 219
GPC, see Gel-permeation chromatography
Grains, vitamin E, 33

## H

Heart cutting, in steric exclusion chromatography, 134, 136, 138
Heptane, 228, 252
Herbicides, 172–174, 177
Hexamethylenetetramine, 249–250, 258, 263
Hexane, 157, 228, 241, 254–255
5-HIAA, see 5-Hydroxyindoleacetic acid
High-performance liquid chromatography (HPLC)
  polar substances on unmodified silica, 223–265
    amine- and carboxylic acid-based separations, coordination of, 257–263
    literature survey, 224–226
    nonaqueous systems containing aliphatic carboxylic acids, 227–238
    nonaqueous systems containing amines, 238–257
    trace-enrichment techniques, 169, 173, 207, 214–216
  vitamins, 1–67
    ascorbic acid, 43–47
    folacin, 55–60
    niacin, 49–52
    riboflavin, 47–49
    thiamine, 48–49
    vitamin A and carotenoid provitamins, 3–17
    vitamin $B_6$, 52–55
    vitamin $B_{12}$, 60
    vitamin D, 17–29
    vitamin E, 29–37
    vitamin K, 37–43
High-performance liquid chromatography–mass spectrometry (LC–MS), 69–109, 264
  applications and conclusions, 101–105
  mass spectrometer as detector, 70–74
  off-line techniques, 74–77, 104
  on-line techniques, 77–97, 104
    aerosol-jet techniques, 94–97
    direct-coupling techniques, 77–83
    GC–MS approaches to interfacing, 93–94
    mechanical transfer techniques, 83–93
    vaporization and ionization, alternative methods of, 97–101
HPLC, see High-performance liquid chromatography
5-HT, see Serotonin
Hydrocarbons, 182–189, 191–196, 217–219, 230–231, 239, 251, 258, 263
5-Hydroxyindoleacetic acid (5-HIAA), 212–214
18-Hydroxyretinoic acid, 16

## I

IEC, see Ion-exchange chromatography
Indole, 258
Insecticides, see Pesticides
Ion-exchange chromatography (IEC), 137
  liquid fuels, 229, 236, 257
  thiamine and riboflavin, 48–49
  vitamin $B_6$, 52, 54–55
Ion gun, moving-ribbon interface, 91
Ionization methods, for LC–MS application, 97–101
Ion-pair chromatography
  folacin, 57–60
  niacin, 51–52

# Index

Ion suppression, in trace enrichment, 161
Isoascorbic acid, 43, 47
Isocratic elution, 153–154, 157–158
Isopyrocalciferol, 22
Isotachysterol, 18–19, 21–23
Isovitamin D, 18–19

## J

Jet separation, LC–MS interface, 93–94

## L

*Lactobacillus casei*, 55
Laser crossed-beam LC–MS interface, 94–97
Laser desorption (LD) ionization, moving-ribbon interface, 91
Laser-ionization mass spectrometry (LIMS), 99–100
LC, *see* Liquid chromatography
LC–MS, *see* High-performance liquid chromatography–mass spectrometry
LD ionization, *see* Laser desorption ionization
LIMS, *see* Laser-ionization mass spectrometry
Liquid chromatography (LC), *see also* High-performance liquid chromatography
 detectors, 160
 liquid fuels, 236
 trace enrichment as specialized form of, 153–160
Liquid fuels, 227–239, 257–263, *see also* Coal liquids; Petroleum; Shale oil
Liquid–liquid extraction, 173, 229
Liquid–solid chromatography (LSC), 224, 226, 239, 264
Lubricating oil, steric exclusion chromatography, 120–121
Lumicalciferols, 17
Lumiflavin, 48–49
Lumisterols, 17–18, 21–23
Lycopene, 7
Lypressin, 214–215

## M

Magnesia
 in high-performance liquid chromatography
  carotenoids, 5–6
  vitamin D, 26
Malathion, 138, 144

Mass spec–mass spec (MS–MS), 99, 101
Mass spectrometer, as detector for liquid chromatography, 70–74
Mass spectrometry (MS), 70
Mass spectrometry and gas chromatography, *see* Gas chromatography–mass spectrometry
Mass spectrometry and liquid chromatography, *see* High-performance liquid chromatography–mass spectrometry
Meat, vitamin D, 19
Membrane separation LC–MS interface, 93
Menadione (vitamin $K_3$), 37–38
Menaquinones, 37–38, 40–43
Methanol, 228, 241
Methotrexate (Mtx), 210–212
2-Methoxyethylamine, 248, 250
Methoxytrate, 60
*N*-Methylaniline, 231
*N*-Methylnicotinamide, 51
Methyl retinoate, 7, 9
Methyl *tert*-butyl ether, 241, 263
Microbiological assay
 folacin, 55–56
 vitamin $B_6$, 52
Microprocessor-based chromatograph, 134, 137
Milk
 high-performance liquid chromatography
  ascorbic acid, 45
  niacin, 50
  vitamin $B_6$, 55
  vitamin D, 19, 26
  vitamin K, 41–42
Monoarylamine, 263
Monochloroacetic acid, 229, 237
Monoethanolamine, 248
Moving-belt LC–MS interface, 87–93
MS, *see* Mass spectrometry
MS–MS, *see* Mass spec–mass spec
Mtx, *see* Methotrexate

## N

Niacin
 liquid chromatography, 49–52
 trace-enrichment analysis, 219
Niacin equivalents, 50
Nicotinamide, 49–51
Nicotinic acid, 49–51

Nonaqueous reversed-phase liquid chromatography, 6

## O

Off-line techniques
   in liquid chromatography–mass spectrometry, 74–77
   in steric exclusion chromatography, 117, 128–133
      problems, 145–147
   in trace enrichment, 163–164, 175–191, 207–208, 217–219
Oil spill, trace enrichment of water sample from, 182–183, 185–187
On-line techniques
   in liquid chromatography–mass spectrometry, 77–97
   in steric exclusion chromatography, 117, 134–144
      problems, 147–148
   in trace enrichment, 164, 191–202, 208–214, 219–220
Organic-gel packings, in steric exclusion chromatography, 116–117
Organic trace analysis, see also Blood; Food; Tissues; Urine; Vitamins; specific compounds
   trace-enrichment techniques, 151–221
Ornipressin, 214–215
4-Oxoretinoic acid, 16
Oxyhydrogen-torch crossed-beam LC–MS interface, 96
5,8-Oxyretinoic acid, 16–17
5,8-Oxyretinoid, 4
Oxytocin, 205–206, 214–215

## P

PAHs, see Polycyclic aromatic hydrocarbons
Paper chromatography, vitamin E, 31
PCP, see Pentachlorophenol
PD–MS, see Plasma-desorption mass spectroscopy
Pentac, 123–124
Pentachlorophenol (PCP), 161, 199–202
Perforated belt LC–MS interface, 87
Pesticides
   steric exclusion chromatography, 120–124, 131, 138, 146–147
   trace-enrichment techniques, 171–177

Petroleum, 236–237, 239, 254–255
Phenanthrene, 188–189
Phenol, 169–170, 236, 244, 246, 248–250, 252–255, 257
N-Phenylbenzylamine, 263
Phthalate esters, 169–170, 178–181, 197–199
Phylloquinone (vitamin $K_1$), 37–43
Plasma chromatography, 83
Plasma-desorption mass spectroscopy (PD–MS), 99–100
Plastochromanol, 30–32
Polychlorinated biphenyls (PCBs)
   steric exclusion chromatography, 120–121, 130
   trace-enrichment techniques, 179, 181–182, 197
Polycyclic aromatic hydrocarbons (PAHs), 143–144, 188–196, 217–219
Polyethylene container, leaching from, 164–166
Polyglutamates, 55–58
Polymers, steric exclusion chromatography, 119–120
Polyoxyethylene surfactants, 125
Polystyrene, 132
Prefractionation/concentration/measurement, in steric exclusion chromatography, 117, 127–128, 145–147
Previtamin D, 17–23, 25–26
Propanoic acid, 227–229, 231, 237, 258, 263
Provitamin A, see Carotenoid provitamins
Provitamin D, 17–18
Psoriasis, 14
Pteridines, 57
Pteroylmonoglutamic acid, 55
Pyridine, 231, 233, 240–241, 244, 252, 254, 257, 263
Pyridine-3-carboxylic acid, see Nicotinic acid
Pyridoxal, 50, 52–55
Pyridoxamine, 50, 52–55
Pyridoxic acid, 52, 55
Pyridoxine, 50, 52–55
Pyr–MS, see Pyrolysis mass spectrometry
Pyrocalciferol, 22
Pyrodoxine, 219
Pyrolysis mass spectrometry (Pyr–MS), 99, 101
Pyrrole, 231, 236, 244, 246, 249–255, 257

# Index

## Q

Quinones, 38

## R

Resins (vitamin D concentrates), 17, 21–23
Retinal, 4, 7, 9–10, 14–15
Retinoic acid, 4, 9, 13–17
Retinoids, 3–10, 13–17
Retinol, *see* Vitamin A
Retinyl acetate, 8, 10, 14, 132
Retinylacetylhydrazone, 14
Retinyl esters, 7, 10, 13–15
Retinyl palmitate, 9–10, 12, 38–39, 132–133
Retroretinol, 3–4
Reverse-phase column (RPC)
  in steric exclusion chromatography, 134–135, 137–138, 140, 143–144
  in trace enrichment, 153–157
Reverse-phase high-performance liquid chromatography
  carotenoids, 5
  folacin, 57, 60
  niacin, 51
  retinoids, 7, 11–12, 14, 16
  thiamine and riboflavin, 48–49
  vitamin $B_6$, 55
  vitamin $B_{12}$, 60
  vitamin D, 20–21, 23–24, 26–27
  vitamin E, 32, 35–36
  vitamin K, 41–43
Riboflavin
  liquid chromatography, 47–49
  trace-enrichment analysis, 219
RPC, *see* Reverse-phase column

## S

Saponification, in analysis of foods and tissues, 10–12, 24–26, 29, 32, 36
SEC, *see* Steric exclusion chromatography
Secondary-ion mass spectrometry (SIMS), 99–100
  moving-ribbon interface, 91, 93
Sennaglycoside, 140–141, 144
Serotonin (5-HT), 212–214
Sewage, 175–176
Shale oil, 233–236
Silica
  high-performance liquid chromatography, 223–265
    amine- and carboxylic acid-based separations, coordination of, 257–263
    carotenoids, 5
    niacin, 50
    nonaqueous systems containing aliphatic carboxylic acids, 227–238
    nonaqueous systems containing amines, 238–257
    retinoids, 7, 10–12, 14, 16
    riboflavin and thiamine, 48–49
    vitamin D, 21–28
    vitamin E, 32, 35–36
    vitamin K, 38, 41–43
  steric exclusion chromatography, 116–117
  trace-enrichment techniques, 157, 160, 217
SIMS, *see* Secondary-ion mass spectrometry
Single-pump ternary chromatograph, 135–136
Skin, human
  retinol, 14
  vitamin D, 27
Sodium lauryl sulfate (SLS), 172
Solanachromene, 31
Solvent-purity test, 154
Soybean oil, 124–125, 131
Spectrofluorometer, 34–35
Steric exclusion chromatography (SEC), 111–150
  advances in SEC columns, 116–118
  advantages, 114–115
  basics of, 112–114
  direct injection, 117–126
  disadvantages, 115–116
  off-line multidimensional chromatography, 117, 128–133
  on-line multidimensional chromatography, 117, 134–144
  prefractionation/concentration/measurement, 117, 127–128
  problems and troubleshooting, 143–149
Straight-phase high-performance liquid chromatography
  retinoids, 7
  vitamin D, 22–23
Sugars, steric exclusion chromatography, 126–127, 144
Suprasterols, 17–18

Surfactants, 125
Synfuels, *see* Coal liquids; Shale oil

## T

Tachysterols, 17–19, 21–22
Tetraalkylammonium hydroxides, 248–255, 257
Tetraalkylammonium salts, 241, 243
Tetrabutylammonium hydroxide, 252
Tetrachloroethylene, 167–169
5,6,7,8-Tetrahydrofolic acid (THF), 56–60
Thiamine
　liquid chromatography, 48–49
　trace-enrichment analysis, 219
Thin-layer chromatography (TLC)
　carotenoid analysis, 5
　vitamin D, 26
　vitamin E, 31–32
Thiochrome, 48–49
Tissues
　high-performance liquid chromatography
　　ascorbic acid, 44–45, 47
　　folacin, 56, 58–60
　　riboflavin, 48–49
　　thiamine, 48–49
　　vitamin A and retinoids, 13–15
　　vitamin $B_6$, 52, 55
　　vitamin D and metabolites, 27–29
　　vitamin E, 33, 36–37
　　vitamin K, 38
　steric exclusion chromatography, 132
　trace-enrichment analysis, 202–214
TLC, *see* Thin-layer chromatography
Tocopherols, 14, 29–39
Tocopheryl acetate, 35–36, 38–39
Tocotrienols, 30–32, 35
Tomato juice, ascorbic acid, 45–47
Tomato plants, steric exclusion chromatography, 138, 144
Toxisterols, 17
Trace-enrichment techniques, 151–221
　analysis of trace-level components above and below detection limits, 153
　clinical applications, 202–214
　detectors, 160
　direct injection, 163–175, 202–207, 214–217
　environmental applications, 164–202

extractability, techniques to increase, 160–161
frontal elution, 152–153
mutual-zone solubility, 161–163
off-line enrichment, 163–164, 175–191, 207–208, 217–219
on-line enrichment, 164, 191–202, 208–214, 219–220
pharmacological applications, 214–220
as specialized form of liquid chromatography, 153–160
Triethanolamine, 248
Triglycerides, 38, 40–41
2,4,5-Trimethylphenol, 255
Tryptophan, 49–50

## U

Ubichromenols, 31
Ultrasonic nebulization LC–MS interface, 97
Urine
　high-performance liquid chromatography
　　ascorbic acid, 45
　　niacin, 51
　　retinoids, 15
　　riboflavin, 49
　　vitamin $B_6$, 52, 55
　steric exclusion chromatography, 130, 137, 144
　trace-enrichment studies, 204–207, 209–210

## V

Vaporization methods, for LC–MS application, 97–98
Vegetable oils
　high-performance liquid chromatography
　　vitamin E, 33–34
　steric exclusion chromatography, 123–125, 131
Vision, biochemistry of, 7, 10, 15
Vitamin A (retinol)
　liquid chromatography, 3–17, 38–39
　　with vitamin E, 36
　metabolism, 15
　steric exclusion chromatography, 125, 132–133
Vitamin $A_2$, 4
Vitamin A acetate, 124–125

Vitamin $B_1$, *see* Thiamine
Vitamin $B_2$, *see* Riboflavin
Vitamin $B_6$
  liquid chromatography, 52–55
  structure, 50
Vitamin $B_{12}$, 60
Vitamin C, *see* Ascorbic acid
Vitamin D
  calculation of potency, 26–27
  liquid chromatography, 17–29
Vitamin $D_2$, *see* Ergocalciferol
Vitamin $D_3$, *see* Cholecalciferol
Vitamin E, 29–37
Vitamin K, 37–43
Vitamin $K_1$, *see* Phylloquinone
Vitamin $K_2$, 37
Vitamin $K_3$, *see* Menadione
Vitamins, *see also* specific vitamins
  high-performance liquid chromatography, 1–67
  steric exclusion chromatography, 140–142
  trace-enrichment analysis, 218–220

## W

Water
  steric exclusion chromatography, 125
  trace enrichment, 155–156, 162, 164–191, 199
Water-soluble vitamins, liquid chromatography, 43–60
White oils, 217
Wire transport LC–MS interface, 84–87

## X

Xanthophylls, 5

/547.30894T759>C1>V2/